International Scientific and Practical Conference "Law and the Information Society: Digital Approach" (LISID-2020)

SHS Web of Conferences Volume 109 (2021)

Volgograd, Russia
6 November 2020

Editors:

S. Cindori
A.M.M. Mostafa
E.Yu. Malushko
T.V. Deryugina
A.A. Mokhov

ISBN: 978-1-7138-3066-5

Printed from e-media with permission by:

Curran Associates, Inc.
57 Morehouse Lane
Red Hook, NY 12571

Some format issues inherent in the e-media version may also appear in this print version.

For additional information, please contact EDP Sciences – Web of Conferences
at the address below.

EDP Sciences – Web of Conferences
17, Avenue du Hoggar
Parc d'Activité de Courtabœuf
BP 112
F-91944 Les Ulis Cedex A
France

Phone: +33 (0) 1 69 18 75 75
Fax: +33 (0) 1 69 28 84 91

contact-edps@webofconferences.org

Additional copies of this publication are available from:

Curran Associates, Inc.
57 Morehouse Lane
Red Hook, NY 12571 USA
Phone: 845-758-0400
Fax: 845-758-2633
Email: curran@proceedings.com
Web: www.proceedings.com

TABLE OF CONTENTS

ABOUT

Welcome to the International Scientific and Practical Conference "Law and the Information Society: Digital Approach" (LISID-2020), which took place on November 5-6, 2020 in a Zoom format in the Volgograd Institute of Management - a branch of the Russian Academy of National Economy and State Services under the President of the Russian Federation, Volgograd, Russia.

LISID-2020 is a good opportunity to share experience, modern knowledge and achievements in the field of legal aspects of the digital society, in the context of the evolution of intellectual property rights in the information society; changing approaches to legal technology in the digital space; law and digital medicine, digital education; socio-cultural studies of information and communication technologies in a digital society, economic models and information society management; security problems in the information society; digital environment, information communication systems and media culture in the information age.

The purpose of the conference is to discuss and forecast future trends in the development of the information society and its legal aspects, exchange advanced knowledge in the field of digital governance and law.

106 manuscripts were submitted to LISID-2020. And 41 manuscripts were approved and accepted by our reviewers. By submitting a document to LISID-2020, the authors agree with the review process and understand the importance of the review process for their work. The manuscripts were reviewed by qualified experts in the fields selected by the organizing committee of the conference, who provided detailed comments. All papers were peer-reviewed using a double-blind review process: authors list their names and affiliations in the manuscript for reviewers, but the reviewers did not know each other's identities and whether the authors received information about who reviewed their manuscript. The organizing committee organized the sessions so that the participants could get the maximum benefit.

We express our deepest gratitude to the organizers, conference participants, LISID 2020 committees and reviewers for their hard work. Without their participation, it would have been impossible for us to successfully hold LISID-2020 and ensure the high quality of conference materials.

We also want to express our gratitude to the large number of authors who submitted and presented the results of their research at the conference.

Special thanks go to the SHS Web of Conferences journal editors and team. We also look forward to furthering fruitful cooperation.

LISID-2020 Organizing Committee

Guest editor(s)

Cindori Sonja, Associate Professor, Faculty of Law University of Zagreb, Croatia

Mostafa Aimoatassembellah Mostafa Mohamed, Deputy Dean of the Faculty of Law, Doctor of Laws, University of El Mansour, Egypt

Malushko E.Yu., Candidate of Pedagogical Sciences, Associate Professor, Deputy Director for Informatization of the Institute of Philology and intercultural communication of the Volgograd State University

Deryugina T.V., Doctor of Law, Professor of the Department of Civil and Labor Law, Civil Procedure of the Federal State Educational Institution of Higher Education “Moscow University of the Ministry of Internal Affairs of the Russian Federation named after V.Ya. Kikotya”

Mokhov A.A., Doctor of Law, Professor, Head of the Department of Medical Law, Moscow State Law University named after O.E. Kutafina (Moscow State Law Academy)

Statement of Peer review

In submitting conference proceedings to *Web of Conferences*, the editors of the proceedings certify to the Publisher that

1. They adhere to its **Policy on Publishing Integrity** in order to safeguard good scientific practice in publishing.
2. All articles have been subjected to peer review administered by the proceedings editors.
3. Reviews have been conducted by expert referees, who have been requested to provide unbiased and constructive comments aimed, whenever possible, at improving the work.
4. Proceedings editors have taken all reasonable steps to ensure the quality of the materials they publish and their decision to accept or reject a paper for publication has been based only on the merits of the work and the relevance to the journal.

Title, date and place of the conference

LISID 2020 Law and the Information Society: Digital Approach
November 6, 2020
Volgograd, Russia

Proceedings editor(s):

Cindori Sonja
Mostafa Aimoatassembellah Mostafa Mohamed
Malushko E.Yu.
Deryugina T.V.
Mokhov A.A.

17, avenue du Hoggar - PA de Courtabœuf – BP 112 - 91944 Les Ulis Cedex A (France)
Tél. : 33 (0)1 69 18 75 75 - Fax : 33(0)1 69 07 45 17 – www.edpsciences.org

Implementation of ECHR decisions on expulsion in a modern digital society

Elvina Ablyakimova[1], *Vladislav* Kostylev[2], *Artem* Rudenko[1], *Kristine* Trifonova[2], and *Sergey* Trifonov[1]

1 Russian State University of Justice, Higher Educational Institution, Crimean Branch, Pavlenko Str. 5, 295000, Simferopol, Russia, e-mail:rudenkocrimea@yandex.ru.
2 Russian State University of Justice, Higher Educational Institution, Crimean Branch, Pavlenko Str. 5, 295000, Simferopol, Russia, e-mail:rudenkocrimea@yandex.ru.
3 Russian State University of Justice, Higher Educational Institution, Crimean Branch, Pavlenko Str. 5, 295000, Simferopol, Russia, e-mail: kri-kazaryan@yandex.ru.
4 Russian State University of Justice, Higher Educational Institution, Crimean Branch, Pavlenko Str. 5, 295000, Simferopol, Russia, e-mail: sergey3fonov@yandex.ru

Abstract. The globalization of world economic relations, political instability and different levels of economic development in different countries predetermine a significant flow of migrants. The problem of illegal migration and application of measures of legal responsibility against migration legislation violators concerns most countries of the world. At the same time, applying such a sanction as expulsion, a number of states, including the Russian Federation, come into conflict with the legal positions of the European Court of Human Rights (hereinafter the ECHR) regarding expulsion of stateless persons and foreigners. The administrative expulsion, unfortunately, has not undergone any significant changes in the digital age, although the urgent need for such reform is obvious. It is emphasized in the legal positions of the ECHR that decisions on expulsion must be justified by pressing social necessity and they must correspond to a legitimate aim, since they may violate the right to respect the private and family life. It is necessary to take into account the legal positions of the ECHR in the framework of proceedings on expulsion and adhere to the principle of individualization of responsibility for an administrative offense, with taking into account the specific circumstances in each specific case.

1 Introduction

The migration policy of the Russian Federation is constantly undergoing reforms. In the policy formation, much attention is paid to regulation of processes related to the illegal stay of a foreign citizen on the territory of the Russian Federation, as well as to prevention of offenses committed by migrants. It is due to the fact that number of foreign citizens on the territory of the Russian Federation is growing rapidly, mainly due to the tense situation in the world. Caused by such factors, migration processes are becoming a negative phenomenon and a threat to the socio-economic and political development of states.

It seems that the migration policy in the context of a modern information society can become more effective through the use of modern digital technologies within the framework of the regulation of migration activities, which will allow to more effectively implement state policy in this area..

2 Problem Statement

Violation of migration rules is characterized by high latency and their sudden onset, and, therefore, it is almost impossible to determine the exact number of migrants appearing on the territory of the Russian Federation. Most foreign citizens and stateless persons enter the Russian Federation legally, but they lose the grounds for staying during their long stay in the Russian Federation and, as a result, they are recognized as illegal migrants later.

Based on the above, the conclusion follows that the control over illegal migration by the state should be strengthened. Expulsion is recognized as one of the most effective and promising mechanisms for combating persons violating the migration regime.

Since adoption of the Code of Administrative Offenses of the Russian Federation [1] in year 2001, the legal regulation of the procedure for the administrative expulsion of foreigners and stateless persons from the Russian Federation has not undergone any significant changes. The very concept of expulsion has not changed during 19 years of the law implementation; it is defined as a forced and controlled movement of foreigners and stateless persons across the State border of the Russian

Federation outside the Russian Federation, and in cases provided for by the legislation of the Russian Federation, it is a controlled independent departure of foreign citizens and stateless persons from the Russian Federation.

3 Research Questions

The main amendments in the legal regulation of the expulsion procedure were introduced by Federal Law No. 410-FZ "On Amendments to the Federal Law On Bailiffs and Certain Legislative Acts of the Russian Federation" [2] dated 06.12.2011. They touched upon the definition of the authority entitled to choose the form of expulsion, the possibility of keeping persons who are assigned for expulsion in special detention centers, as well as determining the conditions where it is possible to apply expulsion in the form of the immigrant's independent controlled exit. Thus, the amendments only specified those provisions that were contained in the Code of Administrative Offenses of the Russian Federation (hereinafter referred to as the Code of Administrative Offenses) earlier.

The procedure for the execution of an administrative penalty in the form of expulsion did not undergo any significant changes during the specified period either.

The problems of imposing an administrative penalty in the form of administrative expulsion, existing at the moment, and the problems of its implementation should be resolved with the adoption of the new Code of Administrative Offenses of the Russian Federation.

In the draft Code of Administrative Offenses of the Russian Federation submitted by the Government of the Russian Federation [3], the concept of "administrative expulsion" is defined similarly to the current version of the Code, which may indicate the legislator's unwillingness to change the current situation. At the same time, Article 3.27 of the Draft, dedicated to the Rules for the appointment of an administrative penalty in the form of administrative expulsion, makes a significant step towards guaranteeing rights of expelled persons. It contains cases when it is impossible to impose an administrative expulsion. They include:

1. When the expelled person has a child who is a citizen of the Russian Federation and permanently resides in the Russian Federation;
2. When the expelled person has a status of a refugee or a person holding temporary asylum;
3. Arrival of a person on the territory of the Russian Federation in an emergency mass order;
4. Recognition of a person as a compatriot.

The emergence of such a provision in the Draft indicates that the Russian authorities are taking into account modern approaches in the migration policy. At the same time, it is worth noting that not all the positions expressed by the ECHR in its decisions were reflected in the Draft.

In addition to changing the procedure for applying administrative expulsion, it is worth paying attention to the composition of administrative offenses containing this type of punishment.

According to a number of scientists, including V.P. Soldatov, it is necessary to amend the wording of the sanction of clause 1.1., Art. 18.8 of the Administrative Code of the Russian Federation at the federal level, providing in it not only the main punishment and additional punishment in the form of a fine and compulsory administrative expulsion, but also a possibility of not applying additional punishment (expulsion) to the offender. They could use the following wording: after the words "... with administrative expulsion outside the Russian Federation", add the phrase "or without it". The scientific research has clearly shown that today there are sufficient grounds not to apply that additional punishment (the international legal act, the judgments of the Constitutional and Supreme Courts), but very few people know about them at the level of city and district courts of law, and judges are not guided by them. How much effort, means and other costs were required for offenders to receive lawful and fair court judgments. In this regard, some other, more ambitious, conclusions and proposals [4] can be done.

The administrative expulsion of a foreign citizen or stateless person outside the Russian Federation can be established and applied as both the main and additional administrative punishment (part 2, article 3.3, Code of Administrative Offenses of the Russian Federation).

The decisions of the Supreme Court of the Russian Federation have repeatedly noted that the application of additional punishment in the form of administrative expulsion from the Russian Federation should be based on data confirming the need to apply that measure of responsibility to the violator as the only possible way to achieve a fair balance of public and private interests within the framework of administrative proceedings.

Within the meaning of the provisions of the Convention for the Protection of Human Rights and Fundamental Freedoms [5], which entered into force for Russia on May 5, 1998, and the Convention on the Rights of the Child [6] (adopted on November 20, 1989), which entered into force for the USSR and for its legal successor, the Russian Federation, on 15 September 1990, and in view of their application by the European Court of Human Rights, the expulsion of a foreign citizen or a stateless person outside the Russian Federation, entailing an interference with the right to respect for private and family life, shall be allowed only in cases where it is necessary in a democratic society and in proportion to public law goals.

Ratifying the Convention, the Russian Federation recognized the jurisdiction of the European Court of Human Rights as binding on interpretation and application of the Convention and its Protocols in cases of alleged violation by the Russian Federation of any provisions of those treaty acts [7].

However, it's necessary to put an emphasis on the latest amendments altered into the Constitution of the Russian Federation. Article 79 of the Constitution of the Russian Federation, set forth in the new edition, indicates that "the Russian Federation may participate in interstate associations and transfer to them part of its powers in accordance with international treaties of the Russian Federation, if this does not entail any restrictions on the rights and freedoms of man and citizen and does not contradict the foundations of the constitutional system of the Russian Federation. Any decisions of interstate bodies adopted on the basis of the provisions of international treaties of the Russian Federation in their interpretation, contrary to the Constitution of the Russian Federation, shall not be subject to execution in the Russian Federation" [8]. Therefore, if any decisions of the ECHR contradict the Constitution of the Russian Federation, they will not be enforced on the territory of Russia.

It is clear from the clarifications of the Plenum of the Supreme Court of the Russian Federation in Resolution No. 23 "On Judgment" dated December 19, 2003, Paragraph 4, that the court should take into account, along with the norms of the substantive law, the decisions of the European Court of Human Rights, which provide an interpretation of the provisions of the Convention for the Protection of Human Rights and Fundamental Freedoms to be applied in the case.

It is necessary to analyze the jurisprudence of the ECHR in cases of administrative expulsion.

The analysis of the decisions taken by the European Court of Human Rights in cases of administrative expulsion, carried out by E.O. Bondar, allows us to speak of the following violations of Art. 3 and 5 of the Convention for the Protection of Human Rights and Fundamental Freedoms of 1950 (hereinafter - the Convention) [5]:

- in some cases, expulsion is replaced by extradition or deportation;
- Art. 3 of the Convention is violated ("forcible return to a country where the applicant is at risk of torture and prohibited treatment or punishment") (also in the context of the execution of court orders that have entered into force);
- the person's freedom is restricted unreasonably (inconsistency with the purpose, where expulsion cannot be carried out; this applies especially to stateless persons) [9].

In our opinion, violations of Article 8 of the Convention should be added to this list.

Let us consider that category of cases in more detail.

The European Court of Human Rights has concluded that the Convention does not guarantee foreigners the right to enter into or to reside in a particular country and not be expelled, having noted that the state's responsibility for ensuring public order obliges the state to control entry into the country; at the same time, decisions in this area must be justified by pressing social necessity, since they may violate the right to respect for private and family life, protected in a democratic society by Article 8 of the named Convention, and they should correspond to a legitimate aim (Case 'Berrehab v. the Netherlands', dated 21 June 1988, §28; 'Boughanemi v. France', dated 24 April 1996, §41; 'El Boujaidi v. France', dated 26 September 1997, §39; 'Uner v. the Netherlands' of 18 October 2006,§54; Liu and Liu v. Russia, dated 6 December 2007, §49; Decision of 9 November 2000 on admissibility of the complaint "Andrey Shebashov v. Latvia" and others.) [10].

With regard to criteria for the admissibility of expulsion taking place in a democratic society, the European Court of Human Rights noted that the importance attached to one or another of them will differ depending on the circumstances of a particular case; bound by the need to establish a fair balance between the competing interests of the individual and society as a whole, the state has certain margin of appreciation; at the same time, the authorities' right to apply expulsion can be an important means of preventing serious and repeated violations of the immigration law, as leaving those violations unpunished would undermine respect for the immigration law.

According to the legal position of the European Court of Human Rights, the totality of social ties between migrants and the society where they live is included in the concept of "private life" in the sense of Article 8 of the Convention for the Protection of Human Rights and Fundamental Freedoms; thus an expulsion of a settled (resident) migrant, regardless of availability or absence of his "family life" could lead to a violation of the right for his "private life" ('Uner v. the Netherlands' of 18 October 2006, §59; 'Alim v. Russia' dated 27 September 2011, § 68).

Assessing the circumstances in which the expulsion of a foreigner gives rise to a violation of Article 8 of the Convention for the Protection of Human Rights and Fundamental Freedoms, the European Court of Human Rights uses the criteria of the necessity of such a measure in a democratic society and its proportionality to the legitimate aim pursued: the nature and severity of the offense committed by the person, and the length of his stay in the country and his behavior, including attempts to legitimize his residence, and citizenship of the persons concerned, and the characteristics of the family life of the immigrant, including availability of children from his marriage and their number, and difficulties that a spouse is expected to face in the country where his spouse, an immigrant, is being expelled, and availability of insurmountable obstacles for a family to live in their home country for one of the spouses, etc.

With regard to the Russian Federation, the European Court of Human Rights noted the difference between what it considers insignificant violation of the rules of registration of foreigners and staying in the country without such a document as a visa or a permit (Resolution in 'Alim v. Russia' dated 27 September 2011, §78, 87 and 89) [11].

Pursuant to the Resolution of the European Court in case 'Alim v. Russia' dated 27 September 2011, (Complaint No. 39417/07, § 81), Article 8 of the Convention does not impose a general obligation on the state to respect the immigrant's choice of the country of residence and to authorize the re-establishment of families on its territory. However, in matters relating to the family life and immigration, the extent of the state's duty to admit relatives to persons staying in its territory will depend on the specific circumstances of those persons and their main interests.

In the judgment of the European Court in the case Zakayev and Safanova versus Russian Federation'

(Complaint No. 11870/03, §41), The Court notes that a number of relevant criteria have been developed in the case 'Boultif v. Switzerland' (Complaint No.54273/00, §48), which should be used in assessing whether the expulsion is a necessary measure in a democratic society and whether the expulsion is proportionate to the legitimate aim pursued. [12]

Those criteria, inter alia, are: citizenship of the interested persons; the applicant's family situation, in particular, the length of their marriage and other factors indicating reality of the couple's family life; whether the spouse knew about such violation at the time of entering into the family relationships.

Within the meaning of Article 8 of the Convention, the aspect of presence or absence of "the family life" is mainly a matter of fact, which depends on actual existence of close family relations. The concept of "family life" is autonomous, i.e. does not depend on how it is understood in the national legislation of any state. According to the case-law of the European Court of Justice, the concept of family is not limited to a registered marriage and it shall also include a de facto marriage relationship. Any child born out of such a relationship shall be ipso jure a part of that family unit from the moment of birth and by the very fact of his birth. Thus, there is a connection between the child and his parents, equal to family life (Judgment of the European Court in the case 'The case of Guliyev and Sheyna v. Russia' (Complaint No. 29790/14) [13].

4 Purpose of the Study

The purpose of the scientific research is, on the basis of the available scientific data of the theory of state and law, international law, constitutional law, administrative law, labor law, the practice of public authorities, to consider topical issues of the implementation of the positions of the ECHR on the expulsion of foreigners in the context of modern digital society.

5 Research Methods

The methodological basis of scientific research consists of general scientific (analysis and synthesis, comparison, description, generalization, modelling) and specific scientific (comparative legal, legal modelling, formal legal, document analysis) methods.

6 Findings

Various approaches have been identified within the framework of the judicial practice of the Supreme Court of the Russian Federation and the ECHR regarding the procedures of administrative expulsion in terms of replacing it with extradition or deportation, violation of a number of norms of the Convention for the Protection of Human Rights and Fundamental Freedoms of 1950. At the same time, in some cases, the similarity of the legal positions of the Constitutional courts of the Russian Federation and the ECHR on research issues has been identified.

The need to strengthen the digital component in the processes of regulating migration activities has been noticed.

7 Conclusion

The position of the European Court is confirmed by its new judgments.

In particular, in the case Akopdzhanyan v. Russia (Application no. 32737/16) [14], the gist of which was that the applicant complained that his expulsion from the Russian Federation, with an eight-year re-entry ban, violated his right for the family life, The European Court decided unanimously that the Russian Government had violated the requirement of Article 8 of the Convention (the right to respect for private and family life) in that case.

In one case, the administrative expulsion from the country may be regarded as an interference with the person's right to family life, which, however, does not entail a violation of this article as long as it is justified within the meaning of paragraph 2, Article 8 of the Convention, and in other case, it can be regarded as a violation of Article 8 of the Convention.

The legal positions of the Constitutional Court of the Russian Federation on similar cases coincide in many respects with the opinion of the European Court of Human Rights. "When it comes to institutions related to termination of the stay or residence of a foreign citizen in the Russian Federation, including administrative expulsion from its territory, the courts of law should not limit themselves to establishing only formal grounds for application of the law, they must investigate and assess the real circumstances in order to recognize that the appropriate decisions in respect of a foreign citizen are necessary and proportionate ... " [15]

There is a significant number of Resolutions of the Supreme Court of the Russian Federation on abolition of judgments on bringing foreign citizens to administrative responsibility with their expulsion outside the country in connection with failure of judges to take into account the actual circumstances of committing those administrative offenses, due to availability of spouses and children of migrants residing on the territory of Russia, who have received citizenship in the manner prescribed by the law.

The Resolutions of the Supreme Court are based on the position of the European Court of Human Rights reflected in the Decision No.39428/12 in the case "Gablishvili v. Russia" of June 26, 2014 [16]. Based on the results of consideration of Gablishvili's complaint, the refusal to allow the foreign citizen to enter Russia was found to be contrary to Article 8 of the Convention for the Protection of Human Rights and Fundamental Freedoms [17], which guarantees everyone the right to respect for his personal and family life in connection with the permanent residence of his wife and children, citizens of Russia, in one of the territories of the Russian Federation.

Based on the foregoing, we should note that the Supreme Court of the Russian Federation has repeatedly made the resolutions cancelling judgments of the lower courts on bringing foreign citizens to administrative responsibility regarding their expulsion from the territory of Russia, thereby calling into question the effective decisions of the authorized bodies not to allow foreigners to enter into the territory of the country.

In our opinion, certain organizational and legal measures are needed to resolve the current problem. It seems necessary to strengthen the state mechanism for protecting the rights of migrants. In particular, one of the measures may be establishment of a high-quality judicial control over defense of migrants' rights, especially in implementation of the provisions of the Convention. It is necessary to focus on the fact that in order to prevent, detect and suppress violations of the migration legislation of the Russian Federation, the application of administrative measures in this area, the use of modern digital technologies and information and technical means is becoming increasingly important, as well as for improving the mechanisms for obtaining by interested state bodies the information on the presence of foreign citizens in the Russian Federation, their movement across its territory, expanding the use of modern biometric technologies for personal identification, ensuring effective control over compliance with the requirements of the labor legislation of the Russian Federation on taxes and fees by employers who engage and make use of foreign workers.

Summing up, it is necessary to note that when exercising their sovereign rights in the field of migration policy, the state authorities should take into account the legal positions of the ECHR. The procedure for bringing to administrative responsibility entailing a penalty in the form of administrative expulsion, must provide substantial procedural guarantees, and must be applied on the basis of the principle of individualization of responsibility for the committed administrative offense, where the competent authorities must take into account the circumstances of each specific case [18].

References

1. Code of the Russian Federation on Administrative Offenses dated 30.12.2001 No. 195-FZ, Collected Legislation of the Russian Federation of year 2002, No.1, Art. 1
2. Federal Law No. 410-FZ "On Amendments to the Federal Law on Bailiffs and Certain Legislative Acts of the Russian Federation", Collected Legislation of the Russian Federation, 12.12.2011, No. 50, Art. 735.
3. Draft of the Federal Law - Code of the Russian Federation on Administrative Offenses, URL: https://regulation.gov.ru/projects#search=02/04/05-20/00102447&npa=102447 (date of access 02.12.2020)
4. A.P. Soldatov, Works of the Academy of the Ministry of Internal Affairs of Russia **4 (48)**, 191-193 (2018)
5. SPS Consultant Plus, Bulletin of International Agreements **3** (2001)
6. SPS Consultant Plus, Convention on the Rights of the Child (approved by the UN General Assembly on November 20, 1989 and entered into force for the USSR on September 15, 1990) (1990)
7. SPS Consultant Plus, Federal Law No.54-FZ "On Ratification of the Convention for the Protection of Human Rights and Fundamental Freedoms and Protocols thereto" dated March 30, 1998 (1998)
8. New text of the Constitution of the Russian Federation with the 2020 amendments, URL: http://duma.gov.ru/news/48953/ (date of access 12.12.2020)
9. Ye.O. Bondar, Bulletin of the Barnaul Law Institute of the Ministry of Internal Affairs of Russia **1(30)**, 58-59 (2016)
10. Judgment of the European Court of Human Rights of July 26, 2011 Case "Liu v. Russian Federation (No.2)" (Application No.29157/09) (First section), Bulletin of the European Court of Human Rights **8** (2012)
11. Judgment of the ECHR dated 27.09.2011 - 'Alim v. Russian Federation' (Application No.39417/07), Bulletin of the European Court of Human Rights **2** (2013)
12. Judgment of the ECHR of 11.02.2010 - Case "Zakayev and Safanova v. Russian Federation" (Application No. 11870/03), Bulletin of the European Court of Human Rights **8** (2010)
13. Judgment of the ECHR of 19.06.2008 - Case "Guliyev v. Russian Federation" (application No. 24650/02), Russian Chronicle of the European Court **2** (2009)
14. Akopdzhanyan v. Russia (Application No. 32737/16), URL:

http://www.echr.coe.int> (date of access: 12.12.2021)

15. RLS Consultant Plus, Resolution of the Constitutional Court of the Russian Federation No.5-P of February 17, 2016 "On Inspection of Constitutionality of provisions of the Federal Law "On Legal Status of Foreign Citizens in the Russian Federation", Article 8, Paragraph 6, and Parts 1 and 3 of Article 18.8 of the Code of Administrative Offenses of the Russian Federation, and the Federal Law "On the procedure for leaving the Russian Federation and entering the Russian Federation", Part one of Article 27, Subparagraph 2, in connection with the complaint of M. Turcan, citizen of the Republic of Moldova (2016)
16. Judgment of the ECHR of 26.06.2014 - Case 'Gablishvili v. Russian Federation' (Application No. 39428/12), Precedents of the European Court of Human Rights **2 (152)** (2015)
17. D. Strazzari, European Journal of Migration and Law, 114-138 (2020)
18. A. Aranguiz, M. Quené, European Journal of Migration and Law, 509-539 (2019)

Consumer protection in the digital environment

Elena Agibalova [1,*], *Igor* Ilovaysky [2], *Yanina* Kayl[3], and *Viktoria* Usanova [4]

[1] Volgograd Institute of Management – Branch of the Russian Presidential Academy of National Economy and Public Administration Volgograd, Russia, agibalovaen76@mail.ru
[2] Volgograd Institute of Management – Branch of the Russian Presidential Academy of National Economy and Public Administration Volgograd, Russia, dominno@rambler.ru
[3] Volgograd Institute of Management – Branch of the Russian Presidential Academy of National Economy and Public Administration Volgograd, Russia, yanika@bk.ru
[4] Volgograd Institute of Management – Branch of the Russian Presidential Academy of National Economy and Public Administration Volgograd, Russian Federation, vau_77@mail.ru

Abstract. The study examines the positive and negative aspects of the spread of digital technologies on legal relations with the participation of consumers. The relevance of the research is due to the complexity of the online environment, in which consumers are vulnerable to risks and challenges that affect their ability to participate in digital transformation. The categories of persons most vulnerable in the digital age are considered; factors contributing to this process; possible counteraction measures by the competent authorities. The purpose of studying these issues is to determine the basis of legal regulation of relations arising in connection with the digitalization of the economy, as well as to justify the improvement of domestic legislation in this area. The main research method was a systematic analysis of the categories "digital environment" and "consumer rights", as well as a general scientific dialectical method, specific scientific and empirical methods of cognition. The study confirmed the hypothesis that consumers of digital goods (services) are less protected than consumers of traditional goods (services), however, the authors conclude that it is necessary to equally protect the rights of consumers of all social groups, while establishing special protection measures for the most vulnerable of them.

1 Introduction

1.1 The digitalization process started with the most significant and extensive areas of society, including areas with weakness, such as consumer rights. Digitalization is having a profound impact on our society and economy, changing the way consumers interact with each other and with the online marketplace. In this context, consumer data became an important economic asset, enabling a wide range of innovative business models, technologies and transactions. With the increasing complexity of the online environment, consumers may be vulnerable to real or potential risks and challenges that could affect their ability to participate effectively in digital transformation.

1.2 The digital transformation of society and the economy brought many new business opportunities to consumers, but also contributed to a number of new risks. It also touched on long-standing consumer policy issues: inaccurate information about goods (works, services); misuse of consumer personal data; cross-border fraudulent and misleading commercial practices; unsafe products; international cooperation on consumer protection, including law enforcement; dispute resolution and redress procedures; sustainable consumption; protection of the most vulnerable categories of consumers.

1.3 The relevance of the topic is due to objective processes of consumer market development and common for many countries tendencies to improve the legal regulation of relations involving consumers.

Every year since 2017, Consumers International (CI) has defined consumer rights in the digital age with the theme of World Consumer Day on 15 March [1].

To protect the rights of consumers, the Government of the Russian Federation approved the Strategy of state policy in the field of consumer protection for the period till 2030 by Order No. 1837-r of 28 August 2017 [2]. Order No. 1197 of 29 December 2018 of the Federal Service for Supervision of Consumer Protection and Human Welfare (Rospotrebnadzor) approved the Concept of Consumer Protection Legislation [3] based on a preliminary analysis of experience in systematizing legal provisions in a number of states (Germany, Italy, Brazil). In 2020, Rospotrebnadzor [4] proposed sweeping amendments to the Russian Federation Law of

* Corresponding author: agibalovaen76@mail.ru

7 February 1992 No 2300-I "On Consumer Protection" [5]. Although these documents do not specifically address the issue of consumer protection in the digital environment, they touch on it.

In Russia, special acts in this area have been discussed in draft form since 2000, but have not yet been adopted. For example, the draft Recommendations on the organization of individuals' activities in the sphere of Internet commerce, designed to harmonize the development of such activities in the Russian segment [6] and the draft Federal Law "On electronic commerce" [7] aimed at protecting the rights of interested parties in selling goods and services using information technologies.

The current version of the 2016 Recommendations on Consumer Protection in E-Commerce of the Organization for Economic Cooperation and Development (OECD) provides a solid basis for guiding global online marketplace policy [8]. However, governments around the world need to do more in consumer protection to keep up with the pace of change inherent in digital technologies and provide consumers with adapted protections and tools to enable them to participate effectively in the digital environment.

2 Problem Statement

2.1 The study aims to substantiate the impact of the digital economy on consumer rights, to consider the positive and negative aspects of spreading digital technologies on legal relations involving consumers, to identify the current status and problems of consumer protection in a digital environment, as well as specific ways to protect consumer rights in the emerging environment, including through digital technology.

2.2 The authors intend to confirm the hypothesis that consumers of digital goods (services) are less protected than consumers of traditional goods (services), and to find out who the most vulnerable consumers are in the digital age, which categories of consumers may be particularly at risk and what can potentially be done to reduce consumer vulnerability in the digital transformation of society.

3 Research Questions

3.1 The research focuses on the impact of digital technologies on legal relations involving consumers and the current state of consumer protection in the digital economy.

3.2 The authors pay special attention to the issue of impermissible conditions in contracts with consumers that infringe their rights, including the unauthorized collection of personal data about consumers, which has become particularly relevant in light of the extensive draft amendments to the Consumer Rights Protection Law prepared by Rospotrebnadzor in 2020 [4].

3.3 As the online environment may affect the ability of consumers to engage in digital transformation, the authors bring up the following issues for discussion:

1. Who is most vulnerable in the digital age (e.g. children, teenagers or the elderly)?
2. What factors can make consumers most vulnerable?
3. What can consumer protection authorities do to address consumer vulnerability in the digital age?

4 Purpose of the Study

4.1 The purpose of the study is to identify the basics of legal regulation of new social relations emerging due to the digitalization of certain social spheres, to substantiate ways to develop legislation on consumer protection in a digitalized economy, considering the balance of interests of society, the state, the online market, as well as consumers of its goods, works and services.

4.2 The introduction of a previously developed institutional model will protect the rights of consumers buying goods and services through information technology, which will improve the institutional protection of the rights of such parties through the creation of a mandatory e-commerce entrepreneurship insurance and e-commerce contract institute, as well as through the optimization of existing institutions by standardizing, strictly selecting and controlling e-commerce entrepreneurship (including intermediation) [9]. However, the authors aim to further develop specific ways to protect consumer rights by shifting the focus of this study to protect the most vulnerable consumer groups in the digital environment.

5 Research Methods

5.1 The research methodology relies on the general scientific dialectical method of knowledge and special scientific research methods: the formal-legal method, the method of interpretation of legal norms and the method of legal modelling. The authors used empirical methods of comparison, description and interpretation. The decisive method of research was a systematic analysis of the categories of "digital environment" and "consumer rights" in terms of their impact on the current state of civil regulation of social relations in the protection of violated civil rights.

5.2 The information and analytical basis for this study included: a) Rospotrebnadzor state reports containing objective systematized analytical information on the results of federal state supervision of consumer protection in the Russian Federation in 2018 and 2019 and serving as the basis for identifying priority areas for consumer protection, developing necessary measures to ensure consumer protection, and improving the legal and regulatory framework in the field [3,10]; b) the experience of foreign countries in the field.

6 Findings

6.1 The impact of digital technology on legal relations involving consumers

The study revealed that the positive aspects of the impact of digital technology on legal relations involving consumers include access to a wider range of goods and services; elimination of human error (minimization of risk) in the production, storage, sale, shipment (shipping) of goods; greater control over the production, storage, sale, shipment (shipping) of goods; elimination of the possibility of digital data tampering associated with the provision of goods and services.

The challenges of digitalization in this area may include: the lack of a realistic opportunity for consumers to inspect and test goods during an online sale; insufficient attention to consumer protection by sellers; inattention by consumers themselves when accepting the terms of a public offer, making it difficult to protect their rights later; an unauthorized collection of personal data about consumers, including by various smart devices; need to build consumer trust in online retailing; lack of legal regulation of online retailing relations in the near future.

The lack of consumer protection in the sale of goods and services using information technology is a serious problem for the development of e-commerce in modern Russia. This is a drawback of e-commerce compared to traditional trade, which reduces the attractiveness of this trade format [9].

6.2 The current state of consumer protection in the digital economy

At this stage of the digital economy, there is a lack of legal protection for consumer rights in electronic transactions. Our analysis of regulatory issues relating to the sale of goods and services using information technologies in Russia has shown that a likely deterrent to the development of electronic commerce is the lack of considering the specifics of this format in existing legislation and the pending draft of a Russian Federal Law "On electronic commerce" [7], designed to reduce uncertainty and guarantee the protection of stakeholders' rights when selling goods and services using information technology [9].

The lack of a specific regulatory framework generates many violations. Some sellers misunderstand that the Consumer Protection Act only applies to transactions made off the Internet. Resolution No. 17 of the Plenum of the Supreme Court of the Russian Federation of 28 June 2012 "On Consideration by the Courts of Civil Cases on Disputes concerning Protection of Consumer Rights" [11] also does not directly address issues of consumer protection in the digital environment. The authors believe that there is a way to remedy this situation by adding to the Act that its provisions apply equally to relations arising in the digital sphere.

At the same time, it is still possible to restore the consumer's violated rights through the Law, in particular Article 26.1 ("Distance selling of goods"), if the unfair seller is a resident of the Russian Federation. But transactions with a foreign seller have not been sufficiently regulated at present. International instruments such as the 1994 General Agreement on Trade in Services (GATS) [12], the Model Law on Electronic Commerce developed and adopted by the United Nations Commission on Trade and Development [UNCITRAL] in 1996 [13], do not contain rules that govern the prosecution of offenders. This is why foreign online hypermarkets (e.g. AliExpress), as well as domestic ones, offer the use of buyer protection programmes developed by them. The relationship in digital trade is increasingly based on self-regulation by its participants. There is an opinion that the development of the digital economy will soon lead to autonomous consumer protection without state intervention [14].

Active changes in the economy, including those associated with the digitalization of sales channels, the transition to remote interaction with the consumer, the emergence of new objects of civil rights (digital products, smart electronics, the Internet of Things, personal data arrays, etc.) lead to an objective need to revise both basic and specific rules that ensure fundamental consumer rights.

6.3 The problem of unauthorized collection of personal data on consumers and other unacceptable contract terms that infringe on their rights

In 2020, Rospotrebnadzor drafted a bill to amend the Consumer Protection Act. The changes will affect Article 16 of the Act, which in the new version will be called "Unacceptable contract terms that infringe on the rights of consumers". The main message of the amendments, according to Rospotrebnadzor, is an attempt to protect citizens from the unlawful collection of personal data by retailers that is not directly related to the transaction. The seller will no longer be able to refuse a consumer into a contract if the consumer does not want to provide the relevant information. "Also to implement the principle of protecting the weaker party to the contract, the draft law provides a list of conditions that infringe on consumers' rights," reads an explanatory note to the draft [4]. It is about conditions included in contracts that the Rospotrebnadzor and, according to its reports, the courts consider to be contrary to the law. As a result, the Law will contain dozens of bans and restrictions on business and, according to some authors, will effectively nullify all attempts by entrepreneurs to prevent consumer extremism and other forms of abuse of rights [15].

Analysing draft Article 16 of the Consumer Rights Protection Law, the authors concluded that Russia is now proposing to introduce, following the European model, a blacklist of unacceptable conditions in a contract with a consumer. However, the European blacklists are much longer and better elaborated. We think that Rospotrebnadzor should have worked with them. In the European Union (hereinafter EU), consumer protection has advanced by dozens of years compared to Russian law. For example, there is the Council of the European Communities Directive on unfair terms in contracts with consumers [16] and the experience of its implementation in all EU countries, as well as numerous publications on the issue. In our view, this was totally overlooked by the drafter of the Consumer Protection Act amendments.

Most importantly, it is very strange that there is no proposal to establish a general rule on the nullity of unfair terms in a consumer contract, giving the court jurisdiction to assess any consumer contract terms for the balance of interests of the parties for protecting the weaker party. This right is available to judges in all European countries and even in England, where ex post control of contractual fairness is generally not developed. The Constitutional Court of the Russian Federation demands the same.

The courts in the Russian Federation are now forced to exercise ex post control over the fairness of consumer contracts under a creative reading of Article 16.1 of the Consumer Protection Law, which literally says exactly the opposite – that conditions that contradict imperative consumer law norms are null and void in terms of worsening the consumer's legal position. In effect, this clause declares all such default rules to be asymmetric-imperative. It is possible to improve the position of the consumer in comparison to such rules, but not to worsen them. However, the problem is that it is hard to write imperative rules for all occasions, and the imagination of traders is limitless. So, our courts do what they can. Europe everywhere has long imposed ex post controls on the fairness of consumer contracts. But in Russia, the courts, including the Supreme Court of the Russian Federation, are forced each time to fictitiously apply the literally unrelated paragraph 1 of Article 16 of the Consumer Rights Protection Law in such a situation to protect the consumer from the unfair conditions imposed on him. For example, in terms of setting fees, claims procedures, etc., there is a need to call things by their proper names and recognise that courts can overturn manifestly unfair terms in consumer contracts rather than pretending that such terms are allegedly contrary to some norm, for there are none. Especially since the doctrine of unfair terms has long developed in our country, with the Resolution of the Plenum of the Supreme Arbitration Court of the Russian Federation on freedom of contract and its limits [17] and Article 428 of the Russian Civil Code [18] in force. The authors of this study suggest that Rospotrebnadzor should take these circumstances into account when drafting the final version of Article 16 of the Consumer Rights Protection Law. It is necessary to enshrine a general rule on the nullity of unfair terms of a consumer contract, giving the court the competence to assess any terms of a consumer contract for the balance of interests of the parties in the new version of Article 16 of the Consumer Protection Law to protect the weaker party.

6.4 The problem of consumers' greatest vulnerability in the digital environment

As the OECD notes, most vulnerable consumers are consumers who are susceptible to harm at a given time because of the market characteristics of a particular product, the quality of the product, the nature of the transaction or the circumstances in which the consumer finds himself [19].

The United Nations Conference on Trade and Development (UNCTAD) highlighted that personal vulnerabilities include characteristics that expose the consumer to a significant risk of harm [20]. For example, a consumer may be most vulnerable because of their age, race, ethnicity or gender; low level of education or literacy; limited language skills; mental health problems; physical disabilities; geographical distance/residence in a low-density region; unemployment or low income [21]. Personal characteristics such as gullibility, impulsivity, risk aversion, poor calculation skills can also make consumers insecure [22]. Vulnerability can also result from personal circumstances such as bereavement, divorce or a period of illness [23].

Certain market characteristics such as lack of competition, insufficient information and/or market complexity can make consumers more vulnerable in particular markets [23-24]. In very complex markets, such as the financial services markets, even the most sophisticated consumer can feel insecure [23]. In such markets, consumers are often guided by simple rules of thumb or heuristic thinking, ignore certain opportunities or choose not to make consumer choices [21].

Certain qualities of the product may also cause the greatest vulnerability to consumers. For example, these could be complex products incorporating the Internet of Things and artificial intelligence technologies [25]. A lack of experience with online services that collect and use consumer data, combined with a lack of user-friendly privacy controls, leave many consumers vulnerable to privacy risks on the information and telecommunications network of the Internet.

Finally, consumers may be vulnerable because of the nature of the transaction. This could be, for example, a transaction involving mobile payments and in-game transactions, which are often made on-the-go, via small screens and may involve limited authentication controls [26-27]. There have been cases of unauthorized purchases made through voice-activated digital assistants [25, 28].

The Russian legal reality is currently actively introducing smart contracts (self-executing contract, self-executing transaction) that use blockchain technology [29]. There is the use of smart contracts in settlement relations, such as the settlement of letters of credit [30]. The use of computer technology in contracting and contract enforcement, in particular smart contracts, can also contribute to consumer vulnerability, as it implies knowledge of scientific and practical provisions related to both the law and technical sciences at the same time.

A number of factors indicate that it is in the digital environment that consumers are most vulnerable. All consumers have no protection from unfair commercial practices in the digital age, as sellers can take advantage of the limitations of a particular medium and transmit misleading advertising or pre-contractual information. The online environment provides new ways in which businesses can prey on behavioural biases, such as the framing effect, loss aversion and overconfidence. When shopping online, consumers suffer the detriment of misleading reference prices, bundled pricing, time-limited offers, scarcity signals and default settings. Since online businesses can collect consumer data and personalize offers and prices, it could be argued that all consumers are potentially vulnerable to digital marketplace manipulation.

However, we believe that the most vulnerable categories of consumers in a digitalized environment are children and teenagers, and the elderly, due to a lack/absence of critical judgement, literacy or relevant skills. We also include consumers with limited access to information and communication technologies (ICT) and little experience with the digital environment as a socially vulnerable group.

1. Children and teenagers. Despite their early experience of the Internet, children and teenagers are vulnerable in the digital environment because they lack the critical judgement necessary to determine the circumstances where they may be exposed to harm. This lack of critical judgement leaves them vulnerable to harm, especially when the commercial nature of the content is not disclosed or masked, as in advertising games, subscription traps, negative publicity, etc. [31-32]. They are unable to resist targeted behavioural advertising when using apps and online games. However, Facebook, Instagram and Snapchat report advertising audience growth in 2020 among users aged 13 to 17 [33].

In addition, many children and young people lack the literacy and/or skills to understand the meaning of the information they encounter online, especially relating to privacy and the use of personal data. We believe that the practice of forcing children and teenagers to consent to the use of their data for marketing purposes to access online services should be prohibited. Children and teenagers cannot give informed consent because they are not required to read and understand the very complex terms/confidentiality notices.

The widespread use of smartphones, digital assistants and other mobile devices by children and teenagers poses additional challenges. Parents or guardians find it more difficult to supervise children and teenagers who manipulate with these devices. For example, when there are no adequate authentication and payment controls, children and teenagers may make purchases without the consent of their legal representatives [27]. Lack of supervision for minors and underage people can lead them to view inappropriate commercial content, such as advertisements for prohibited items [35].

2. Older consumers. Compared to children and teenagers, older consumers tend to have less experience with the Internet and have lower ICT skills. As a result, they are more likely to face security risks. In the UK, for example, consumers aged 65 and over often fail to check the reliability of an online website before posting their credit card details [23].

Older people are more susceptible to misleading marketing practices and often fall victim to online dating scams, romance scams, lotteries, inheritance, investments [23]. After all, they tend to be lonely, more trusting and more likely to suffer from vulnerabilities such as health conditions, physical disabilities, mental illness or cognitive impairment, making them vulnerable to scammers [23].

Research shows that older people are most vulnerable in mainstream markets (such as energy, water, telecommunications, financial services) and markets for special services for older people (e.g. care homes, funeral homes, pension funds) [23]. While these markets are not new to the digital age, digital innovation has transformed many of them so that older consumers can only get their benefits online. In addition, given that some older consumers have limited ability to use online services, business policies that provide access to and/or discounts on basic services only to online consumers are infringing on the rights of older people.

According to the Federal State Statistics Service (Rosstat), the elderly are twice as likely as young people to use a computer. Half of the older generation uses the Internet, and only a third of them are active users. They communicate on social media (57%), search for information related to health topics (49%), use messengers to make calls and video calls (46%) and analyse data about products and services (45%). Only 10% of older consumers use the internet to order goods or services online. Their most popular services include financial transactions, including banking services, money transfers, insurance services, securities transactions (46%), clothes and shoes (36%), and household goods (26%). 73% of older people surveyed pay for their purchases using bank cards. Only just over a quarter of people over 55 actively use online access to public services. The most demanded services are those related to health care and medicine (35%), payment of taxes and fees (17%), and housing and utilities obligations (14%) [36].

3. Consumers with limited access to ICTs. To participate effectively in e-commerce, consumers need access to quality communication infrastructures as well as the appropriate skills to navigate the online environment. Lack of access to ICT and experience with the digital environment affects consumer vulnerability.

At the beginning of 2020, more than 4.5 billion people are using the Internet. Almost 60% of the world's population is already online. In Russia, the number of internet users in 2020 is 118 million. This means that 81% of Russians use the Internet. However, barriers remain that prevent people all over the world from having fair and equal access to the digital environment. Today, more than 40% of the world's population, some 3.2 billion people, are not yet connected to the Internet. More than a billion "unconnected" people live in South Asia (31% of the total). African countries account for 27%. There is a correlation between Internet access and the age of users in these regions: more than half of Africa's under-20s and more than 460 million people under 13 in South Asia are not online.

There is an inequality between the number of men and women on the Internet. Women are less likely to be online than men. The gap is particularly wide in developing countries. For example, women in South Asia today are three times less likely to use social media than men. More than half of the women now living in India are not aware of the existence of mobile Internet at all. This inequality results from a range of factors, including access, cost, lack of education, skills and technological literacy, as well as innate gender biases and socio-cultural norms. The main reason for this imbalance lies in deeply rooted social norms and traditions. Regardless of the cause, the number of "unconnected" people will depend to a large extent on increasing digital accessibility for women, especially in developing countries [33].

There is a gap in Internet access between rural and urban areas, with rural areas tending to lag urban areas in being able to use the Internet at high speeds.

Thus, lack of access and/or ICT skills leads to a digital divide, which is significant for people in rural areas, older people, women, people in developing countries, less educated people, and people with disabilities and the poor [23]. As businesses do not always develop digital goods and services tailored to the specific needs of the listed minority groups, it is necessary to classify them among the most vulnerable categories of consumers in the digital environment.

6.5 Ways to address consumer protection in the digital age

Analysing the possible ways of protecting consumer rights in the digital environment, the authors conclude that it is necessary to equally protect the rights of consumers of all social groups, including through digital technologies, while establishing special protections for socially vulnerable groups.

The Russian Federation has now developed an effective mechanism for consumer protection authorities to apply enforcement measures. As of 1 October 2019, a new form of judicial protection for consumer rights has become class action lawsuits [37]. Rospotrebnadzor is developing online dispute resolution mechanisms - new effective forms of consumer protection.

However, we believe that targeted policies to encourage consumer-oriented business practices, investment in ICT access, and education and awareness campaigns can help protect and empower consumers in the digital age.

1. The incentive policy way. Some foreign countries developed and successfully implemented specific policies targeting vulnerable consumers to protect them in the online environment. The Children's Online Privacy Protection Act of October 21, 1998 [COPPA] [38] prohibits the collection, use and disclosure of personal information about children on the Internet without parental consent. The law applies to operators of commercial websites and online services (including online advertising) aimed at children under the age of 13. The UK encourages entrepreneurial policies focused on markets for older people and other innovations for socially vulnerable consumers [23].

Improved business processes, better business practices by entrepreneurs will contribute to consumer protection when problems arise in a particular market or for a particular group of vulnerable consumers. In recent years, abroad there were several initiatives to self-regulate internet marketing aimed at children and other disadvantaged consumers [31]. Entrepreneurs themselves have proposed effective tools to help vulnerable consumers. In the UK, for example, some banks have allowed consumers to block dubious transactions [23]. There are free online tools that consumers can use to manage their online spending. Tools have been introduced to prevent children from making unauthorized purchases and viewing prohibited content online, which provide authorization and/or age verification of the user [26].

We believe that the state should encourage entrepreneurs who consider the challenges faced by the most vulnerable categories of consumers in the digital environment. And entrepreneurs should take the lead in making online goods (services) available to as many consumers as possible [23]. Market monitoring and law enforcement need to be improved to identify and combat unfair commercial practices related to consumer vulnerability in the digital environment.

2. Increasing access to ICTs and developing digital competencies. We believe the first harbinger of digital consumer empowerment is adequate ICT skills and literacy as well as accessibility to ICTs. There is worldwide recognition of the need for ICT literacy. Most countries teach children and young people ICT literacy in schools. While building digital competences is largely the responsibility of education authorities, we believe that consumer protection agencies can also play a role in developing targeted education and ICT literacy programmes.

Russia is currently implementing a programme of new digital professions from the government as part of the federal programme "Human Resources for the Digital Economy". Free training is available in over 20 areas that are in demand in the digital economy. In 2020, residents from 48 regions of the country participated in the programme and received 33,000 personal digital certificates [39].

3. Targeted information and awareness-raising campaigns. In some foreign countries, consumer protection agencies have developed and implemented information and awareness-raising campaigns for older consumers. For example, the Federal Trade Commission https://ru.wikipedia.org/wiki/%D0%90%D0%BD%D0%B3%D0%BB%D0%B8%D0%B9%D1%81%D0%BA%D0%B8%D0%B9_%D1%8F%D0%B7%D1%8B%D0%BA(FTC) is implementing projects to protect older people from online fraud [40]. It also has a number of educational resources for children and young people on the Internet. Similarly, the European Commission developed a Children's Internet portal with numerous resources for children and teenagers in the digital age *(EC, n.d.)*.

When running information and awareness-raising campaigns, the state should consider the importance of making them available in formats that meet the needs of vulnerable consumers. For example, in 2018, the Canadian Competition Bureau published the second edition of its educational guide on how to avoid cheating in the eight different languages spoken by new Canadians [41].

Russia also runs information and awareness-raising campaigns. For example, the Rospotrebnadzor website contains more than 75,000 materials, including 25,000 about products that do not comply with mandatory requirements, 360 information leaflets and guides, such as a leaflet to prevent deception of older people purchasing goods and services[42].

7 Conclusion

7.1 The study proves the hypothesis that consumers of digital goods (services) are less protected than consumers of traditional goods (services) and shows that, at this stage of the digital economy, consumer rights in digital transactions are not sufficiently protected, primarily at the legislative level. A huge number of irregularities result from the lack of a specific regulatory framework governing e-commerce. In this context, there is an emerging trend towards autonomous consumer protection on the basis of self-regulation by e-commerce players without state intervention.

7.2 It was revealed that although Rospotrebnadzor takes an active part in the discussion of international recommendations and approaches at the WTO, OECD, UNCTAD, etc., the work on improving domestic legislation in consumer protection field does not always consider the best international practices. The Rospotrebnadzor's 2020 draft amendments to the Consumer Rights Protection Law need improvement, as they do not consider the EU experience in regulating unacceptable terms of the contract with the consumer that infringe on their rights. By analogy with the practice in European countries, we propose that the new version of Article 16 of the Consumer Rights Protection Law should contain a general rule on the nullity of unfair consumer contract terms, giving the court jurisdiction to assess any consumer contract terms for the balance of interests of the parties to protect the weaker party.

7.3 We substantiate that all consumers are potentially vulnerable in the digital economy. A number of factors point to the greater vulnerability of consumers in the digital environment, as it implies not only legal literacy but also certain technical skills: the emergence of non-traditional digital sales channels and remote interaction, new objects of civil rights (digital products, smart electronics, Internet of Things, personal data sets, etc.), the use of smart contracts and blockchain technology in contract negotiation and enforcement. However, the most digitally disadvantaged categories of consumers are children and teenagers, the elderly, consumers with limited access to ICTs and lack of experience with them.

Toward addressing consumer protection in the digital economy, the authors propose to ensure equal consumer protection for all social groups, including through digital technologies, while establishing special protections for socially vulnerable groups through public policies to encourage entrepreneurship, increase access to ICT and develop digital competencies, as well as targeted information and awareness-raising campaigns. The main objective of consumer protection in the digital environment is to reduce consumer risks.

References

1. World consumer rights day 2017 calls for protecting consumer rights in the digital age (2017), URL: https://zppdon.ru/news/9234/ (date of access: 16.03.21).
2. Government of the Russian Federation. *The decree of the Government of the Russian Federation "On approval of Strategy of state policy of the Russian Federation in the field of consumer protection for the period till 2030"* (August 28, 2017 No. 1837-r). Moscow, Russia (2017)
3. Federal service for supervision of consumer protection and human welfare [Rospotrebnadzor]. Consumer protection in the Russian Federation in 2018: State report. Moscow, Russia (2019a)
4. Federal service for supervision of consumer rights protection and human welfare [Rospotrebnadzor] (2020a), URL: http://www.consultant.ru/cons/cgi/online.cgi?req=doc&base=PRJ&n=199634#07191586993096075 (date of access: 16.03.21).
5. Russian Federation. Law of the Russian Federation "On consumer protection" (February 7, 1992 No. 2300-1). Moscow, Russia (1992)
6. V. Naumov, Recommendations for organizing the activities of individuals in the field of Internet Commerce in the Russian Federation (2000), URL: https://www.osp.ru/ecom/2000/07/13031267 (date of access: 16.03.21).
7. Draft Federal law No. 310163-4 "On electronic Commerce" (2006), URL: www.consultant.ru/cons/cgi/online.cgi?req=doc&base=PRJ&n=42158#028127057741049954 (date of access: 16.03.21).
8. Organization for Economic Cooperation and Development [OECD]. (2016), URL: https://dx.doi.org/10.1787/9789264255258-en
9. I.B Ilovaysky, Y.Y Kayl, D.A. Tokarev & V.A. Usanova *Institutional Model for the Protection of Rights of the Parties Concerned in the Sale of Goods and Services with the Use of Information Technologies.* (Switzerland, Cham: Springer Science + Business Media, 2019)
10. Federal service for supervision of consumer protection and human welfare [Rospotrebnadzor]. *Consumer protection in the Russian Federation in 2019: State report.* Moscow, Russia (2020b)
11. Plenum of the Supreme Court of the Russian Federation. *The resolution of Plenum of the Supreme Court of the Russian Federation "About consideration by courts of civil cases on disputes on protection of consumer rights"* (June 28, 2012 No. 17). Moscow, Russia (2012)
12. World trade organization [WTO] (1994), URL: http://ivo.garant.ru/#/document/2541542/paragraph/650:0 (date of access: 16.03.21).

13. United Nations Commission on international trade law [UNCITRAL]. (1996), URL: https://base.garant.ru/2555780/ (date of access: 16.03.21).
14. Y.A. Vasilyeva, *Risks for consumers of logistics services in the conditions of digitalization of society* (Saint Petersburg, Russia: Kult-inform-press publishing house, 2019)
15. E. Krivtsov The law "On consumer protection" is being changed to suit judicial practice and the views of officials. Legal. Repert. (2020), URL: ://legal.report/zakon-o-zashhite-prav-potrebitelej-izmenyajut-pod-sudebnuju-praktiku-i-vzglyady-chinovnikov/ (date of access: 16.03.21).
16. Council of the European Communities [EEC] (April 5, 1993 No. 93/13/EEC) (1993), URL: https://base.garant.ru/2565749/ (date of access: 16.03.21).
17. Plenum of the Supreme Arbitration Court of the Russian Federation. (2014). *Resolution of the Plenum of the Supreme Arbitration Court of the Russian Federation "Freedom of contract and its limits"* (March 14, 2014 No. 16, Moscow, Russia)
18. Russian Federation. (1994). *Civil code of the Russian Federation (part one)* (November 30, 1994 No. 51-FZ, Moscow, Russia)
19. Organization for Economic Cooperation and Development [OECD]. (2014a), URL: http://www.oecd.org/sti/consumer/Toolkit-recommendation-booklet.pdf (date of access: 16.03.21).
20. United Nations Conference on Trade and Development [UNCTAD]. (2018), URL: https://unctad.org/meetings/en/Presentation/WG%20Vulnerable%20and%20Disadvantaged%20Consumers%20.pdf (date of access: 16.03.21).
21. Organization for Economic Cooperation and Development [OECD]. (2010), URL: https://dx.doi.org/10.1787/9789264079663-en (date of access: 16.03.21).
22. European Commission [EC] (2016), URL: https://ec.europa.eu/info/sites/info/files/consumer-vulnerabilityfactsheet_en.pdf (date of access: 16.03.21).
23. Competition and Markets Authority UK [UK CMA]. Consumer vulnerability: challenges and potential solutions. (2019), URL: https://www.gov.uk/government/publications/consumer-vulnerability-challenges-andpotential-solutions/consumer-vulnerability-challenges-and-potential-solutions(date of access: 16.03.21).
24. Consumer Affairs Victoria [CAV]. What do we mean by vulnerable and disadvantaged consumers? (2004), URL: https://www.consumer.vic.gov.au/library/publications/resources-andeducation/research/what-do-we-mean-by-vulnerable-and-disadvantaged-consumersdiscussion-paper-2004.pdf (date of access: 16.03.21).
25. United Nations Conference on Trade and Development [UNCTAD]. Working Group on Vulnerable and Disadvantaged Consumers Contribution, Intergovernmental Group of Experts on Consumer Law and Policy. (2018), URL: https://unctad.org/meetings/en/Presentation/WG%20Vulnerable%20and%20Disadvantaged%20Consumers%20.pdf (date of access: 16.03.21).
26. Organization for Economic Cooperation and Development [OECD]. Report on Consumer Protection in Online and Mobile Payments. Paris, France: OECD Publishing. (2012), URL: https://doi.org/10.1787/20716826(date of access: 16.03.21).
27. Organization for Economic Cooperation and Development [OECD]. Consumer Policy Guidance on Mobile and Online Payments. *OECD* (2014b), URL: https://dx.doi.org/10.1787/5jz432cl1ns7-en(date of access: 16.03.21).
28. Organization for Economic Cooperation and Development [OECD] (2019a), URL: https://one.oecd.org/document/DSTI/CP(2019)10/FINAL/en/pdf/(date of access: 16.03.21).
29. E.N. Agibalova, *Scientific and Technical Revolution: Yesterday, Today and Tomorrow. ISC 2019. Lecture Notes in Networks and Systems* (Switzerland, Cham: Springer, 2020).
30. E.N. Agibalova, I.B. Ilovaysky, Y.Y Kayl & V.A. Usanova, *Scientific and Technical Revolution: Yesterday, Today and Tomorrow. ISC 2019. Lecture Notes in Networks and Systems* (Switzerland, Cham: Springer, 2020)
31. Organization for Economic Cooperation and Development [OECD]. *Good practice guide on online advertising.* (2019b), URL: https://doi.org/10.1787/20716826 (date of access: 16.03.21).
32. United Nations Children's Fund [UNICEF]. *Children and Digital Marketing: Rights, risks and responsibilities.* (2018), URL: https://www.unicef.org/csr/css/Children_and_Digital_Marketing_-_Rights_Risks_and_Responsibilities.pdf (date of access: 16.03.21).
33. Sergeeva Y. *All Internet statistics for 2020 – figures and trends in the world and Russia.* (2020), URL: https://www.web-canape.ru/business/internet-2020-globalnaya-statistika-i-trendy/ (date of access: 16.03.21).
34. Organization for Economic Cooperation and Development [OECD]. Consumer Policy Guidance on Mobile and Online Payments. OECD Digital Economy Papers, 236. (2014b), URL: https://dx.doi.org/10.1787/5jz432cl1ns7-en (date of access: 16.03.21).
35. UK Council for Child Internet Safety [UKCCIS]. Children's online activities, risks and safety: A literature review by the UKCCIS Evidence Group. (2017), URL: https://assets.publishing.service.gov.uk/govern

ment/uploads/system/uploads/attachment_data/file/650933/Literature_Review_Final_October_2017.pdf (date of access: 16.03.21).
36. Federal state statistics service [Rosstat]. Results of a sample Federal statistical survey on the use of information technologies and information and telecommunications networks by the population. (2019), URL: https://rosstat.gov.ru/folder/313/document/63640 (date of access: 16.03.21).
37. Russian Federation. Federal law of "On amendments to certain legislative acts of the Russian Federation" (July 18, 2019 No. 191-FZ, Moscow, Russia)
38. The Children's Online Privacy Protection Act of October 21, 1998 [COPPA]. URL: https://www.govinfo.gov/content/pkg/PLAW-105publ277/pdf/PLAW-105publ277.pdf (date of access: 16.03.21).
39. New digital profession from the state (n.d.). URL: https://цифровойсертификат.рф/?utm_source=fb_wp&utm_medium=cpc&utm_campaign=teacher&utm_term=stat&fbclid=IwAR065dBxOY_lPfwEm9Cd7s0jPN4Um_i0imcdb1O1-Ey9BNUBszibGMp44BY (date of access: 16.03.21).
40. Federal Trade Commission US [US FTC]. Protecting older consumers 2017-18: A report of the Federal Trade Commission. (2018), URL: https://www.ftc.gov/system/files/documents/reports/protecting-olderconsumers-2017-2018-report-congress-federal-tradecommission/protecting_older_consumers_-_ftc_report_10-18-18.pdf (date of access: 16.03.21).
41. Competition Bureau Canada. The Little Black Book of Scams 2nd edition. (2018), URL: https://www.competitionbureau.gc.ca/eic/site/cb-bc.nsf/eng/04333.html (date of access: 16.03.21).
42. Federal service for supervision of consumer rights protection and human welfare [Rospotrebnadzor]. Recommendations to citizens: how to avoid becoming a victim of fraud? (2019b), URL: https://www.rospotrebnadzor.ru/about/info/news/news_details.php?ELEMENT_ID=10961&sphrase_id=1821993 (date of access: 16.03.21).

SHS Web of Conferences **109**, 01003 (2021) https://doi.org/10.1051/shsconf/202110901003
LISID-2020

Transformation of the principles of international private law in the digital age

Mariia Aleksandrina[1,*]

[1] Kulikova st. 36/2-21, Astrakhan, Russia, e-mail: aleks9999.72@inbox.ru

Abstract. The article focuses on the problem of transformation of the content of the principles of international private law in the context of digital development of the state, society and economy. The implementation and transformation of the principles of international private law in relation to the main types of cross-border public relations related to digital technologies are studied. It is established that in the context of the development of digital technologies, the most obvious transformation of the principles can be traced precisely in the process of concluding and executing smart contracts. A smart contract does not fit into the traditional criteria for determining either a civil contract or a foreign economic transaction. There is no uniform legal understanding of this phenomenon. It is noted that the content of the norms regulating modern cross-border private law relations related to the use of digital technologies should be built (along with other industry categories) on the basis of the principles of international private law, which, in turn, are transformed in connection with the needs of modern development of digital technologies and correspond to them.

1 Introduction

The rapid development of digital technologies has predetermined the formation of a new so-called digital information society, which requires appropriate changes in legal regulation. The structural and substantive reconstruction of the whole legal system involves not only the adaptation of existing legal norms to the features of public relations in a digital society (including an extended interpretation of their content), but also the further formation of digital law as an independent branch. Since the digital space and relations in a digital society often go beyond the borders of a single state, the problem of international legal regulation in this sphere is in the foreground. International private law is able to adapt earlier than other branches to the new conditions of digital development of the state, society, and economy, as it is characterized by more flexible methods of regulation, the lack in most cases of strict binding to national law, and the possibility of non-state regulation of relations in accordance with the principles of freedom of contract and autonomy of the will of the parties. In this regard, the research of the problem of transformation of the principles of international private law in the context of the development of the state, society and economy is of great theoretical and practical importance.

2 Problem Statement

In the course of this research, the following tasks are to be solved:

Determination of the system of principles of international private law in the context of the development of digital technologies;

Identification of changes in the content of the principles of international private law;

Study of the results of the transformation of the system of principles of international private law in accordance with the needs of practice;

Highlighting the features of the implementation of the principles of international private law in the process of regulating cross-border private law relations related to the use of digital technologies.

3 Research Questions

In the context of the development of digital technologies, the most obvious transformation of the principles can be traced in the process of concluding and executing smart contracts. A smart contract ("intellectual" contract, "clever" contract) is a self-executing contract based on the blockchain platform, concluded in electronic form, in which a mathematical code is used to enter all the terms of the contract into the blockchain. A smart contract does

* Corresponding author: aleks9999.72@inbox.ru

not fit into the traditional criteria for determining either a civil contract or a foreign economic transaction. There is no uniform legal understanding of this phenomenon. The question of the mechanism for forming the terms of a smart contract remains is not clear. The place of conclusion, the personal law of the participants, the place of execution is difficult to determine. The problem of determining the applicable law to such contracts is related to the adaptation of the known conflict of laws principles exclusively to traditional civil law contracts. The problems of determining the place of conclusion of a smart contract, the place of execution, and the law of the country with which it is most closely connected are especially characterized for such contracts. The fundamental postulates, and, above all, the principles of international private law, do not find their expression in the traditional understanding. The problems of doctrinal formulation, legal consolidation of the principles of international private law, and the use of digital technologies in the field of international private law relations are paid attention both in domestic science [1-3]and in foreign studies [4-6].

4 Purpose of the Study

The purpose of the research is to study the problem of transformation of the content of the principles of international private law in the context of digital development of the state, society and economy.

5 Research Methods

The article uses the following methods: dialectical method, comparative legal method of research, method of system analysis.

6 Findings

The principles of international private law are the doctrinally formulated basic ideas that are formed in the course of the development of international private law relations and follow from the content of industry sources and their general meaning, which contribute to the achievement of the goals of legal regulation of international private law relations and in accordance with which the functions of international private law are performed.

The branch principles of international private law in various sources include: firstly, the principle of autonomy of the will of the parties to international private law relations (lex voluntatis)[7-8], the principle of the closest connection of legal relations [7,9] (at the same time, they are one of the main principles of conflict-of-laws regulation) and the principle of freedom of contract (the latter is concretizing in relation to the principle of autonomy of will), the territorial principle (the principle of applying the court's own law (lex fori) [10] the principle of conflict-of-laws regulation of private law relations with a foreign element (Tolochko, 2001), the principle of priority of the norms of international treaties over national legal norms [5] and the obligation of their execution [9] traditionally recognized in the scientific literature and enshrined in legal norms.

And also among the principles, for example, N. V. Oksyutchik highlights the principle of duality of sources of international private law, that is, the complex regulation of private law relations with a foreign element based on sources of international law and national law (in the literature, the principle of combining international and national regulation is highlighted) [4]; the principle of dispositivity (the possibility fixed in international legal sources for the parties to private-law relations with a foreign element to refuse to use them as regulators of these specific relations); the principle of combining state and non-state forms of regulation of international private law relations (that is, the "established order" that operates over a long period (lex mercatoria)) (for example, N. V. Oksyutchik formulates the principle of the application of non-state forms of regulation of legal relations) [4], however, the recognition of *lex mercatoria* as a legal regulator has not yet received mass distribution and is controversial [11]; fourthly, the principle of equality of the state as a participant in international private law relations and other subjects of international private law in accordance with the concept of functional immunity.

In addition, the literature highlights the principle of applying the norms of foreign law on the basis of conflict of laws rules or the agreement of the parties, taking into account the restrictions established in the imperative norms of national law and the public order clause [12]. Baryshev V. A. formulates two independent principles: the principle of mandatory application of foreign law, if it is required by conflict of laws rules or the parties have chosen it to regulate legal relations, and the principle of protecting the national law and order from the norms of foreign law that are incompatible with the established public legal consciousness and morality [9]. Akimova Yu. M. names two interrelated principles: the principle of application of foreign law and the principle of protection of domestic law and order [13].

In addition to the above-mentioned principles, the literature points to such principles of international private law as the principle of granting certain regimes and the principle of reciprocity [7]. In addition, it seems appropriate to add the principle of ensuring the execution of international private law obligations (ensuring the freedom of movement of goods and financial resources; compliance with guarantees to foreign investors) and the principle of conscientious execution of international private law obligations (conscientious competition and compliance with antimonopoly legislation), as well as the principle of personal data protection and information protection.

In the context of the development of digital technologies, the most obvious transformation of the principles can be traced in the process of concluding and executing smart contracts.

A smart contract ("intellectual" contract, "clever" contract) is a self – executing contract based on the blockchain platform, concluded in electronic form, in which a mathematical code is used to enter all the terms of the contract into the blockchain. Thus, the blockchain contains a special algorithm that allows you to automatically track the execution of participants (signatories) of their obligations.

A smart contract does not fit into the traditional criteria for determining either a civil contract or a foreign economic transaction. There is no uniform legal understanding of this phenomenon. For the first time, the term "smart contract" was legislated in the Decree of the President of the Republic of Belarus of December 21, 2017 No. 8 "On the development of the Digital Economy" [14]. The practical application of smart contracts remains outside the legal field due to their specifics, and it is impossible to distinguish the legal features of a computer program. Therefore, the original agreement of the parties containing the basic conditions, which has acquired the form of a digital code, is subject to regulation. Therefore, based on the analysis of foreign legislation and practice in the literature, it is proposed to consolidate the status of a smart contract as a way to formalize contractual obligations in the normative legal acts on contracts at the same time as making changes to the procedural legislation and normative legal acts on information technologies [15]. Federal Law of the Russian Federation No. 259-FZ of 31.07.2020 "On Digital Financial Assets, Digital Currency and on Amendments to Certain Legislative Acts of the Russian Federation" does not define a smart contract [16]. However, in the draft of this regulatory legal act, a smart contract was defined as "an contract in electronic form, the execution of rights and obligations under which is carried out by automatically performing digital transactions in a distributed register of digital transactions in a strictly defined sequence of such an contract and upon the occurrence of circumstances determined by it" [17].

Smart contracts do not originate from the parties, but as a result of the creation of a program based on digital technologies. Third parties with special names and special functions participate in their creation, control, execution and confirmation of transactions. The parties only conclude a smart contract, while simultaneously confirming their identity (and sometimes even this is not required). The question of the mechanism for forming the terms of a smart contract remains is not clear. The place of conclusion, the personal law of the participants, the place of execution is difficult to determine. The problem of determining the applicable law to such contracts is related to the adaptation of the known conflict of laws principles exclusively to traditional civil law contracts. The problems of determining the place of conclusion of a smart contract, the place of execution, and the law of the country with which it is most closely connected are especially characterized for such contracts. The distributional nature of the obligation contained in the smart contract implies the application of the personal law of each of the participants equally, as well as the participant of the blockchain [2]. Therefore, the traditional principles (the principle of conflict of laws regulation; the principle of autonomy of the will of the parties, based on the conflict of laws principle of determining the applicable law lex voluntatis; the principle of dispositivity, which implies the possibility of excluding the application of certain norms of international treaties and international customs; as well as the principle of combining state and non-state forms of regulation of international private law relations, which consists in the possibility of using business practices in order to regulate contractual relations, that is, the so-called "routine order" that operates over a long period (lex mercatoria)) as the basis of legal regulation in relation to smart contracts, they are transformed into one general principle of automated determination of a more favorable right for a smart contract and autonomous supranational regulation. And in this form, these principles are preferably laid down in the algorithm for determining the applicable law when formulating the terms of a smart contract. At the same time, the so-called lex electronica system of self-regulation of relations on the Internet, based, among other things, on blockchain technology, is allowed to exist [1].

Since, first of all, just as in the case of concluding a traditional complex or unnamed contract, a conflict of laws problem may arise in connection with its qualification as an international commercial (cross-border) transaction [18], a similar problem may accompany smart contracts. In practice, the parties to smart contracts, in accordance with the principle of autonomy of will, often choose the most preferred legal systems that are most favorable to smart contracts. But if the nodes of the distribution registry network are located in the zones of different jurisdictions, there is a splitting of the treaty statute, which makes it much more difficult to choose the applicable law [2]. Building a strict algorithm for formulating the terms of a smart contract makes this process easier and faster, and also allows you to solve the conflict of laws problem. But to do this, the algorithm should be based on the principle of automated determination of a more favorable right for a smart contract (including in view of the possible use of smart contracts to ensure compliance with legal contracts). However, first "many technical and legal issues need to be resolved, which will require time and interdisciplinary discussion between lawyers and software developers" [16].

The quality of the contract is determined primarily by the use of clear and unambiguous terms, phraseological constructions that exclude the possibility of ambiguous interpretation.Arbitrary interpretation of certain regulatory requirements, on the basis of which the content of the contract is based, and its terms become impossible in the case of using smart contracts [18].

Qualification of smart contracts as accession contracts made under a suspensive condition [19], does not deprive them of fundamental differences from traditional civil law contracts. In this case, due to the lack of a single text expression of the contract, which does not allow us to reproduce the content of the transaction in the usual form, the formulation and fixing of the terms of the contract within the usual limits do not seem real. Smart contracts are more appropriate to use when concluding mass standard contracts. But if it is necessary to individually agree on the terms of the contract and, at the same time, the transfer of values expressed in bitcoins is not required, the creation of a smart contract is not substantiated [20]. Thus, the fundamental postulates, and, above all, the principle of freedom of contract, do not find their expression in the traditional understanding. This explains the need for a more specific definition of the boundaries of contractual regulation of relations arising on the basis of digital technologies, as opposed to traditional contractual relations [21].

The global nature and, in general, the unity of digital technologies used, including for the creation of smart contracts, can contribute to the formation of the principle of unified international legal regulation of relations in the sphere of digital space on the basis of the following principles of international private law: the principle of priority of the norms of international treaties over national legal norms and the obligation of their execution, as well as the principle of duality of sources of international private law, that is, the principle of comprehensive regulation of private law relations with a foreign element based on sources of international law and national law (in other words, the principle of combining international and national regulation).

The implementation within the framework of the digital space of the principle of equality of the state as a participant in international private law relations and other subjects of international private law in accordance with the concept of functional immunity does not cause any special problems. The program does not provide a different legal regime for users, and sometimes it depersonalizes them.

The principle of ensuring the execution of international private law obligations (especially in terms of ensuring the freedom of movement of goods and financial resources) and the principle of conscientious execution of international private law obligations in smart contracts receive the most guaranteed implementation due to the features of their execution. Smart contracts are self-executing contracts. If the conditions specified in the contract are met, an automatic transaction is made. The blockchain platform is built on the distribution principle. And even if there was a loss of data about the transaction for some technical reasons, the information does not completely disappear, and the payment can be confirmed by other participants. This is achieved by combining transactions into groups-blocks, from which a continuous chain is subsequently formed. At the same time, a distribution register is most acceptable for record-keeping of assets and their movement [23]. Provided that smart contracts are properly implemented, they can provide transaction security that surpasses traditional contract law. The costs of coordinating audit and ensuring compliance with such agreements are thus reduced [6] Therefore, the above-mentioned principles are implemented, among other things, through the self-fulfillment of obligations characteristic of digital law, based on a distributed registry system.

Regarding the payment method, the traditional contract provides for its possible options: payment is made by transferring cash to the seller (contractor, etc.); by depositing funds in a notary's deposit; by bank transfer (by payment orders, by letter of credit, by collection, by cheques, etc.). In addition, the contract may specify the time when the payer executes the obligation to pay for goods (works, services): the time when funds are credited to the correspondent account of the recipient's bank, the time when funds are debited from the payer's current account, and the time when funds are credited to the recipient's current account. This is one of the differences from smart contracts, which can only work with assets located in the digital ecosystem. And the lack of legal regulation of the turnover of cryptocurrencies is the reason for limiting their turnover. For the same reason, the corresponding uniform judicial practice has not developed [23]. In fairness, it should be noted that the first steps in this direction are being taken at the national level. On January 1, 2021, Federal Law No. 259-FZ of 31.07.2020 "On Digital Financial Assets, Digital Currency and on Amendments to Certain Legislative Acts of the Russian Federation" [16] had been enacted. However, it does not contain the concept of a smart contract. And in this regard, the above-mentioned principles of ensuring the execution of international private law obligations (especially in terms of ensuring the freedom of movement of goods and financial resources) and the conscientious execution of international private law obligations are expressed and implemented, including in the norms that fix the requirements for the activities of information system operators and operators of the exchange of digital financial assets [16].

Cryptocurrencies are used to make payments when executing the terms of smart contracts. But recently, the use of cryptocurrencies has been questioned, and bitcoin has been severely criticized due to the constant increase in the cost of maintaining the underlying blockchain system [24]. Other disadvantages of cryptocurrencies include: the lack of official exchange procedures, the impossibility of canceling transactions and returning funds in case of fraud, the vulnerability to hacking of cryptocurrency payment systems, and the lack of guarantees for the execution of transactions [25]. In addition, as practice has shown, there are no criteria on the basis of which it is possible to at least approximately calculate the possible exchange rate of cryptocurrencies. But, on the other hand, this payment method is currently more convenient and faster in relation to smart contracts. Its attractiveness is especially evident in the case of the supply chain implementation.

Payment systems based on blockchain technology continue to develop. The most well-known bank settlement system is SWIFT. However, recently it has shown vulnerability to hacker attacks. We can name the GPSG system, which is considered as a competitor to SWIFT, the new CLS Not platform, etc.[26]. In this regard, the implementation of international settlements on the basis of the principle of freedom and security of the movement of financial resources becomes relevant.

In the digital age, problems related to the protection of personal information and the identification of the subjects of relations arising through the use of the Internet also come to the foreground. At the international level, the principle of personal data protection is implemented in the provisions of the Convention for the Protection of Physical Persons in the Automated Processing of Personal Data of January 28, 1981. (as amended on 15 June 1999) [27], which enshrines that any person should be given the opportunity to know about the existence of an automated personal data file and the name, place of residence or location of the file controller, to obtain confirmation of whether his data is stored in an automated file, to seek, if necessary, correction or destruction of such data, to resort to legal remedies in cases of violations of rights in this sphere. In general, the Convention aims to protect physical persons from abuse in the collection and processing of personal data and to regulate the cross-border flow of personal data. Russia joined the Convention by signing its Protocol on October 10, 2018. And at present, the Convention on the Protection of Physical Persons in the Automated Processing of Personal Data of January 28, 1981. (as amended on June 15, 1999), known as "Convention 108", remains the only global international treaty in the field of personal data protection.

On the basis of the principle of personal data protection, the problem of identifying subjects should be solved: in which cases identification is possible and necessary; the right to use a pseudonym (network name); the right to oblivion and "digital death". The literature suggests two solutions to this problem: to recognize a virtual person who cannot be identified as an independent subject of private legal relations arising on the Internet; to recognize personal data as an object of civil law [28].

7 Conclusion

Modern needs determine the creation of a completely new system of legal norms regulating relations that are based on digital technologies. The content of such norms is built (along with other industry categories) on the basis of the principles of international private law, which, in turn, are transformed in connection with the needs of modern development of digital technologies and correspond to them. The system of principles of international private law, the implementation and transformation of individual principles in accordance with modern trends in the development of digital technologies and their application in various private law areas require special research.

References

1. N.V. Lukyanov, Legal Studies **11**, 28-35 (2018)
2. M.V. Mazhorina, Lex Russica 7, 93-107 (2019)
3. N.V. Mamaeva, Science and education today **2 (25),** 6-7 (2018)
4. N.V. Oksyutchik, Journal of International Law and International Relations **1-2 (84-85)**, 13-17 (2018)
5. O.N. Tolochko, Belarusian Journal of International Law and International Relations **2** (2001), URL: https://elib.bsu.by/bitstream/123456789/29310/1/2001_2_JILIR_tolochko_r.pdf (date of access: 16.03.21).
6. Sherman, Voshmgir. Token Econoimy: How the Web 3 reinvents the Internet (English) Taschenbuch – 5, (2020) URL: https://blockchainhub.net/blog/blog/token-economy-book-shermin-voshmgir/ (date of access: 16.03.21).
7. I.V. Getman-Pavlova, International private law: a short course of lectures (Moscow: Publishing house Yurayt, 2014)
8. D.P. Strigunova, Bulletin of the Moscow University of the Ministry of Internal Affairs of Russia **1,** 47-51 (2017)
9. V.A. Baryshev, International private law. (Minsk: "Vysheyshaya shkola", 2019)
10. A.V. Asoskov, Fundamentals of Conflict of Laws (Moscow: Logos, 352, 2017)
11. A.N. Novikova, Social Sciences **12 (24)**, 59-63 (2013)
12. D.P. Strigunova, Modern law **4**, 153-158 (2014)
13. Yu.M. Akimova, Actual Problems of Russian Law **5 (30),** 600-606 (2013)
14. President of the Republic of Belarus Decree "On the Development of the Digital Economy" (21 December 2017 No 8). Minsk, Belarus. URL: https://www.economy.gov.by/uploads/files/sanacija-i-bankrotstvo/Dekret-PrezidentaRespubliki-Belarus-ot-21-12-2017-N-8-O-r.pdf (date of access: 16.03.21).
15. N.B. Krysenkova, International public and private law **5**, 28-30 (2019)
16. State Duma of the Russian Federation Federal Law "On Digital Financial Assets, Digital Currency and on Amendments to Certain Legislative Acts Of the Russian Federation" (July 31, 2020 No. 259-FZ). Moscow, Russia. URL: http://www.consultant.ru/document/cons_doc_LAW_358753/ (date of access: 16.03.21).
17. State Duma of the Russian Federation Federal Law "On Digital Financial Assets, Digital Currency and on Amendments to Certain Legislative Acts Of the Russian Federation" (May 17, 2018). Moscow, Russia. URL: https://sozd.duma.gov.ru/bill/419059-7(date of access: 16.03.21).
18. M.V. Mazhorina, Journal of the Higher School of Economics **2**, 143-159 (2016)
19. M.L., Davydova, & M.Yu. Kozlova, Bulletin of the Nizhny Novgorod Academy of the Ministry of Internal Affairs of Russia **3 (47),** 113-118 (2019)
20. M., Aleksandrina, & M.Y. Kozlova, (2021) "Smart Contracts' vs Legal Technology in Contract Practice. Lecture Notes in Networks and Systems **155**, 1204-1212
21. A.V. Efimov, Journal of Russian law 7, 52-63 (2020)

22. E.V. Voblenko, *Accounting of digital transactions: dissertation for the degree of Candidate of Economic Sciences:* (St. Petersburg State University of Economics, S.-P., , 2020)
23. O.S., Grin, E.S Grin & A.V. Soloviev, Lex Russica **8 (153),** 51-62 (2019)
24. D.V. Lipnitsky, Industrial Economics **1,** 82-100 (2019)
25. S. Yu. Pertseva, State Councillor **2 (26),** 58-63 (2019)
26. V.E. Ponomarenko, Legal Science **5,** 113-121 (2017)
27. Council of Europe Convention for the protection of physical persons in the automated processing of personal data (28 January 1981, 15 June 1999) Strasbourg, France. URL: http://www.consultant.ru/document/cons_doc_LAW_121499/637cbbf46b43d8a53a38d1319556d0beae26d7de/ (date of access: 16.03.21).
28. E.V. Talapina, Journal of Russian Law **2,** 5-17 (2018)

Peculiarities of development of digital financial ecosystem

Aleksandr Almosov [1], *Svetlana* Bondareva [2*], *Oksana* Darelina [3], *Sofya* Potomova[4] and *Elena* Timacheva[5]

[1] Russian Presidential Academy of National Economy and Public Administration, Volgograd Institute of Management, Gagarina Street, 8, Volgograd, vsamol@yandex.ru
[2]Russian Presidential Academy of National Economy and Public Administration, Volgograd Institute of Management, Gagarina Street, 8, Volgograd, s.a.bondareva@mail.ru
[3] Russian Presidential Academy of National Economy and Public Administration, Volgograd Institute of Management, Gagarina Street, 8, Volgograd, darelina@yandex.ru
[4] Russian Presidential Academy of National Economy and Public Administration, Volgograd Institute of Management, Gagarina Street, 8, Volgograd, potomova@inbox.ru
[5] Russian Presidential Academy of National Economy and Public Administration, Volgograd Institute of Management, Gagarina Street, 8, Volgograd, borisova_e@mail.ru

Abstract. The development of information society based on creating digital information channels fosters the expansion of virtual interaction of the participants of social relations. It gives rise to technological transformation of financial infrastructure and fosters related changes in other spheres of economy and social life. This research is aimed at finding out peculiarities of forming and developing digital financial microsystems and determining possible scenarios of the development of the Russian national financial ecosystem. The analysis of the statistical data has shown that developing digital technologies leads to the transformation of financial services consumer behaviour. They move from off-line environment with cash payments to the on-line service with non-cash settlements predominance and distant interaction. The trust in fintech-companies and digital financial services is constantly increasing. The caried-out analysis has shown that digitalization of financial sector increases its financial accessibility and fosters poverty reduction (Wang ,Yang, Chiu & Lin, 2020) and provides economic growth (Siddik & Kabiraj, 2020). It forms the basis for the transformation of financial architecture and the expansion of digital financial ecosystems in Russia. On the basis of world experience as well as Russian practice, the authors determined the elements and principles of digital financial ecosystem development, identified types of forming and levels of ecosystems functioning.The authors considered two variants of forming the national digital financial ecosystem that differ in the form of interaction of state and private ecosystems functioning. The authors considered two variants of forming the national digital financial ecosystem that differ in the form of interaction of state and private ecosystems. The obtained conclusions are important for making project decisions by the authorities, financial organizations and technological companies.

1 Introduction

At present country's competitiveness and economic development are mostly determined by the level of economy digitalization. The new institutional environment is being formed, so called digital environment. It is based on the transformation of the information and channels of its transmission into digital form that fosters the expansion of virtual interaction of social relations participants.

Dynamic development of new technologies and their penetration in all spheres of human activity served as the driver of the digital transformation of institutional environment. It served as the driving force for fast fundamental changes in world economy and structural industrial reorientation. Such large-scale changes were interpreted as the fourth industrial revolution – Industry 4.0.

Simultaneous significant changes in social sphere are based on the reassessment of social values. Contemporary generation actively uses new technologies creating permanent demand for innovative personalized goods, services and technologies. These conditions cause the development of economic sharing at the same time the market is becoming the subject to mass personalization. It implies that human activity on the one hand is adapted to technological changes and on the other hand differences in individual preferences are the source of current changes.

* Corresponding author: s.a.bondareva@mail.ru

Thus, new industrial environment implies forming a digital ecosystem by means of digitalization of all physical assets, information and financial resources aimed at integration of the participants of economic relations into digital space for interaction, decision making and regulating. These trends influence the development of the financial sector.

2 Problem Statement

Technological transformation of financial infrastructure initiates and fosters related changes in other spheres of economy and social life. Development of digital technologies leads to the transformation of behaviour of financial services consumers. They move from off-line environment with cash payments to on-line service with predominance of non-cash settlements with distant interaction. This tendency can be observed in Russia in the context of transactions by means of payment cards. Over the last decade the share of "card transactions" related to payment for goods and services has almost tripled reaching 91% in the total volume of card transactions. At that the number of card transactions has increased over 10 times mostly due to their use for paying for goods and services (Figure1).

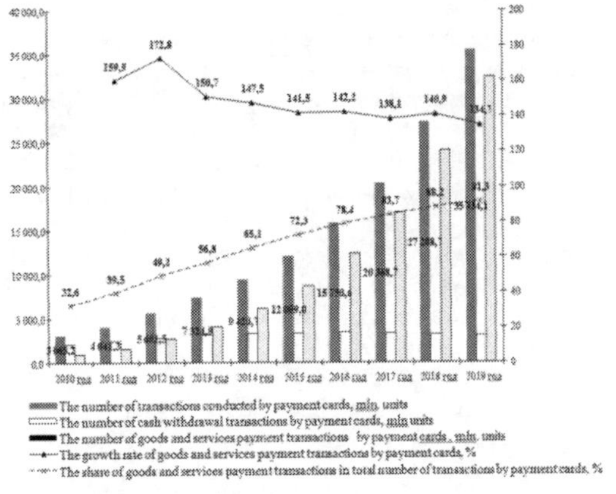

Fig. 1. Dynamics and structure of card transactions in Russia over the period 2010-2019.
The Source: Bank of Russia

Gradual refusal from cash is caused by the development of payment infrastructure, increasing financial literacy of the population and penetration of digitalization in all spheres of activity. According to the research made by National Agency for Financial Studies over a half of Russians (56%) use online banking. In 2020 the share of users of mobile banking has increased up to 51% (in 2018 - 26%) [1]. The experts point to a rather high level of digital technologies penetration into the Russian financial market [2].

One of the determinants of the structural changes in the financial sector is changing consumers' trust. The demand for traditional banking service is reducing among young generation that actively consumes services and goods of technological companies. As it has been shown by the researchers the consumers of the platform products Apple, Amazon, Alibaba will agree to receive banking services from these companies (if they provided them) hoping to get a related level of service [3]. In many respects it can be explained by increasing digital financial literacy of the population, forming skills of online interaction, scales of using and expanding of smartphones' functionality, developing e-commerce.

3 Research Questions

This research is aimed at answering the following questions:

1. How does the banking sector change in the context of digital expansion?
2. What elements does digital financial ecosystem include?
3. What principles are fundamental for forming and developing digital financial ecosystem?
4. What types of digital financial ecosystems exist?
5. What variants of development of Russian national digital financial ecosystem are possible?

4 Purpose of the Study

The relevance of the study is determined by the fact that digital financing and providing financial accessibility aimed at poverty reduction and economic growth are becoming high priorities of the governmental, business and scientific focus. In this respect the authors set the goal to study peculiarities of forming and developing ecosystem as the form of interaction in digital environment and expanding financial accessibility. In order to reach the set goal the following objectives were set and achieved; we studied the factors of digitalization; considered the factors of digitalization of financial services; found out advantages of digital mode of interaction in the financial market; researched the models of creating digital banks; identified elements of digital financial ecosystem and principles of its development; characterized different types of digital financial ecosystems; predicted scenarios of development of Russian national financial ecosystem.

5 Research Methods

Methodology of the research is determined by the complexity and diversity of the challenges met that requires using different methods and tools. The present research has been conducted by means of integrating forms, methods and logical operations of the research, using analyses and synthesis abstraction and concretization comparison, generalization and systematization, methods of descriptive statistics and statistical analysis, data analysis and interpretation in graphs and figures.

6 Findings

In conditions of increasing demand in the market on-line service is being developed, digital financial innovations including digital financial products, digital channels and business-models are introduced, As a result digitalization of financial relations can be observed: increasing use of digital financial innovations along with the consuming of financial services and managing financial resources.

New technologies in financial sphere can provide accessibility, effectiveness and safety of the services rendered, increase intuitive ability of services, decrease transaction costs. Platform solutions for digital transactions can be used by users with different levels of income without any time and territory restrictions. Digital financial services significantly reduce transaction costs, raise effectiveness of risk-oriented pricing and risk management, reduce information asymmetry, expand the range of financial products and increase transparency of activity of financial organizations as well as consumers of financial services. In addition using technological innovations in inclusive financing give rise to competition between financial organizations that fosters quality improving and reducing the cost of financial goods and services [4].

Creating new models of expanding financial services (external agent networks, banks without the network of branches), new opportunities for consumers' access and personal online control are some of the directions of developing the financial sector which is still in the process of constant innovation search and adaptation. The phenomenon of financial technologies has created new opportunities for financial organizations as well as for consumers of financial services. on the basis of speed, flexibility, accessibility and diversity.

Financial innovations have a significant impact on sustainable development and can be considered as an instrument for economic integration [5]. Mobile payments play a vital role in promoting financial accessibility and are important instruments providing the growth of the standard of life even among the disadvantaged population [6]. At the same time the researches prove that developing financial technologies is significantly dependent on the growth of economy and opportunities of the financial sector [7].

In current conditions banks actively implement technological service novelties and develop online and mobile banking trying to save their positions in the market. However, renewing the old-fashioned IT-infrastructure of a classical bank and meeting all the requirements of the regulating authority demand significant costs that in its turn is reflected on the cost of services. The practice shows that the fastest and the most costless project is launching start-ups of fintech-companies or digital banks (neobanks).

1. Developing by the classical bank additional digital brand. Such model was used by Singapore bank OCBC having established FRANK, by Spanish CaixaBank having established LKXZ, Financial corporation "Otkrytie" having established Roketbank and Tochka Bank.

2. Using a digital channel for resaling traditional banking products on the platform of the company independent from the bank. The examples of such model are Simple (USA), Moven (USA), Talkbank (Russia).

3. Creating of social digital financial organization as subcompany of a classical bank. This variant was chosen for example for creating Hello Bank (subcompany of French BNP Paribas) and Touchbank (included in the financial group OTP Group).

4. Creating initially independent digital bank using only digital channels of service, the pattern of Fidor Bank (Germany), Tangerine (Canada), Tinkoff Bank (Russia), Modulbank (Russia).

5. Creating neobank by a popular fintech-company for making additional profit by means of using payment services and financial platforms with a wide client base. Such practice can be observed in the activity of the companies dealing with e-commerce (Amazon, Alibaba), mobile communication (Megafon), software providing (Apple).

It should be noted that development of any digital ecosystem is determined by the extent of integration of financial technologies. It can be explained by the fact that modern economy is significantly "financialized" [8] and differs by its high level of monetization of social relations and BigData. In modern ecosystem financial technologies are a compulsory component, consequently the ecosystem based on the combination of digitalization and financed should be considered as a digital financial system.

Digital financial ecosystem is a whole complex of interrelated elements of organization of functional and institutional financial interrelations of economic subjects. Its creation can be initiated by some economic subjects: state structures (governmental authorities, central bank, institutions), business-structures (fintech-companies, banks, corporations), non-commercial organizations, communities and social groups. Financial ecosystem plays a pivotal role in allocation of capital, reaching an agreement by depositors and investors that is especially important in case of limited sources of financing [9]. However, it should be noted that often financial services and applications are taken by the consumers for granted and don't appear to be a competitive advantage and don't foster the inflow of new participants into the ecosystem.

The world experience shows that without providing conditions for development of finances at nano-economic level it is impossible to provide sustainable rates of economic growth. Digital financial ecosystems facilitate realization of resource potential of population in economy that is made up by human capital, entrepreneurial activity, and personal investments [10].

A digital financial system is developed on the basis of a number of principles.

"One stop-one-shop principle" which characterizes the ecosystem as a specific integral complex where all necessary things for comprehensive functioning of economic subjects are concentrated in one place. This principle facilitates satisfying different individual needs by means of one online resource (multifunctional digital platform). The services and applications can be provided by the developer of the digital platform as well as integrated by other partner companies. On the basis of mutually favourable cooperation the ecosystem accumulates financial (in the first turn payment) services as well as Internet-trade services, public services, social interaction, research and obtaining information and etc.

The principle of mutually favourable cooperation («win-win» principle) implies that all participants of the ecosystem get a positive effect as e result of their interaction. The owner besides direct economic effect (profit growth and costs reducing) has an opportunity to get additional information about participants which enables him to improve a business-model of the ecosystem as well as goods and services. The users of the ecosystem obtain extended financial, information and social accessibility, higher speed of transactions, effectiveness of decision making and lower transaction costs. The partners providing third-party applications and services obtain an expanded client base and increased market for trading goods and services.

The principle of openness derives from the above-mentioned principles of ecosystem and implies the opportunity of its integration with other services, digital products, data bases. In practice digital ecosystem can be closed as well with limited functionality and number of participants. The more open is the ecosystem the more opportunities it has for development including global scale.

The principle of ecosystem's flexibility determines opportunities for meeting individual needs. The essence of digital financial ecosystem implies personalization of services and meeting the needs of every actor. Every actor is unique and ecosystem constantly adjusts to specific needs. Moreover, modern society as well as institutional environment are permanently transformed that is why development of the ecosystem depends on its ability of fast adaptation to changing values, technologies, institutes.

The principle of distant identification is determined by the specifics of digital technologies enabling distant access to resources and interaction. Different identification methods can be used. At present the most popular ones are unique codes and access passwords, but in perspective the following methods will become popular: active using biometric data, telemetry, technologies of faces recognition with the help of artificial intelligence.

Principle of trust in digital environment is the basis for forming and developing a digital financial ecosystem. It is trust based on individual approach and effective system of security that creates competitive advantages and opportunities for developing a digital ecosystem. Long ago economists highlighted trust as a key factor in decision making [11, 12]. The individual's trust to other participants of financial relations facilitates expanding new financial instruments, increasing speed and functionality of financial transactions by means of active usage of new financial technologies [13]. In the era of digital technologies new cost is created due to the trust and the last can be considered as a special currency.

The principle of simplicity is determined by human inherent motivation to minimize efforts aimed at reaching the intended effect. Simplicity of using a digital platform enables consumers to make research, compare and assess different variants of supply in the financial market. Due to it they find themselves in a more favourable position based on empirical knowledge. Simplifying financial relations including understandable financial services results in reducing risks and more effective control. [13, 14].

Digital financial ecosystem has a complex and dynamically developing structure. The compulsory elements of a digital financial ecosystem are presented in Figure 2.

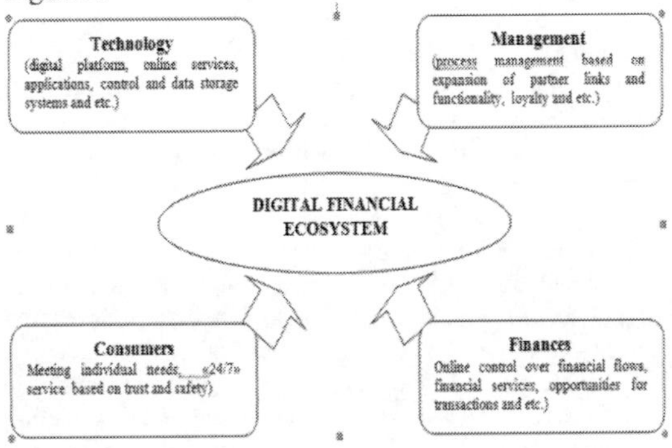

Fig. 2. Elements of digital financial ecosystem
The source: provided by the authors.

In the contemporary world there are many digital financial ecosystems. Ecosystem as business model lay the foundation for the development of such companies Sber. Yandex, Apple, Amazon, Alibaba and others.

In the world market ecosystem of the group Alibaba is leading by the number of users. Created in 2003 [15] payment service of Alibaba company Alipay for payment in industrial e-commerce at present has over 1 billion users all over the world and continues its international expansion [16]. To the Kapron, Z. [17] this company that has changed the character of retail consumption in China can take on a global market of digital payments [17]. The peculiarity of development of the ecosystem of Alibaba Group is the inflow of users due to world expansion of e-commerce platforms. The basis of this ecosystem is a world famous service of mobile payments Alipay which includes opportunities of paying for purchases, money transfers, communal payments including payments by QR-code. The directions of development of Alibaba Group is on the one hand are connected with broadening its functionality by means of providing a full range of financial services, mobile system of organizing medical aid, charity projects in which participants can follow how their money was used and other applications. On the other hand the ecosystem of Alibaba is developing due to technical innovations such as the system of detecting images with the help of artificial intelligence which enables to evaluate the damage of the car, the cost of car repairmen and provide information about the nearest automobile service centers and the cost of their services with the help of photo send to the insurance agent [16].

By the diversity of its key spheres including telecommunications, financial services, FoodTech, e-commerce telemedicine, education, transport, cloud services, media and entertainments, estate, IT-services, analytics and others. The prominence can be given to

Russian ecosystem Sber. However, this ecosystem doesn't have a unified digital platform at the moment ("one-stop-shop approach" doesn't work). Many consumers even don't know that DocDoc, Delivery, Okko, Citymobil, Domclick are elements of one ecosystem, and Sberb should still work in this direction.

The whole complex of digital financial ecosystems functioning in the country makes up the national digital financial ecosystem. In Russia further development of this sphere will be connected with active development of ecosystems of public organizations. (governmental authorities) as well as with increasing competitiveness of private ecosystems. The positive examples of public ecosystems are the "Portal of State Services" and the system of quick payments of the Bank of Russia. In order to expand payment transactions the Bank of Russia announced its decision to issue national currency in digital form – digital ruble [18]. It proves the increasing activity of the public sector in the digital environment.

Private digital financial ecosystems functioning in Russia can be divided into universal and niche. Forming universal ecosystems is aimed at providing consumer with a whole range of services for meeting existing and potential needs in one application. Yandex, Tinkoff can be referred to such ecosystems. Niche ecosystems develop a definite range of functions necessary for a particular group of consumers or for a particular sphere of activity. For instance, AlfaBank creates an ecosystem for representatives of small and medium business, Rosbank has concentrated its efforts on the perspective in their opinion niche - buying residential estate, Rosselkhozbank made a presentation of the ecosystem for farmers. Private ecosystems are created in order to attract the consumer and to make additional profit as well as to create entrance barriers for new participants.

In practice all ecosystems intersect, use external channels for promoting and rendering their services. Development of Russian national digital financial ecosystem will be based on generating the model of interaction of public and private ecosystems. There are at least two possible variants of interpenetration (Figure 3).

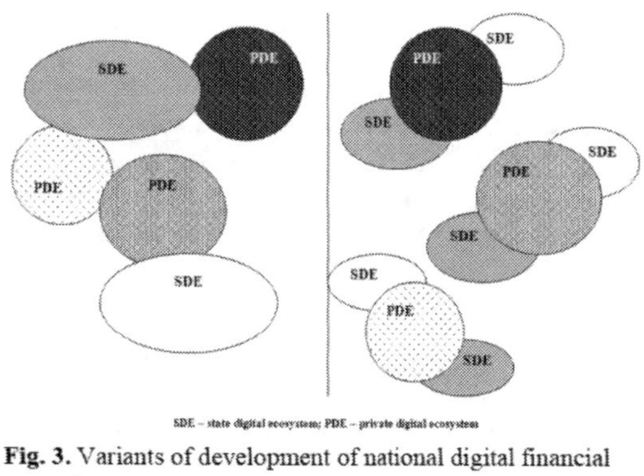

Fig. 3. Variants of development of national digital financial ecosystem
The source: Provided by the authors

The first variant implies autonomous functioning of state and private ecosystems with expanded functionality of them and different extent of interpenetration. At that state ecosystems will dominate in the sphere of public service. The advantages of this variant for state administration are that it saves the unity of the system of data in public sector, providing safety, control and managing BigData in public sector. However, in case of this approach state platforms will significantly stand behind private ecosystems by its functional and technological state. Besides, there will be duplicating of functions by a number of ecosystems [19].

The second variant is possible if ecosystems of state structures become applications of the largest private ecosystems. In this variant public services will be dissolved in the services of private ecosystems. The consumer will have an opportunity of receiving public services at any place and any time without leaving the private ecosystem. The main disadvantages of such variant are the loss of an unified channel of service and obtaining information as well as increasing information security risks.

As the world experience shows if the ecosystem is aimed at consumers values and interests it succeeds and develops permanently. If the digital financial ecosystem has the priority of fast accumulation of profit striving to increase the number of consumers at the expense of consumes' interests it has no potential for development and it doesn't facilitate financial accessibility and consequently development of economy and society as shows Kenya's experience [20].

Within the crisis caused by COVID-19 the advantages and opportunities of digital ecosystem only increase. Functional and technological specifics facilitate mitigating of negative consequences of social and economic crises caused by the pandemic as it gives an opportunity to solve urgent problems to use different services and purchase in social distancing [21]. In this respect the strategic priorities in development of Russian national financial ecosystem in conditions of a new reality are the following: (1) providing sufficient liquidity for market functioning and demand maintaining; (2) intensification of the exchange of information related to healthcare; (3) inclusive financing; (4) using digital channels for obtaining services and making payments for reducing contacts between people, and in case of necessity supporting people having no access to digital technologies; (5) directing financial resources to improve digital infrastructure in order to provide qualitative and permanent work of digital channels of communication.

Conclusion

Success and effectiveness of the Russian national digital financial ecosystem directly depend on new and productive forms of integration of public and private ecosystems for the benefit of consumers. Development of the digital ecosystem at the macrolevel increases economic stability and financial accessibility. Innovations in the sphere of digital finances have a long-term positive impact on financial market. Digital transactions increase transparency of economy reducing informal sector of economy, unofficial payments and circulation of false banknotes. Speed, simplicity and accessibility of financial services positively influence social welfare reducing the level of poverty and expanding opportunities for the realization of entrepreneurial initiative.

References

1. NAFI. Think Tank. Analytics and predictions. (2020), URL: https://nafi.ru/analytics/bolee-poloviny-rossiyan-polzuyutsya-tsifrovym-bankingom/ (date of access: 16.03.20)
2. RBC. Neobanks and lifetime-advisers: banking innovations of the nearest decades (2020), URL: https://trends.rbc.ru/trends/futurology/5f4d3aa99a794781e4b9ae68 (date of access: 16.03.20)
3. E. Fiyaksel, I. Sokolov, A. Solokhin, I. Kutepov, E. Privalova, E. Rozhkova, & L. Lyakisheva Centre for Strategic research (2018), URL: https://publications.hse.ru/mirror/pubs/share/direct/219492205.pdf (date of access: 16.03.20)
4. Md. Hasan, Y. Morshadul, Lu, Khan, & Shajib. Global Business Review (2020)
5. D. Salampasis, & A.-L. MentionFinTech: Harnessing innovation for financial inclusion. In Handbook of blockchain, digital finance, and inclusion 1st ed., **2** (2018)
6. G-24. Fintech for financial inclusion: A framework for digital financial transformation. AFI Special Report. (2018), URL: https://www.g24.org/wp-content/uploads/2018/09/G-24 (date of access: 16.03.20).
7. F. Guo, S. Kong & J. Wang, China Economic Journal **9(3),** 253–271 (2016)
8. S. Dubinin, Economic theory **21 (4)** 6-11 (2017)
9. S. Bose, G. Dong, & A. Simpson, The Financial Ecosystem. Palgrave Studies in Impact Finance **1**, 19-46 (2019)
10. V. Slepov, S. Bondareva, Finances and Credit **48** 2-17 (2016)
11. J. Keynes, *Sellection* (Moscow: Eksmo 2007)
12. F. Fukuyama, *Trust: social moral virtues and the way to prosperity* (Moscow, 2004)
13. V. Slepov, & S. Bondareva, *Personal finances: textbook / edited by prof.* (Magistr: INFRA-M Moscow, 2019).
14. R. Feniks, *Financial services: reloding* (Ivanov and Ferber, Mann, 2012)
15. Vedomosti. How Alibaba designed a model of smart business. URL: https://www.vedomosti.ru/management/articles/2018/11/01/785430-alibaba (2018) (date of access: 16.03.20).
16. Z. Kapron, Journal of Payments Strategy and Systems **12,** 68-73 (2018)
17. Bank of RussiaReport for public consulting "Digital ruble" (2020), URL: https://cbr.ru/eng/analytics/consultation-paper/the-consultation-paper-a-digital-ruble/ (date of access: 16.03.20).
18. O. Peterson, Research and Business Development Department **18**, 329-340 (2018)
19. Q. Wang, Y. Jinbao, Chiu Yung-ho & Lin Tai-Yu. Managerial and Decision Economics **41 (7)** 1225-1236 (2020).
20. D. Arner, J. Barberis, J. Walker, R. Buckley, & D. Zetzsche, Digital Finance & Crisis. SSRN Electronic Journal, **17** (2020)

Employment structure transformation through economic activity types perspective

Alexandr Almosov[1], *Yulia* Brekhova[2], *Elena* Malysheva[3], *Viktor* Molokanov[4], and *Ekaterina* Chumakova[5*]

[1] Volgograd Institute of management – branch of The Russian Presidential Academy of National Economy and Public Administration, 8 Gagarin street, Volgograd, Russia, almosov-ap@vlgr.ranepa.ru
[2] Volgograd Institute of management – branch of The Russian Presidential Academy of National Economy and Public Administration, 8 Gagarin street, Volgograd, Russia, brekhova-yv@ranepa.ru
[3] Dean of the faculty of Economics Volgograd Institute of management – branch of The Russian Presidential Academy of National Economy and Public Administration, 8 Gagarin street, Volgograd, Russia, decanatec@vlgr.ranepa.ru
[4] Associate Professor at the Department of Economics and Finance Volgograd Institute of management – branch of The Russian Presidential Academy of National Economy and Public Administration, 8 Gagarin street, Volgograd, Russia, vikmol@inbox.ru
[5] Associate Professor at the Department of Economics and Finance Volgograd Institute of management – branch of The Russian Presidential Academy of National Economy and Public Administration, 8 Gagarin street, Volgograd, Russia, chumakova-ea@vlgr.ranepa.ru

Abstract. The present study was conducted to identify the reasons in modern information society that caused significant changes in the labor market particularly in its industrial structure through the perspective of individuals employed in real and service sectors. The analysis of the employed population's statistical sample by economic activity type in enlarged groups of industrial and non-industrial spheres of activity over the past 20 years has shown that scientific and technological transformations have led to a significant shift in the employment structure. The research used statistical data comparison of economic activity types and the ratio of the regions employed in real and service sectors. New types of economic activity are emerging and start prevailing inherent exclusively in the current state of economy in the context of digitalization. A significant structural shift has been revealed in the course of the analysis of the employed population by the enlarged industry affiliation, which demonstrates diametrically opposite positions in the overall structure of employment. The resulting ratio allows us to propose and apply the coefficient of the employment structure, demonstrating a multiplier effect on identifying such jobs creation and preservation of which leads to further employment provisioning. The obtained tool allows managing population employment as well as related social and economic processes as migration outflow and population growth in all age groups, population reproduction, expansion and improvement of educational and health services in a particular territory. In addition, application of the developed employment structure tool allows estimating the amount of investment required to solve these issues.

1 Introduction

Labor market issues cannot leave on an arm's length neither federal nor regional or municipal authorities and citizen, since the changes at this "testing ground" are so tangible that they affect all components of the market from the amount of individual citizens' wages to macroeconomic employment indicators across both region and country [1].

It has become very interesting recently to study the labor market situation through economic transformation perspective as it was not yet the case several decades ago. In addition, the study of the labor market through the population employment perspective is always relevant since it is one of its main components in the workforce.

The study of the Russian Federation recent past reveals a radical change in approaches to the study of employment, its interpretations, legislative consolidation, and structure in various dimensions [2]. The authors believe that it makes sense to change the existing research vectors for new ones that take into account many factors of shifting reality. In the course of transition from total and universal employment typical for the Soviet economy to currently used employment assessment with unemployment as its constant component.

Therefore, nowadays in the context of significant changes in society with information priorities and dominants through the digitalization of the economy perspective, the scientific community cannot ignore employment and unemployment phenomena that

* Corresponding author: chumakova-ea@vlgr.ranepa.ru

influence the very basic characteristics positioning in labor market [3], [4].

2 Problem Statement

The interest of researchers and economists in employment study through various perspectives and contexts is obvious today. A number of current studies illustrate the following aspects:

- correlation between social and labor relations development and labor and entrepreneurship transformations with a focus on displacing a large number of the working-age population due to economy modernization [5], [6], [7];
- numerous barriers to the labor market from a regulatory and legal point of view and the corresponding instruments for its regulation [8], [9];
- statistical data on changes in the sex-age structure of the labor market [10], [11].

Assessment of population employment and its structure by industry, the scientific community only states statistics in dynamics, indicating its significant change, but in this direction the structural changes in employment by type of economic activity by segments of the production and non-production sectors of the economy as a result of its transformation which is undoubtedly an urgent area within the framework of solving a wide range of tasks related to labor market issues.

3 Research Questions

The study was aimed at getting answers to the following questions:

1. What structural changes in labor market have occurred from the transformation of the economy over the past decades?

2. What allows to assess the current ratio of the employed by type of economic activity?

3. What additional group of employed people can be identified in the service sector where jobs are financed by the government?

4 Purpose of the Study

The scientific inquiry made it possible to determine the goal of the study which centers on the development of the coefficient of the structure of population employment which makes it possible to assess how many jobs in the real sector of economy lead to the creation of work places in the service sector, as well as to assess the multiplier effects on the labor market of a particular territory using the example of the Volgograd region. In order to accomplish this goal the following tasks were set: to analyze employed population change of the Volgograd region by enlarged groups of types of economic activity; to propose an indicator describing the structure of population employment; to distinguish between the proposed coefficient of population structure of employment in certain types of economic activity, the creation of jobs in which is created by the state; to offer a lever of employment management in the labor market of a separate territory.

5 Research Methods

In the framework of this study, classical general scientific research methods were applied: economic and statistical methods of collecting and processing information, comparison, analogy, analysis, synthesis, induction and deduction, and graphic method.

6 Findings

In this paper, the analysis of employment of the population is carried out at the level of the Russian Federation as a whole and at the regional level in particular (using the example of the Volgograd region) through the perspective of the impact of economic modernization on the structure of employment, namely on significant changes in the structure of employment by industry for enlarged economic activity types.

Such changes are illustrated by the graphic cross which indicates that the last few decades the shares of individuals employed in the real sector of the economy and those employed in the service sector in the overall structure of employment of the population have taken diametrically opposite positions.

For the analyzed period of time a variety of completely new types of services and, as a result, new types of economic activity appeared. The production modernization has led to the release of able-bodied citizens from their jobs while the growth rates of labor productivity for different types of activities in the manufacturing sector of the economy are very different, but the obvious leader here is the agriculture sector where labor productivity has increased 20 times [12]. In these conditions, the transformation of the labor market and the structure of employment by industry seem logical.

Within the framework of this study, the assessment of the structure of employment of the population was carried out for enlarged subgroups of types of economic activity in accordance with the all-Russian classifier of types of economic activities of the Russian Federation - the real sector of the economy (from "Agriculture" to "Construction") and the service sector (from "Wholesale trade" and retail to "Activities in the field of culture and sports").

This approach is justified since the study of the structural components of the employed labor force according to the enlarged groups of economic activity in the Russian Federation as a whole from 1990 to 2019, and in the Volgograd Region from 1995 to 2019 in the graphical representation has a diametrically opposite relationship with building up the existing trend.

Let us denote those employed in the real sector of the economy by A, and those employed in the service sector by B and demonstrate the dynamics of employment of the population in the manufacturing and non-manufacturing sectors in the Russian Federation as a whole and in the Volgograd region (Figures 1, 2).

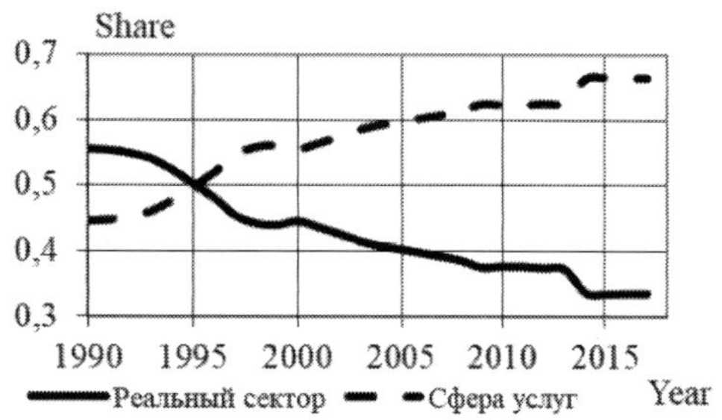

Figure 1. Change in employment ratio in manufacturing and non-manufacturing sectors in Russia [13]

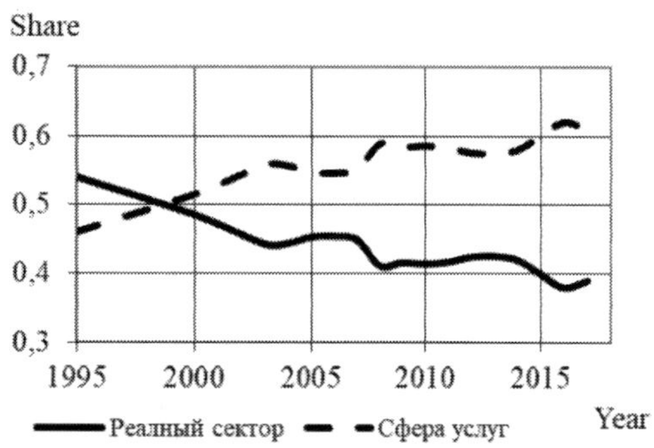

Figure 2. Change in the ratio of employment in manufacturing and non-manufacturing sectors in the Volgograd region [13]

Due to the formation of such a "cross" it can be recommended not only to preserve the change in the shares of employment in the real sector of the economy and in the service sector against the background of each other, but to offer a tool that allows one to combine two structural elements into one.

Let us denote the proposed tool through the coefficient of the structure of employment (the ratio of employed in "B" to those employed in "A" showing the final ratio of employed in the manufacturing and non-manufacturing sectors.

According to official statistics, in 1990 across the country as a whole 41,794 thousand people worked in the manufacturing sector and 33,531 thousand people worked in the non-production sector with an employment structure ratio of 0.8, then by the end of 2019, 25,071 thousand people worked in the manufacturing sector and in the non-productive sector there are 47 593 thousand people with an employment structure coefficient of 1.9 [14].

To put it another way, today there are 2 jobs in the non-production sector per worker in the manufacturing sector whereas 27 years ago, the value of this coefficient on the contrary was 1.2 jobs in the manufacturing sector per one job in the service sector.

Employment data by industry for 1995-1996 show approximately equal shares, 0.99 and 1.07, respectively, when one job in one consolidated group of types of economic activities of a production nature also had 1 job in the group of non-production sectors.

With a decrease in the total number of employed people in Russia by 2,661 thousand people (minus 3.5%) over the analyzed period from 1990 to 2019, the values in absolute and relative terms for employment in the real sector of the economy and for employment in the service sector have fundamentally changed.

Thus, in the real sector of the economy there was a decrease in jobs from 41,794 thousand people to 25,071, the retirement amounted to 16,723 thousand people or 59.99%, and in the service sector, the growth of employment was 14,062 thousand people or 41.94% increasing from 33 531 thousand people in 1990 to 47 593 thousand people in 2019.

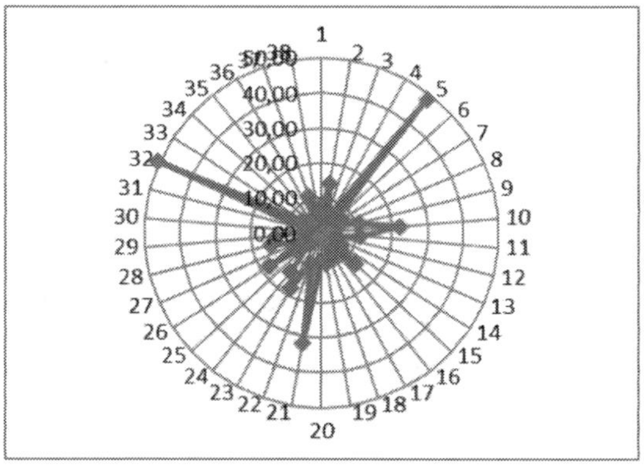

Figure 3. Employment structure ratio of municipalities of the Volgograd region [13]

The additional detailing of the structure of employment in the service sector is an important direction of this research as it highlights in a separate group "B" those employed in the spheres, 95 percent financed by the government is education; activities in the field of health and social services; public administration and military security and social security. In such conditions, the layout of the employment structure coefficient for the Volgograd region indicates that there are 4 jobs in the service sector per worker in the real sector of the economy, 2 of which are financed by the state.

As a result of the development of the structure of employment coefficient as well as its detailing and the allocation of a separate segment, an effective lever for managing the employment of the population appears both in the municipality, region, and country as a whole. The proposed tool is applicable to any territory which is its versatility.

This multiplier effect makes it possible to identify in which particular area of economic activity it is worth attracting investments in order to maintain employment in the main industry and related industries [15].

7 Conclusion

When making a management decision using the coefficient of the employment structure of the population and the calculated indicators of the cost of creating one job in each separate type of economic activity, it is possible to carry out using simple mathematical calculations - how much investment is required to create (maintain) jobs in "A" in order to automatically jobs were created in "B" with government funding of jobs in "C".

In the context of economy modernization, it is important to follow rapid changes both in general and in particular, therefore an important tool in the modern labor market is the identification of relevant tools, such as monitoring the structure of employment of the population by industry through the indicator of the coefficient of the structure of employment of the population which makes it possible to offer effective directions labor market management of the relevant territory.

References

1. H. Wang, L. Ding, R. Guan, Y. Xia, Technological Forecasting and Social Change, 161 (2020) https://doi.org/10.1016/j.techfore.2020
2. M.A. Ortskhanova, M.I. Kitieva, F.Ya. Polonkoeva, Competitiveness in the Global World: Economics, Science, Technology **4-4 (40)**, 111-117 (2017)
3. V.S. Dubrovina, E.A. Chumakova, *Collection of materials of the 18 annual open competition of research works of students and young scientists in the field of economics and management "Green Sprout"*, 5-6 (2018)
4. O.A. Kolennikova, M.S. Toksanbaeva, Population **3 (77)**, 90-104 (2017)
5. E. Cruz, X. Raurich, Review of Economic Dynamics **38**, 198-219 (2020) https://doi.org/10.1016/j.red.2020.04.004
6. M. Metin Başbay, C. Elgin, O. Torul, Economic Systems **42**, (2018) 556-568 https://doi.org/10.1016/j.ecosys.2018.03.003
7. T. Tsekeris, Cities, 95 (2019) https://doi.org/10.1016/j.cities.2019.102472.
8. J. Clemens, M. Wither, Journal of Public Economics **170**, 53-67 (2019)
9. G. Van den Broeck, T. Kilic, World Development **119**, 81-99 (2019) https://doi.org/10.1016/j.worlddev.2019.03.008
10. V.N Arkhangelsky, Yu.V. Zinkina, S.G. Shulgin, Population **21-2**, 18-34 (2018)
11. I. Haaland, C. Roth, Journal of Public Economics **191** (2020) https: //doi.org/10.1016/j.jpubeco.2020.104256
12. G. Gozgor, Structural Change and Economic Dynamics **45**, 77-83 (2018)
13. Official site of the Federal State Statistics Service, URL: http://www.gks.ru (date of access: 01.12.2020)
14. A.V. Molostova, E.A. Chumakova, *Materials of the national scientific-practical conference dedicated to the memory of Professor Z.N. Boschaeva*, 19-20 (2019)
15. O.A. Aleksandrova, Yu.S. Nenakhova, A.V. Yarasheva, Population **1 (75)**, 34-46 (2017)

Analysis of knowledge assessment results of civil servants regarding anti-corruption issues

Olga Astafurova[1*], *Ada* Golomanchuk[2], *Tatyana* Omelchenko[3], *Anna* Borisova[4], and *Julia* Kayushnikova[5]

[1] Volgograd Institute of Management – Branch of the Russian Academy of National Economy and Public Administration under the President of the Russian Federation, 8 Gagarina str., Volgograd, 400066; Volgograd, Russian Federation. astoa@vlgr.ranepa.ru
[2] Volgograd Institute of Management – Branch of the Russian Academy of National Economy and Public Administration under the President of the Russian Federation, 8 Gagarina str., Volgograd, 400066; Volgograd, Russian Federation. golomanchukav@mail.ru
[3] Volgograd State University, 100 Prospect Universitetsky, Volgograd; Volgograd, Russian Federation
[4] Volgograd Institute of Management – Branch of the Russian Academy of National Economy and Public Administration under the President of the Russian Federation, 8 Gagarina str., Volgograd, 400066; Volgograd, Russian Federation. borisova-as@mail.ru
[5] Volgograd Institute of Management – Branch of the Russian Academy of National Economy and Public Administration under the President of the Russian Federation, 8 Gagarina str., Volgograd, 400066; Volgograd, Russian Federation. kayshnikova@yandex.ru.

Abstract. The article presents the program shell developed by the authors to estimate the knowledge of civil servants regarding anti-corruption issues. This development is a part of an information and analysis system, which is aimed not only at automating the analysis of data on knowledge control but it also includes both educational functions and possibilities for automation of the activity of the personnel divisions on the prevention of corruption offenses, formation of maps of corruption risks, observation, study and forecast of deviant corruption-related changes in the conduct of civil servants, assignment of testing, organization of training, etc. Due to the introduction of this program, the personnel divisions will obtain both systemic organization of awareness-raising measures (lectures, seminars) involving servants and the possibility to estimate their psychological condition and the general base of knowledge regarding prohibitions and restrictions in the civil service, practical and theoretical nuances of providing data about income, expenses, assets, liabilities, etc. The authors consider the issues of the introduction of the programs in the state authorities of both federal and regional levels. They point out the need to take into account the peculiarities of the targeted information system of the organization : the compatibility of the development being introduced with the current version of the operating system, as well as the security policy of the targeted system. The study presents the first results of the testing of the software product and includes an assessment of their main factors.

1 Introduction

Corruption processes taking place in society are one of the urgent challenges to social well-being and national security as a whole [1], [2], [3]. Manifestations of corruption at the level of state authorities are the most dangerous as they interrupt the normal functioning of both these structures and the entire state system [4]. There are thematic seminars, lectures, testing organized to raise the legal awareness and the level of legal literacy of civil servants regarding anti-corruption issues and to prevent corruption offenses in the civil service [5], [6], [7].

Modern information technology can considerably facilitate the development of this area of activities. For example, the information and analysis system aimed at informational support of the process of gaining anti-corruption knowledge and skills on the basis of modern information technology [8], [9], [10], [11], [12], [13], [14].

This information and analysis system is developed within the framework of the grant of the Russian Foundation for Basic Research "Formation of an Anti-Corruption Environment in State and Municipal Authorities through Introducing the Information and Analysis System 'Anti-Corruption Methods and Tactics for Civil and Municipal Servants'" . Any information system implies the automation of particular functions. The system under consideration ensures an integrated approach to solving the problems of anti-corruption education, stage-by-stage control of the implementation of measures, objective assessment of their results, analysis of the anti-corruption environment and forecasting of its development. The use of the software product will make it possible to cease to process information manually and reduce the possibility of errors during data processing and compilation of analytical reports [15].

2 Problem Statement

* Corresponding author: astoa@vlgr.ranepa.ru

The organization of using any software is associated with certain difficulties of introducing new technologies. It requires all kinds of settings and often enhancement of specific functions and integration with other systems in compliance with the reality of the organizational structure and management procedures.

It is necessary to take into account the peculiarities of the targeted information system of the organization during the deployment of software products. In particular, not only the compatibility between the versions of operating systems of one family but also the compatibility of the development being introduced with the current update version of each given operating system can turn out critical.

In addition, certain difficulties and restrictions can follow from the security policy of a particular targeted system. In this regard, the deployment of the software product in each given information system gives rise to problems which are impossible to consider at the stage of testing of the software. As a result, the settlement of such problems falls on engineers and administrators. Consequently, development and adjustment of the software product do not provide any guarantee that it will be applied effectively.

The user's qualification is a significant aspect of this problem. Programs with considerable functional possibilities, as a rule, have a complex interface capable of providing users with a huge number of tools. They are intended for professional users, i.e. experts whose main activity is the operation of a particular program. Software with a simpler set of tasks can be managed by ordinary users that refer to it only from time to time, if needed, for example, to take tests. In this case the efficiency of the software operation depends on the availability of the interface ensuring an intuitive understanding of the principles of interaction with the program and the promptness of obtaining the result.

3 Research Questions

The main objective of software developers is to bring their development as close as possible to the needs of the customer. Even at the design stage it is important to involve potential users in the development process, to form their requirements.

In general, the introduction of the information system turns out more difficult than the use of small program packages. This is due to particular organizational difficulties and the complexity of adjustment of the software product. Moreover, large software products do not always ensure fast information processing. As a rule, this is caused by an incomplete understanding of how to use the possibilities of the new system.

Another argument in favor of the development of a separate program shell for testing is the request of users to have the possibility to take tests (to watch video lectures and other materials) using mobile devices.

Based on the above, it was decided to develop a simple program shell that includes a testing subsystem of the information and analysis system being developed with its further transfer to the mobile version.

The testing system is a program module performing the functions of user authorization, selection of a test from the list of available tests, the examination of users in terms of the selected test with the results saved in the database for further analysis, as well as functions related to the analyst account, which are limited in this version of the program.

This software product was designed for Windows 7 and Windows 10 operating systems by means of MS Visual Studio 2019 in the C# programming language. The list of system requirements has two main components: NetFramework – ensures the functioning of the developed software product, and MSSQLlocalDB – ensures interaction with databases. These components are included in the installation package, which checks their availability and compliance of their versions at the beginning of the installation. As a result of the check, in case the mentioned components are not available or the versions are not adequate, they are downloaded and installed. A computer restart may be required to complete the installation of some components.

The stage of identification and deployment of the components is followed by the installation of the software product itself. During the installation, it is necessary to select the folder for the program files. It is strongly recommended not to install the software product into system folders, such as Program Files or Windows, to avoid problems during work with the software database. After the installation is complete, a shortcut to run the Tests of Corruption program appears on the desktop.

Every launch of the software product opens the login form, in which the user is asked to enter the name or any other sequence of characters (hereinafter "login") that uniquely identify the person taking a test. Every user of the program must use his or her own login to ensure correct processing of the testing results. This requirement is due to the fact that the first time when users enter their logins, they are simultaneously registered in the list of accounts stored in a special database intended for this purpose. Afterward , the use of a formerly registered login causes addition or update of data of the tests taken under the given account. It is necessary to consider the case sensitivity – a difference in one capital letter of the login will result in the registration of a new account. This simplified authorization system is applied to reduce the time for introduction and testing at this stage of studies. The program contains an integrated analyst (personnel officer) account protected by a password. In this version of the software product the analyst account makes it possible to access information about all accounts registered in the course of operation of the program and the testing results. To implement these functions, the login into the analyst account launches the corresponding interface that is different from the interface of interaction with other users.

After the login into a standard account, the software product opens a window that provides the possibility to select any test from the suggested list. The corresponding testing form is created and opens after the test is selected. The following form components can be

 https://doi.org/10.1051/shsconf/202110901006

defined as common for all the tests: the output area for the current question, the output area for the answer options for the current question, the timer and the confirm button for the selected answer. As a rule, using the mouse is enough to interact with the test forms, but it is also possible to use hot keys to save time and provide more comfort to the person taking the test. Every test should be completed within a particular time period, after which the test closes automatically and the current test result is saved in the database. During the testing the user can track the elapsed time with the timer located in the form.

The result is saved in the database after each testing. The saved information includes the login of the person taking the test, the date and time of the testing. The number of scored points is recorded as the number of correct answers or according to the scales approved on the basis of the assessment technique, which depends on the type of testing .

4 Purpose of the Study

The purpose of this study is not so much to develop a software product and test it on the basis of executive authorities as to analyze the results of knowledge assessment, to enhance the methods and approaches related to the processing of the obtained data.

5 Research Methods

Working on the article, we applied various principles of psychology, in particular, the principle of the unity of consciousness and activity by S.L. Rubinstein, the concept of system determination of mental phenomena by B.F. Lomov, the principle of development. The theoretical framework of the study includes the basic provisions of V.N. Myasishchev's psychological concept of relations and V.A. Yadov's dispositional concept with regard to the study of psychological peculiarities of money relations, as well as the theory of personality development by A. Maslow and L.I. Antsyferova.

Research methods. The Questionnaire for Assessment of Money Behavior of Civil Servants was created to achieve the set objectives. It is based on a set of diagnostic techniques, which includes as follows:

1 – Technique by Yu.M. Semenov,

2 – Technique by A. Furnham,

3 – Technique by E.B. Fantalova, Level of Correlation between Value and Availability (the authors' version),

4 – Technique by A.M. Poleev,

5 – Dembo-Rubinstein self-assessment technique (the authors' version),

6 – Authors' scale of monetary satisfaction.

The reliability of the research results is ensured by the methodological substantiation of the applied approaches and research methods, use of complementary techniques, special selection of persons for tests among civil servants and check of correctness of the selection of persons for tests, control of the empirical research procedure, comparative analysis of the results.

6 Findings

In the course of testing of the software, the findings were obtained according to four techniques for tests: the program for assessment of the psychological prerequisites for employees' corrupt practices (hereinafter "test 1"), the program for assessment of employees regarding anti-corruption issues (hereinafter "test 2"), the program for assessment of the civil servants' awareness regarding the provision of data about income (hereinafter "test 3"), the program for assessment of employees' deviations (hereinafter "test 4").

The technique of test 1 implies the assessment of the answers according to seven set scales, in a similar way, the technique of test 4 assesses the answers of the persons taking tests according to six set scales. Each scale has a threshold value – the norm the excess of which indicates particular deviations in the behavior. As a result of the conducted tests, the average indices for each of the scales of the tests as well as the indices of maximum deviation from the norm among the persons taking tests were calculated.

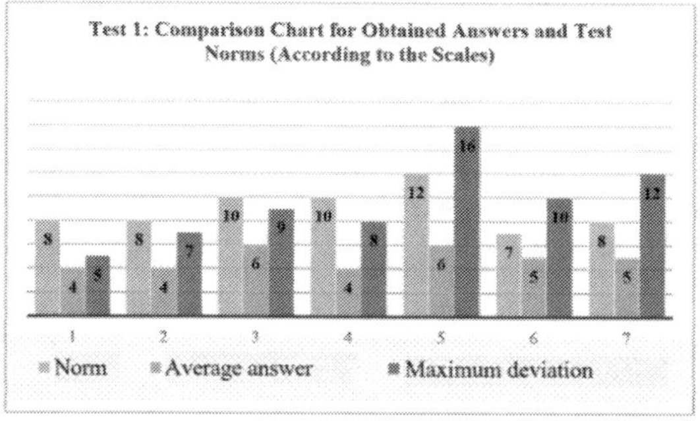

Figure 1. The results of the assessment of the psychological prerequisites for employees' corrupt practices ("test 1")

Table 1. The technique of test 1

Scale No.	Name of Scale	Description of Deviations
1	Social desirability attitude	Excess of the norm – dishonest answers
2	Tendency to overcome norms and rules	Excess of the norm – unwillingness to obey laws and norms
3	Tendencies to addictive behavior	Excess of the norm – incapability of accepting reality and resisting obstacles (partially – tendency to corruption)
4	Tendency to self-injurious and self-destructive behavior	Excess of the norm – dissatisfaction with oneself, tendency to act deliberately in a "bad", "wrong" way
5	Tendency to aggression and violence	Excess of the norm – excessive aggression

6	Volitional control of emotional reactions	Excess of the norm – high degree of self-control
7	Tendency to corrupt behavior	Excess of the norm – tendency to corruption and other deviations

According to the results of the first test, it is possible to assume that the average indices of all the scales are within the norm, and even the maximum deviations exceed the threshold value only of three of seven scales (Figure 1, Table 1).

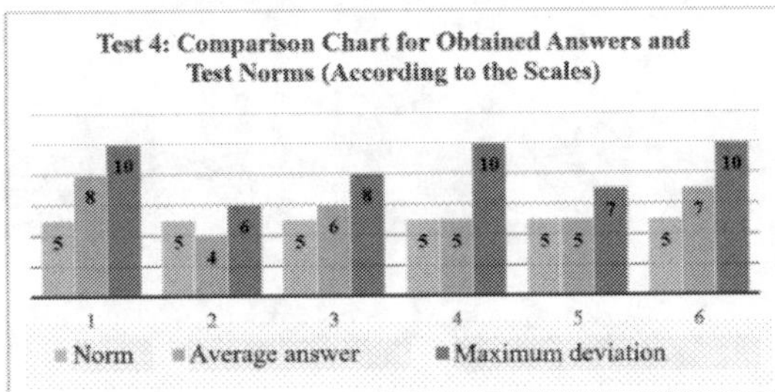

Figure 2. The results of the assessment of the employees' deviations ("test 4")

Table 2. The technique of test 4

Scale No.	Name of Scale	Description of Deviations
1	Social desirability	Excess of the norm – dishonest answers
2	Tendency to violation of norms and rules	Excess of the norm – unwillingness to obey laws and norms
3	Stress tolerance	Excess of the norm – low resistance to stress, high exposure to stress
4	Control of emotional reactions	Excess of the norm – high degree of self-control
5	Tendency to risk	Excess of the norm – high tendency to risk
6	Tendency to aggression and violence	Excess of the norm – excessive aggression

The results of the fourth test demonstrate a greater number of deviations from the norm. The average indices and maximum deviations of most scales border or exceed the norm (Figure 2, Table 2).

The technique of test 2 and test 3 implies calculation of points by the number of correct answers. The maximum number of points for test 2 is 20 (20 questions), and for test 3 – 45 (45 questions).

The subject of testing is the knowledge of legal rules in the area of corruption. To analyze the results of tests, the same ranges were selected according to the number of correct answers, and the number of persons taking tests within each range was calculated. The number is expressed as a percentage.

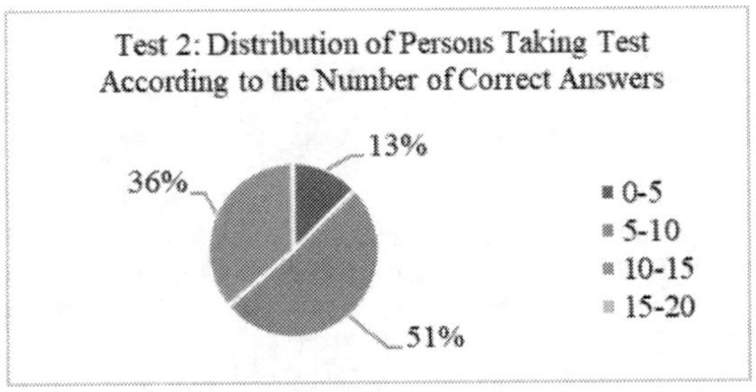

Figure 3. The results of the assessment of the civil servants' awareness regarding provision of data about income ("test 2")

The results of test 2 are indicative of insufficient training of the persons taking tests in respect of anti-corruption issues (Figure 3).

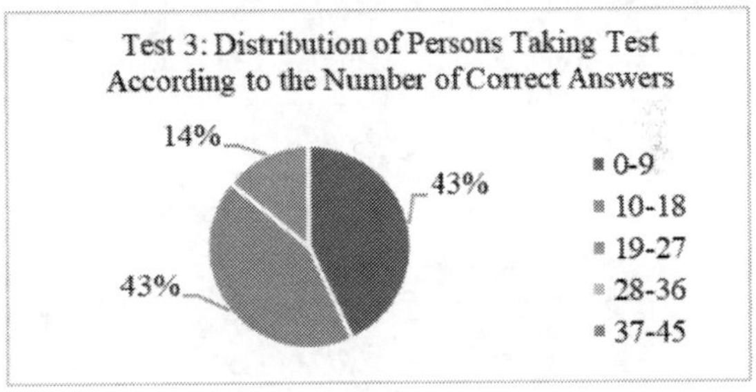

Figure 4. The results of the assessment of the civil servants' awareness regarding the provision of data about income ("test 3")

The results of test 3 are indicative of insufficient training of the persons taking tests in respect of the provision of data about income (Figure 4).

7 Conclusion

The introduction of the information and analysis system in the activity of divisions countering corruption and other offenses will make it possible to develop the field of corruption prevention and enhance anti-corruption education. The proposed software product will automate the assessment of the knowledge of civil servants regarding their possession of the fundamentals of the anti-corruption legislation, provision of data about income, analytics of the psychological condition, money behavior and its prerequisites, evaluation of the tendency of officers to corruption. We are sure that the implementation of the program will significantly simplify the work of personnel divisions, help them to effectively exercise their authorities to prevent corruption offenses among civil servants.

Acknowledgment

The study was carried out with financial support of the Russian Foundation for Basic Research within the framework of scientific project No. 18-29-16119 "Formation of an Anti-Corruption Environment in State and Municipal Authorities through Introducing the Information and Analysis System "Anti-Corruption Methods and Tactics for Civil and Municipal Servants".

References

1. T. Gulchehra, T. Albina, International Journal of Innovative Technology and Exploring Engineering **9 (1)**, 3219-3221 (2019) doi:10.35940/ijitee.A9159.119119
2. H. Li, T. Meng, Journal of Chinese Political Science **25 (3)**, 431-456 (2020) doi:10.1007/s11366-020-09672-w
3. H. Zhang, Y. Song, S. Tan, S. Xia, H. Zhang, C. Jiang, ... Y. Lv, Cities **90**, 64-73 (2019) doi:10.1016/j.cities.2019.01.042
4. S.V. Kuznetsov, R.R. Magizov, Y.V. Radzevanovskaya, V.K. Alexey, I.G. Mukhametgaliev, M. Andino, International Journal of Recent Technology and Engineering **7 (6)**, 99-102 (2019)
5. C. Hauser, Journal of Business Ethics **159 (1)**, 281-299 (2019) doi:10.1007/s10551-018-3808-3
6. O.F. Lobazova, International Journal of Innovative Technology and Exploring Engineering **8 (12)**, 4667-4672 (2019) doi:10.35940/ijitee.L3867.1081219
7. J.-H. Meyer-Sahling, K.S. Mikkelsen, Review of Public Personnel Administration (2020) doi:10.1177/0734371X20949420
8. O.A. Astafurova, A.S. Borisova, A.V. Golomanchuk, T.Y. Yagotinsteva, International Journal of Information and Education Technology **10 (3)**, 209-214 (2020) https://doi: 10.18178/ijiet.2020.10.3.1365.
9. R. Baguma, F. Mutungi, T. Janowski, *Paper presented at the ACM International Conference Proceeding Series*, 484-494 (2019) doi:10.1145/3325112.3325266
10. E. Denisova-Schmidt, M. Huber, E. Leontyeva, A. Solovyeva, Empirical Economics **60 (8)** (2020) doi:10.1007/s00181-020-01827-1
11. T.K. Mackey, R.E. Cuomo, Global Health Action **13 (1)** (2020) doi:10.1080/16549716.2019.1695241
12. A. Mouna, B. Nedra, M. Khaireddine, *Transforming Government: People, Process and Policy* (2020) doi:10.1108/TG-03-2020-0040
13. C. Strand, M. Hatakka, *Anti-corruption efforts in national ICT policies: A study of policy environments in sub-saharan Africa* (2019) doi:10.1007/978-3-030-18400-1_43
14. C. Strand, M. Hatakka, *Existing and promising theoretical approaches to understanding ICTs contribution to anti-corruption efforts: A literature review* (2020) doi:10.1007/978-3-030-39296-3_6
15. O.A. Astafurova, A.S. Borisova, A.V. Golomanchuk, T.A. Omelchenko, International Journal of Information and Education Technology **10 (2)**, 90-94 (2020) https://doi: 10.18178/ijiet.2020.10.2.1345.

SHS Web of Conferences 109, 01007 (2021) https://doi.org/10.1051/shsconf/202110901007
LISID-2020

Data and factors analysis of anti-corruption environment components of Russian public authorities

Olga Astafurova[1*], and *Ada* Golomanchuk[2]

[1] Volgograd Institute of Management – Branch of the Russian Academy of National Economy and Public Administration under the President of the Russian Federation, 8 Gagarina str., Volgograd, 400066, Volgograd, Russian Federation astoa@vlgr.ranepa.ru
[2] Volgograd Institute of Management – Branch of the Russian Academy of National Economy and Public Administration under the President of the Russian Federation, 8 Gagarina str., Volgograd, 400066; Volgograd, Russian Federation. golomanchukav@mail.ru

Abstract. The paper is devoted to the examination of functional features of anti-corruption activities in Russian public government bodies and authorities using the cognitive modeling method . Currently, a contradictory situation is observed in corruption related assessment practice. At present, the leadership of public authorities reports on a decrease in acts of corruption whereas the statistics testify rather to the persistence of the previous level of corruption offenses than to the decline. Employees are concerned about the increasing requirements related to compliance with anti-corruption restrictions, as well as income, expenses and property obligations reporting. Citizens continue to express distrust in government bodies and state apparatus as a whole. These contradictions could be resolved if the interests of both employees of state bodies and citizens of the country are taken into account. It is assumed that when assessing anti-corruption environment effectiveness, it is advisable to take into account not only the risks of citizens, but also the risks of public officials and state authority's employees. We applied cognitive modeling as it is difficult to formalize most stages of government bodies' activities, whereas we need to take into account many factors and conflicting goals that have complex weakly structured relationships. Fishbone Diagram or Cause-and-Effect-Diagram and the cognitive map were basic initial stage of cognitive modeling which allowed conducting the analysis of external and internal factors influence on the anti-corruption environment development in government bodies. The authors of the paper have shown benefits of expediency transferring of some personnel department functions into digital format.

1 Introduction

The performance of the existence as a system, as an integral "organism", directly depends on the orderly functioning of all its structures, which are to implement the powers of the state via its authorities. Corruption at the level of these structures disrupts the normal order of their functioning which entails disruptions in the entire state "organism" and hinders the achievement of public goals and objectives.

Necessary conditions for effective work to combat corruption have been introduced into the Russian legislation, but in order to create sufficient conditions [1] , it is necessary to activate anti-corruption enforcement in all the constituent entities of the Russian Federation.

There are some critical challenges in the development of possible transformational options in anti-corruption environment in state authorities including difficulties to formalize most of its stages; uniqueness of the emerging tasks; need to take into account many factors and goals that have a complex structure of interconnection, and often contradict each other. These are weakly structured positions [2]. For these reasons, we believe that combinations of various factors to which we include personal expert experience, assessment based on it, systematized knowledge, confirmed by practice and even intuition, combined with the use of modern technologies, are the most successful for the development of a truly effective automated decision-making program in anti-corruption issues [3].

Cognitive analysis or cognitive modeling is an effective approach to solving poorly structured problems [4]. It is based on the concept of a cognitive map which is a directed graph which edges are assigned with weights [5]. The vertices of the cognitive map correspond to the factors or concepts which determine the situation and the horizontal edges correspond to the causal relationships between the factors [6].

In order to synthesize effective control strategies, methods of analytical processing implemented in decision support systems are applied to the cognitive map [7], aimed at studying the structure of the system and obtaining predictions of its behavior under various control actions [8].

The main advantage of the proposed apparatus of cognitive maps is the ability to systematically identify effects that can help or hinder the implementation of seemingly obvious solutions and which are difficult to

* Corresponding author: astoa@vlgr.ranepa.ru

assess intuitively with a large number of factors and the variety of numerous ways of interaction between them.

The forming and using cognitive maps process includes the following procedural sequence:

- Defining a list of concepts;
- Determining the influence between each concept pair;
- Building a cognitive map;
- Stability analysis;
- System characteristics analysis on the cognitive map.

We find it most appropriate to use Silov Fuzzy Cognitive Maps as a basis for cognitive model constructing of the anti-corruption environment. This type of cognitive maps provides higher adequacy of the developed models in comparison with classical, symbolic cognitive maps, since it allows using qualitative and quantitative characteristics, taking into account several experts opinion, generating and selecting control strategies and carrying out their dynamic modeling [9].

2 Problem Statement

PEST and SWOT analyses of the factors influencing the processes of combating corruption in government bodies in the context of systemic transformation were applied in order to identify the main concepts.

Using PEST analysis at the beginning of the study, we examined the influence of environmental factors on anti-corruption processes in public authorities (Table 1). As a result, some main groups of factors including political, economic, social, cultural and technological were identified.

Table 1. PEST-analysis of external environment influence on anti-corruption processes in public authorities

Political (P)	**Economic (E)**	**Social (S)**	**Technological (T)**
Current legislation in the field of public anti-corruption regulation Domestic policy of the State. State regulation of anti-corruption processes. Personnel policy in government bodies. Maintaining government prestige and authority among individual citizens and businessmen.	The state of national economy. Digitalization of economy. Low level of salaries of municipal authorities. Possession of real estate. Low level of competition. Demand in qualified personnel	Lack of publicity and transparency in decision-making by the authorities. Legal under-protection of government officials. Availability of employment benefits. The prestige of the profession. Imperfect mechanisms of direct and indirect feedback between citizens and government authorities. Legislation that determines social factors. Societal Basic values. Ethnic, cultural, religious factors. Lifestyle trends, quality of life. High society stratification, low social consciousness.	Redundancy of state and municipal functions. Low efficiency of external and internal control. Development of innovative technologies in Russia and globally. The state of information technology in the regions.

In order to draw conclusions on the analysis of the internal and external environment affecting the development of the anti-corruption environment in state authorities, a SWOT analysis was conducted. The purpose of the SWOT analysis was to identify the main factors through the systematization of available information on the strengths and weaknesses of anti-corruption policies, as well as on the potential opportunities and threats from the external environment, including those related to digitalization.

A set of significant factors and conditions affecting the development of the anti-corruption environment in state authorities, obtained as a result of SWOT analysis, is given in Table 2.

Table 2. SWOT-analysis of factors and conditions affecting anti-corruption environment development in public authorities

Strengths (S)	**Weaknesses (W)**	**Opportunities (O)**	**Threats (T)**
Emergence of new	Underdeveloped civil	Development of a	Negative manageme

forms of staff management. Informational and analytical systems use for corruption combating. Social responsibility and awareness increase.	society institutions. Wages gap in government and municipal employees. Insufficient provision of civil service personnel with the necessary competencies in information and communications technologies. Educational institutions` activities are not sufficiently coordinated with the government bodies` needs. Low level of remuneration. Negative social and psychological factors: poor socialization, unstable life position, the decline in the prestige of civil servant profession.	system of anti-corruption activities. Reduction of acts of a corruption orientation due to increased awareness of employees and the formation of an appropriate attitude towards the facts of corruption. Improving the activities of personnel departments. Maintaining social stability.	nt tendencies. Migration of qualified personnel outside the regions. Laying off workers and reducing the number of jobs in personnel departments. Changing family traditions.

As a result, we identified three groups of factors:

- a number of factors consistent with the initial data on anti-corruption environment [10] (employees involved in corruption liability, data on non-compliance with existing regulations, data on HR department services for anti-corruption work, social and psychological factors, etc.);
- a number of factors reflecting a positive impact on anti-corruption environment [11] (transparency, revealed facts of corruption, educational work of experts and HR services, etc.);
- a number of factors characterizing the main destabilizing effects on the anti-corruption environment (corruption offenses, corruption crimes, training of specialists without taking into account the competencies in organizing personnel work to combat corruption, lack of civil servants with sufficient informational and communicational skills, etc.).

3 Research Questions

The authors clarified the list of basic concepts and determined cause-and-effect relationships between them which were achieved through the use of the Fishbone Diagram (or Cause and Effect Diagram or Ishikawa diagram). The diagram displays all the factors influencing the problem under consideration. The complex application of the above methods (approaches) allows one to get the greatest effect when building a cognitive model of the influence of the main external and internal factors on the transformation of the anti-corruption environment in government bodies.

Further analysis of the factors influencing the transformation of the anti-corruption environment in state authorities under the influence of the internal and external environment was carried out using the Fishbone Diagram (Figure 1).

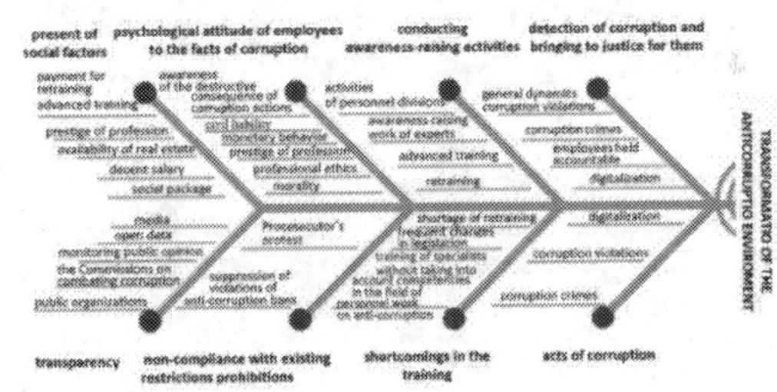

Figure 1. Fishbone Diagram of anti-corruption environment transformation in state authorities and government bodies.

The factors of the 1st order in their turn are influenced by smaller bones which include such factors as:

Admissible acts of a corrupt nature are the corruption offenses result [12]; corruption crimes; digitalization which in turn depends on information and communication technologies development (ICT), the development of digital employee competencies are of the 3rd order factors. Besides, the factors of the 3rd order influencing corruption offenses are administrative, material and disciplinary responsibility.

The factors "revealing corruption facts and bringing responsibility for them" include the following influence: the general dynamics of corruption offenses, corruption-related crimes, bringing to responsibility employees (federal civil servants, civil servants, municipal employees, employees holding municipal positions, other officials) and digitalization. The "digitalization" factor as well as development of information and communication technologies (ICT), the development of digital competencies of employees is of the 3rd order [13].

The "awareness-raising work on anti-corruption activities is influenced by educational work of experts explaining typical situations; the activities of personnel departments, the workload of employees of these departments with routine processes (filling in magazines, assigning and conducting testing, analyzing test results ...) can be automated. These functions will be assumed by the information and analytical system developed under the Russian Foundation for Basic Research grant "Formation of the Anti-Corruption Environment in State and Municipal Authorities by introducing information and analytical system "Methods and tactics of combating corruption for state and municipal employees" [14].

Deficiencies in personnel training are the result of training specialists without taking into account competencies in the field of organizing personnel work to combat corruption [15]; shortcomings in retraining, untimely passing of advanced training, as well as frequent changes in legislation.

Official statistics on corruption offenses of various departments are very diverse and contain numerous indicators in the reporting forms. Sufficiently deep knowledge of statistics and mathematics is required to be able to operate with these indicators. In addition to criminal law information on corruption, knowledge in the field of criminology, psychology, sociology, economics, political science, and other branches of knowledge related to the study of crime is required. Unfortunately, the state educational standards on jurisprudence do not provide as compulsory those disciplines which knowledge is so necessary for professionals to study criminal or judicial statistics and combating crime [16].

Non-compliance with existing restrictions and regulations is influenced by prosecutor protest of the; presentation of the prosecutor; suppression of violations of anti-corruption prohibitions [17];

Psychological attitude of employees to the facts of corruption depends on civic responsibility; prestige of the profession; morality, moral principles, personal qualities and individual motivation. This factor can be influenced by explaining the basics of monetary behavior and professional ethics. The employee must be aware of the destructive consequences of acts of corruption.

Transparency is influenced by the media; the need to upload open data to appropriate information systems; public organizations; anti-corruption commissions; public opinion monitoring.

Changes in social factors are facilitated by decent wages; affordable real estate; available social employment benefits which include payment for medical services, gym or swimming pool subscription, free meals, etc.; further education, retraining or advanced training covered by the employer; occupational prestige.

4 Purpose of the Study

The purpose of data and factors analysis of the anti-corruption environment components of state authorities and government bodies determine mutual influencing indicators in a systematic study of theoretical and practical problems arising in the course of this process and determining their solution perspective.

The object of inquiry is social relations that develop in the course of individual components formation of the anti-corruption environment of state authorities and government bodies.

The subject being analyzed is Russian legislation norms in the field of combating corruption and modern technologies for the formation of individual components of anti-corruption environment of state authorities and government bodies..

5 Research Methods

In the course of the study, we applied various methods including the universal scientific method of historical materialism; general scientific methods of cognition (comparison, ascent from the abstract to the concrete, analysis, synthesis, induction, deduction, generalization, classification and abstraction), partially scientific methods (formal legal, comparative legal, method of interpreting legal norms, dogmatic method), interdisciplinary method of formalization.

6 Findings

The results of trends analysis served as the basis for identifying the cognitive map concepts that have the greatest impact on the anti-corruption environment of state authorities and government bodies. The structure of the cognitive map is presented in Figure 2.

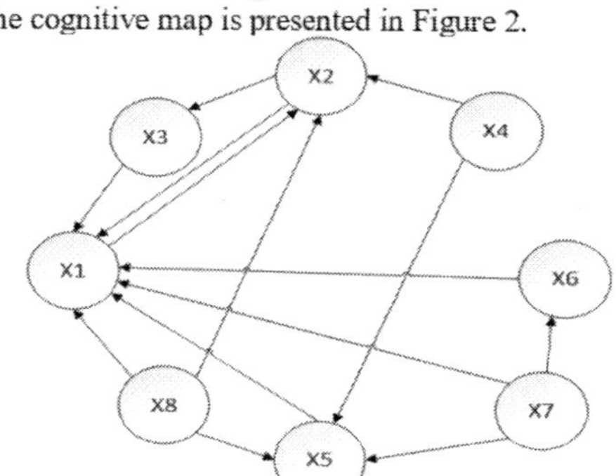

Figure 2. Cognitive map of Internal and External factor influencing anti-corruption environment of state authorities and government bodies

As the first and second concepts we defined X1 which is the Index of Admissible Corruption Offense and X2 which is the Index of Corruption Facts Revealing; these two indices have a mutual influence on each other. In addition, we identified the Social Index X3 which is in turn affects the Index of Admissible Corruption Offense.

Transparency Index X4 is the fourth index that influences the Index of Corruption Facts Revealing and the Index of Psychological Attitude of Employees to corruption facts (X5). The index of the psychological attitude of employees to Corruption Facts affects the Index of Permitted Corruption Offenses.

The sixth index is the Personnel Training Index X6 which also affects the Index of Admissible Corruption Offense.

The seventh index is the Anti-Corruption Education is X7 which affects the Index of Admissible Corruption

Offense, the Index of the Psychological Attitude of Employees to corruption facts and the Index of Personnel Training.

The eighth index X8 is the Index of Existing Restrictions and Prohibitions which influences the Index of Admissible Corruption Offense, the Index of Employees Brought to Justice for Corruption offenses and the Index of the Psychological Attitude of Employees to the Corruption Facts.

The mutual influence of the concepts on each other and on the problem under consideration at the initial stage of the study was determined primarily by experts, but the further development of the model involves the construction of econometric relationships.

Practical outcomes can be used for cognitive modeling of possible scenarios for anti-corruption environment development in government bodies.

This model is scalable, that is it makes it possible to detail large factors into smaller components. The factors were grouped to map the model and determine the interrelationships of the model. The connections between factors show their influence as one-sided (a change in one factor leads to a change in another), mutual (both factors affect each other), as well as the influence of a factor on itself (a change in a factor at a previous moment leads to its change at a later moment). If the influence is direct (the growth of the initial factor leads to an increase in the associated factor), then the relationship is considered positive, if the opposite influence (the growth of the factor leads to a decrease in the associated factor), then the relationship is negative. Links in the model diagram appear as influence arrows. An important quantity for creating a model is the bond strength. The value characterizes the intensity of influence of one factor on another.

7 Conclusion

The anti-corruption system consists of many different elements, including measures and means of preventing and combating corruption.

The aim of the study is to identify the complex effect of external and internal factors influencing and conditioning the development of anti-corruption environment in government bodies, namely by reducing uncertainty in predicting personnel shortage with the necessary up-to-date work skills and qualifications; determining possible scenarios for anti-corruption environment development; identification of necessary retraining and advanced training programmes. The paper presents the results of the cognitive modeling first steps, namely the construction of a cognitive map of mutual influencing concepts and factors which have an overall impact on the problem under consideration. The assessment of each concept importance is conducted at the first stage of cognitive map structure formation. Assigning connections between concepts (factors) can be made in the form of values (terms) of a linguistic variable (weak; strong; very strong) or in numerical form in accordance with the approximate value of the connection strength between the concepts.

Acknowledgment

The study has been carried out with the financial support of the Russian Foundation for Basic Research within the framework of the Research project No. 18-29-16119 "Formation of the anti-corruption environment in the state and municipal authorities through the introduction of the information and analytical system 'Methods and tactics of countering corruption for state and municipal employees'".

References

1. C. Baez-Camargo, P. Bukuluki, R. Sambaiga, African Studies **79 (2)**, 232-249 (2020)
2. Z.K. Avdeeva, S.V. Kovriga, D.I. Makarenko, Large Systems Management **16**, 26-39 (2007)
3. E.V. Golomanchuk, O.A. Astafurova, D.A. Mehrabyan, Business. Education. Law **3 (52)**, 309-316 (2020) doi: 10.25683 / volbi.2020.52.378.
4. S.M. Brown, Journal of Management Studies **29**, 287-307 (1992)
5. W.R. Zhang, S. Chen, R.S. King, IEEE Transactions on Systems, Man and Cybernetics **22**, 103-113 (1992)
6. Yu.G. Fedulov, A.B. Yusov, *Social Policy: Formalization, Measurement, Forecasting*, 384 (Direct Media, Moscow, 2015)
7. C.D. Stylios, P. Groumpos, Journal of Intelligent & Fuzzy systems **8 (2)**, 83-98 (2000)
8. M.Yu. Mizulin, G. Fedulov, A.B. Yunusov, *Application of Systems Theory to Politics and Economics: monograph*, 312 (Moscow, RASS, 2011)
9. O.A. Astafurova, A.S. Borisova, A.V. Golomanchuk, T.A. Omelchenko, International Journal of Information and Education Technology **10 (2)**, 90-94 (2020) https://doi: 10.18178/ijiet.2020.10.2.1345.
10. H. Kim, International Journal of Urban and Regional Research **45 (22)**, 1057-1071 (2020)
11. C.L. Atkinson, International Journal of Public Administration **43 (13)**, 1169-1182 (2020)
12. A. Castro, N. Phillips, S. Ansari, Academy of Management Annals **14 (2)**, 935-968 (2020)
13. K. Veeramani, S. Jaganathan, Ksii Transactions on Internet and Information Systems **14 (9)**, 3693-3711 (2020)
14. K.-H. Yu, S-D. Kang, C. Rhodes, Business & Society **59 (7)**, 1377-1409 (2020)
15. V. Vukovic, Journal of Comparative Economics **48 (1)**, 124-143 (2020)
16. V.I. Shulga, Bulletin of the Vladivostok State University of Economics and Service **1 (28)**, 28-33 (2015)
17. Z.Q Wang, Advances in Education Research **85**, 101-103 (2015)

The structure of the e-state and public administration at the digital age

Alexey Barkov[1], *Aleksey* Sokolov [2*], and *Alexander* Kiselev[3]

[1] Military University of the Ministry of Defense of the Russian Federation, Leningradsky Prospekt str., 55, Moscow, Russia; Financial University of the Russian Federation, Volochayevskaya str., ¾, Moscow, Russia; barkov_a_v@mail.ru
[2] Military University of the Ministry of Defense of the Russian Federation, Volochayevskaya str., ¾, Moscow, Russia, srrpj@mail.ru
[3] Military University of the Ministry of Defense of the Russian Federation, Volochayevskaya str., ¾, Moscow, Russia; e-mail: pain068@yandex.ru

Abstract. The introduction of information technologies into public life in all spheres of the life of the state and society has set the scientific community the task of scientific understanding of the theory and practice of the introduction and application of these technologies, modernization of the legal policy of the state in the information society. The consequence of this was the formation at the turn of the XX-XXI centuries. In the political and legal though of views on the electronic state, embodying ideas about how the state should function in the era of high technologies. Management in modern conditions begins to be built through the use of electronic communication systems at all levels. In the modern legal doctrine, the structure of the electronic state has ambiguous content. And this expresses not just the vision of the structure of the idea by various scientists, moreover, the consolidation of such provisions in scientific research is a trigger for the institutionalization of the idea, its development at the theoretical level and obtaining its expression in building its own infrastructure. The idea of an electronic state as its main goal is aimed at modernizing the public administration system through information and communication technologies and the capabilities of the global Internet. It must solve the accumulated problems and meet the criteria set by modern society.

1 Introduction

The structure of the e-government includes such elements ase-government, e-justice and e-democracy. But such a vision of the structure is not unambiguous, it can be refined, modified. In connection with the development of the idea of an electronic state, there is a transformation not only of management approaches in the administrative sector, but also in the corporate sphere, new approaches to management appear. The electronic state is not only a new model of interaction between the state and citizens; it is also a way of rethinking Russian statehood and Russian society through the prism of a yet unexplored, but objectively necessary system of views. For development, the state needs a new vision of its existence, taking into account the preservation of the eternal values and ideals of law. It is quite possible that it is possible to achieve de-bureaucratization precisely due to the implementation of the idea of an electronic state, which changes not only the approach to management, but also the very culture of interaction, mental foundations.

2 Problem Statement

1. Clarification of the structural elements of the e-government system. The peculiarities of the system as a philosophical concept lie in the fact that the elements, being combined into complexes, are forced to change. Each of the presented structural elements represents a new direction in scientific research.
2. To determine the main directions of improving public administration and the era of the electronic state. Accordingly, it is necessary to identify and characterize the ongoing changes, to make a forecast of the consequences of these changes.

3 Research Questions

3.1 Reforming the bureaucratic apparatus.
3.2 Providing the state with conditions for citizens to choose the form of participation in democracy.
3.3 Training of qualified personnel, responsible and servicing the resources and workflow of the electronic state.
3.4 Further implementation of the "one window" principle, incl.

* Corresponding author: srrpj@mail.ru

3.5 Increasing the computer literacy of the population and the creation of consulting, training and accompanying centres.
3.6 Inclusion of non-state structures in the infrastructure of the electronic state.
3.7 Creation of specialized state portals, acting on the principle of an aggregator of all services provided by the state.
3.8 Development and implementation of mechanisms of state legal regulation of the sphere of implementation of the idea of an electronic state. within the framework of the integration of state and municipal standards for the provision of services.
3.9 Creation of a secure electronic document management system.
3.10 Full transfer of authorities to electronic document management.

4 Purpose of the Study

The purpose of the study is to determine the structural elements of the electronic state, to characterize them, as well as to identify the main directions for improving public administration in the context of digitalization.

5Research Methods

The main methods used in the work are methods of empirical knowledge - comparative legal, historical and legal, formal logical methods, description, systemic method, as well as analysis, synthesis, induction and deduction methods.

6 Findings

1. Against the background of these processes, there are qualitative and quantitative changes in the content and essence of the idea of an electronic state. Quantitative changes are reflected in the revision of the norms of positive law and program acts, in law enforcement practice, including e, on the organization of the functioning of the electronic state. Qualitative changes are associated with changes in systems of a higher order in relation to the electronic government. At the same time, the ideological core of the electronic state remains unchanged, which must maintain its stability even with strong opposition of legal and rational, scientific and practical, legal and corruption factors.
2. Gradually, in the Russian Federation, government officials and society are developing a value attitude towards the electronic state. In addition, it is the content of the value attitude towards it that will determine and clarify the content of the electronic state. At the moment, we believe that such a relationship has not yet been fully formed, which acts as a serious constraining factor both for further scientific research of the electronic state, and for the practice of implementing this idea in Russian conditions.
3. Now the e-government is either a way to save your time for obtaining various kinds of documents (for those citizens who have the appropriate level of computer literacy), or something inaccessible, which presents a serious problem, both temporal, intellectual, and technical (for those citizens who do not have an appropriate level of computer literacy). It seems that for state bodies and local self-government bodies, this is the sphere of the use of labor resources; these are the problems of the availability of qualified personnel, the problems of ensuring the filling of sites.

7 Conclusion

The structure of the electronic state can include additional elements, both inherent to it and alien, as well as - elements dictated by political and other considerations. Some researchers, such as E.G. Vasiliev and D.V. Kononenko, believe that the e-state is an integral part of e-democracy [1]. Understanding the structure largely depends on what is meant by the term "e-government".

We believe that the idea of an electronic state, being a system, is at the same time a part of a system of a higher order. Such, in our opinion, the system for the idea of an electronic state is the theory of the rule of law. Consequently, the development of the idea of an electronic state will be carried out in coordination with the development of larger-scale ideas and theories.

8 Turning to history, you can see that the key and most basic element of the idea of e-government is e-government. P.U. speaks about this in his work. Kuznetsov, noting “the phrase“e-government” was initially used in the international Purpose of the Study

The purpose of the study is to determine the structural elements of the electronic state, to characterize them, as well as to identify the main directions for improving public administration in the context of digitalization.

9 Research Methods

The main methods used in the work are methods of empirical knowledge - comparative legal, historical and legal, formal logical methods, description, systemic method, as well as analysis, synthesis, induction and deduction methods.

10 Findings

1. Against the background of these processes, there are qualitative and quantitative changes in the content and essence of the idea of an electronic state. Quantitative changes are reflected in the revision of the norms of positive law and program acts, in law enforcement practice, including e, on the organization of the functioning of the electronic state. Qualitative changes are associated with changes in systems of a higher order

in relation to the electronic government. At the same time, the ideological core of the electronic state remains unchanged, which must maintain its stability even with strong opposition of legal and rational, scientific and practical, legal and corruption factors.

2. Gradually, in the Russian Federation, government officials and society are developing a value attitude towards the electronic state. In addition, it is the content of the value attitude towards it that will determine and clarify the content of the electronic state. At the moment, we believe that such a relationship has not yet been fully formed, which acts as a serious constraining factor both for further scientific research of the electronic state, and for the practice of implementing this idea in Russian conditions.

3. Now the e-government is either a way to save your time for obtaining various kinds of documents (for those citizens who have the appropriate level of computer literacy), or something inaccessible, which presents a serious problem, both temporal, intellectual, and technical (for those citizens who do not have an appropriate level of computer literacy). It seems that for state bodies and local self-government bodies, this is the sphere of the use of labor resources; these are the problems of the availability of qualified personnel, the problems of ensuring the filling of sites.

11 Conclusion

The structure of the electronic state can include additional elements, both inherent to it and alien, as well as - elements dictated by political and other considerations. Some researchers, such as E.G. Vasiliev and D.V. Kononenko, believe that the e-state is an integral part of e-democracy [1]. Understanding the structure largely depends on what is meant by the term "e-government".

We believe that the idea of an electronic state, being a system, is at the same time a part of a system of a higher order. Such, in our opinion, the system for the idea of an electronic state is the theory of the rule of law. Consequently, the development of the idea of an electronic state will be carried out in coordination with the development of larger-scale ideas and theories.

Turning to history, practice of introducing information technologies into the activities of government institutions" [2]. The use of new technologies was envisaged in the context of improving the productivity of the state apparatus, then the process of providing services began to transform in accordance with existing computer technologies and the growing needs of society. Further, changes took place in the structure of the judicial authorities, for example, in Russia this period fell at the beginning of the 21st century, when the first sessions of the courts of general jurisdiction and arbitration courts began to be held using video conferencing. This event marked the development of the second element of e-government - e-justice.

The last element of the e-government is e-democracy, which provides for the implementation of both representative and direct democracy. Yu.V. Nikulina points out that "through e-democracy, citizens determine what services and how they will be provided by the government, mutual responsibility is established within the framework of interaction" [3].

The embodiment of democratic principles in an electronic state is possible with the achievement of a high level of legal culture of citizens and the formation of an information society. Nevertheless, the question arises as to whether the establishment of an electronic state in a country with a monarchical form of government is possible. Note that in states such as Qatar, the United Arab Emirates, Kuwait, Saudi Arabia, Oman, Jordan, Sweden, Norway, Liechtenstein, Luxembourg, Belgium, Denmark, the Netherlands, Monaco, in which the main form of government is monarchy, the state is at a high level [4].

To date, no international rating includes an indicator reflecting the level of development of e-democracy from the standpoint of the realization of human rights and freedoms.

Considering these circumstances, we note that the development of e-democracy becomes possible in a state governed by the rule of law under a republican form of government with a democratic regime. A.A. Korobov, T.S. Melnikova, and N.V. Mityaeva support this position: "Unlike e-government, which can be implemented under both democratic and authoritarian regimes, the idea of e-democracy can find application in a state with developed democratic institutions" [5].

In states governed by the rule of law, it is advisable to maintain a balance between the need to maintain sovereignty in the information environment and preserve a democratic regime. Otherwise, very serious consequences will arise for the national statehood. Therefore, the search for a new level of management of society while preserving its national, historical and other characteristics and traditions is the most important task. In this regard, let us highlight the main directions of development of the idea of an electronic state on the example of the Russian Federation.

1. Reforming the bureaucratic apparatus.

Every year the volume of work and the load on state and municipal employees in all spheres of management is growing rapidly. Many Russian scientists pay attention to this problem. There is no doubt that this problem is urgent both in Russia and throughout the world. Accordingly, scientists, managers and legislators are trying to find the optimal mechanisms for de-bureaucratization.

Human participation in the managerial activity in history is of fundamental importance. However, O.N. Sinitsyn and D.A. Rakitin draws attention to the fact that "in the Russian state system, citizens are accustomed to the fact that all decisions come «from above" [6].

Accordingly, with the transition to a new management model, in which there is significant democratization of the administrative system, many Russians get the opportunity to independently communicate with government bodies without intermediaries in the face of numerous officials. E-government, in fact, provides a simplified way of interacting with authorities in comparison with personal direct contact.

2. The provision by the state of conditions for citizens to choose the form of participation in democracy.

A significant innovation for the development of an electronic state, in our opinion, is the realization of the potential of electronic democracy. This is due to the peculiarities of the state structure and management, the degree of development of civil society, the level of preparedness of state bodies and the competence of employees, as well as a number of other factors.

Of particular interest is the optimization of certain functions of government bodies, carried out through information and telecommunication technologies and computer equipment with Internet access. Taking into account the significant role of informatization of society, a natural question arises: is it possible to talk about the implementation of the idea of an electronic state, in particular, the introduction of such an element as electronic democracy, with the achievement of a complete transfer to remote interaction, or is it worth considering additional components that contribute to the modernization of lawmaking, legislative, electoral and other processes.

It is important that in the future e-voting systems be not imposed by the state as the only possible form of participation in political processes. The population should have a choice giving the right to switch to electronic participation while maintaining the traditional one. Only when the population independently chooses electronic voting as the main way to participate in elections, it will be possible to refuse paper ballots.

3. Training of qualified personnel, responsible and servicing the resources and workflow of the electronic state.

The electronic state not only relies on existing legal norms, but also aims at improving the forms and mechanisms for the exercise of citizens' rights. It is for this reason that at present, special attention is being paid to improving the qualifications of personnel who exercise their powers in government bodies and ensure the functioning of the electronic state.

Currently, according to V.Yu. Turan in, "a university graduate must have civil, professional, business, spiritual, moral, organizational, strong-willed, and intellectual qualities" [7]. Graduating from an educational institution, in fact, a specialist must be trained who meets a set of skills, knowledge and qualities that will allow him to immediately delve into the process of functioning of the electronic state.

We believe that it is necessary to create educational programs in higher educational institutions aimed at training people who will fill positions in the relevant authorities. They must have a sufficient level of knowledge to work with electronic systems, ensure their functioning, know the legislation in the field of civil, constitutional, information and other branches of law, and have the skills to work with citizens' appeals.

Thus, an e-government specialist must have knowledge and skills in the field of modern technologies and jurisprudence. This will help ensure a better work of government bodies, increase the level of accessibility of public services for citizens in order to ensure their constitutional rights and freedoms.

4. Further implementation of the "one window" principle, including.

Within the framework of the integration of state and municipal standards for the provision of services.

The process of evolution of the e-government goes from simple informing citizens about services to complex two-way interaction, in which citizens get access to all types of services of the state administration.

D.V. Kononenko points out that in modern conditions, in order to achieve the goal of communication between the state and society, it is necessary "to develop means of providing remote access for organizations and citizens to information about the activities of public authorities; provision of public services using multifunctional centers and the Internet; creation of a secure interdepartmental electronic document management system; introduction of unified departmental information systems for planning and management reporting; the formation of a regulatory legal framework "[8].

Now, in Russia, state bodies of each branch of government have their own websites created on platforms, the interaction between which is either not developed or not provided at all. However, getting services on the principle of "one window" from a single website of the electronic government in Russia remains difficult.

As noted by A.F. Sokolov, "state authorities must be integrated in such a way that their information systems can exchange with each other, and all information is available from one point. This principle has proven its effectiveness in Canada, Singapore and Australia, and it is he who underlies the concept of an "electronic representative" "e-Enjoy" in the UK" [9]. M.S. Yakubov and U.R. Kurbaev believe that "the structure of public services should be produced in accordance with "life events" and rganized according to the principle of"one window ". The main point is to provide citizens with the opportunity to receive the entire range of public services in one place" [10].

In this regard, we propose to create a single platform for the "Electronic State of the Russian Federation", which will combine the judicial system (websites of courts of general jurisdiction, commercial courts, magistrates' courts), a system of executive authorities (the State Service portal, as well as websites of regional and municipal authorities providing services) and the system of legislative authorities (consists not only in the provision of services, making an appointment, but also in the elements of e-democracy - the possibilities for the participation of the population of the municipality, the region in the adoption of laws and regulations governing legal relations in these territories).

5. Increase of computer literacy of the population and creation of consulting, training and accompanying centers.

This should also be carried out taking into account the following means:

- Provision of state and municipal services through terminals, cable TV, as well as through other means of communication;

- Development of centers for the provision of services in electronic form in all administrative-territorial units, the creation of their electronic nationwide map;

- The formation of a new culture of the electronic state, including the informational, legal culture of the population and the management culture of employees of state and municipal authorities.

It is necessary to form a culture of interaction between citizens and the state in the new electronic space. Otherwise, due to a lack of understanding of the ongoing processes, legal nihilism will prevail, which, as N.I. Matuzov is characteristic of the Russian mentality [11].

6. Inclusion of non-governmental structures in the infrastructure of the electronic state.

It is important that the idea of an e-government extends not only to the sphere of public administration, but and covers public relations. Unfortunately, non-state structures that provide services to citizens are not included in the e-government system. For example, the Bar is a public institution that does not belong to state authorities. It is impossible to apply through the portals of state services in order to obtain the services of a lawyer. In addition, the notary, which originated as an institution of civil society, is currently not connected to the system of providing electronic services. We believe that the designated institutions, state authorities and local self-government bodies should form a single e-government system in order to provide the entire range of services for the population. Thus, the content of the concept of a structural element of the electronic state - electronic government can be expanded.

7. Creation of specialized state portals, acting on the principle of an aggregator of all services provided by the state.

In accordance with Art. 29 of the Constitution of the Russian Federation: everyone is guaranteed freedom of thought and speech; everyone has the right to freely seek, receive, transmit, produce and distribute information in any legal way; freedom of the media is guaranteed. In accordance with this provision, timely and full provision of up-to-date data to the population on what decisions are made in municipal and state authorities and what types of services are provided by these bodies should be supported. The determining criterion for the implementation of this right is the presence of feedback from the population. Without the active participation of citizens in the activities of state and municipal authorities, it is difficult to talk about building an "electronic state" as a whole [12].

Trust and understanding on the part of the population can be achieved if the authorities themselves seek to provide the society with objective and reliable information about their decisions and actions. Citizens should receive all the information on the portal and, if they have any questions, directly contact a specialist consultant via the electronic service. Only in this case can we talk about the implementation of this mechanism in practice.

Ensuring easy adaptation and transparency of public services will significantly facilitate the transfer of interaction between the state and society to a new level. One of the main and most optimal forms of interaction between public authorities and society in the framework of the development of the idea of an electronic state is the creation of specialized state portals, acting on the principle of an aggregate of all services provided by the state. Understanding what service a user can order, what actions need to be taken, the degree of citizens' confidence in new communication methods will increase significantly.

The introduction of new technologies requires not only consistent and balanced state policy, but also the availability of special knowledge, reconstruction of the thinking style of a modern person in a rapidly progressing information space. The problems of translating the relationship between the state and society into electronic form were written by Yu.V. Soustin, D.V. Kononenko, A.A. Shanin, A.G. Madoyan, L.L. Popova, O.A. Kirekov and others.

D.S. Mikheev emphasizes, "The service delivery system is still at the stage of formation, until a verified and successful practice has taken shape. The advantages include the formed regulatory framework aimed at wide openness and publicity of procedures for the provision of municipal services" [13]. Nevertheless, problems often arise when citizens want to get a specific result, but, being confused in the wording and the actual meaning of the operation, order a different service. Of course, we are talking about special cases and over time, there should be fewer such situations. In addition, in order to avoid these problems, serious qualifications must be presented to specialists working in the field of electronic services.

Accordingly, the following are important here:

- Focus on the real needs of citizens, determined through social surveys in electronic means of communication;

- Taking into account the specifics of the needs and capabilities of various social strata and groups of the population (first of all, socially unprotected segments of the population and people with disabilities);

- The use of advanced ways of cooperation with the business sector.

8. Development and implementation of mechanisms of state legal regulation of the sphere of implementation of the idea of an electronic state.

Legislation must develop dynamically in order to keep up with new technologies introduced into the processes of interaction between government bodies and the population. Innovation is becoming increasingly important.

To date, there is an insufficiently unified state of the Russian legal framework for the electronic state, which is because the authorities have not yet come to a single legal standard that ensures the generality of the principles of government.

9. Building a secure electronic document management system.

V.P. Kirilenko and G.V. Alekseev conclude that "it is necessary to refine the legal and administrative mechanisms for ensuring security in the area under consideration, since traditional administrative algorithms and regulations aimed at ensuring national security do not work effectively enough in the information sphere" [14]. The flexibility and efficiency of adaptation to new threats to information security, population demands, changes in technologies and legislation should also be ensured.

The adverse consequences of hacker attacks can cause serious damage to citizens and the state, legal institutions and constitutional principles. The existence of an effective method of punishment for crimes in the field of information technology becomes a preventive security measure and a measure to protect the rights and legitimate interests of citizens, as well as restore social justice.

The very fact of criminal law protection of public relations governing the production, use, distribution and protection of computer information has a preventive, preventive value. The United States has the Computer Systems Protection Act (1979), and information on cybercrimes is contained in the United States Code. The French Penal Code provides for liability for violations of human rights associated with the use of filing cabinets and data processing on computers. The Spanish Criminal Code establishes responsibility for the receipt and use of information constituting personal or family secrets in information, electronic or television filing cabinets. This gives grounds to conclude that the protection of information relations is relevant all over the world in the era of the development of the idea of an electronic state.

In addition to hacker attacks, it is possible for the population to commit crimes by submitting deliberately false information to the authorities in electronic form. In addition, on the part of municipal and state officials, the possibility of violation of the rules of law regarding personal data and other information protected by law is not excluded. Accordingly, in the electronic state, it is necessary to develop and implement a universal law providing for liability for all types of cybercrimes: both existing and possible in the future. The possibility of adopting international laws or conventions in this area is not excluded.

10. Complete transfer of authorities to electronic document management.

An important direction of improving the electronic state is the transition to electronic document management and the rejection of paper information carriers. Nevertheless, today this system is of an auxiliary nature, since when confirming legal facts, original documents with seals and signatures of officials are still required. Of course, an electronic signature is being actively implemented now, but it is not yet possible to completely abandon material carriers.

An electronic document can be recognized as eligible if the following conditions are met: there is its material analogue - a paper equivalent; the document contains the appropriate details - personal data about the person who issued it; the requirements and rules for tangible media in accordance with the law are observed.

The process of transition to electronic document management is developing in connection with the implementation in Russia of the Information Society and Electronic Russia programs, where a key element is the communication of authorities using electronic document management. Therefore, the attention of specialists in the field of public administration, politicians, legal practitioners and theoretical scientists is focused on confirming the legitimacy of electronic copies, which are much easier to counterfeit if they have sufficient skills in working with computer programs. E.V. Lapteva, D.M. Sokolova, A.V. Ermolaeva, G.A. Dvoenosova, E.R. Sultanova and others, actively pursued the study of this topic. The development of electronic document management in Russia presupposes a complete transition to electronic interaction in government by 2019.

All these directions of development of the electronic state were initially formed in theoretical works and industry studies, then, based on the developed theoretical base, the implementation processes begin. This can be seen from the changes in the current legislation of Russia and foreign countries

References

1. E.G. Vasileva, D.V. Kononenko, Bulletin of the Volgograd state University, **1 (30)**, 9-16 (2016)
2. P.U. Kuznetsov Russian law journal, **3**, 37-47 (2014)
3. Yu.V. Nikulina Scientific works of the Republican Institute of higher education, **15**, 97-104 (2016)
4. UNE-Government Survey 2016. URL: https://publicadministration.un.org/egovkb/en-us/Reports/UN-E-Government-Survey-2016 (Date of access: 20.12.2020).
5. A.A., Korobov, T.S. Melnikova, N.V. Mityaeva *Formation of a public need for the development of e-government and e-democracy systems in modern Russia.* (Saratov socio-economic Institute of Plekhanov Russian University of Economics, 2016).
6. O.N. Sinitsyna, D.A Rakitina. Traditional national-cultural and spiritual values as the Foundation of innovative development of Russia, **1(7)**, 67-68 (2015)
7. V.Yu Turanin. The science. Art. Kultura, **2**, 247-251 (2015)
8. D.V Kononenko, Bulletin of Volgograd state University **2**, 34-38 (2013)
9. A.F. Sokolov. Regional Economics: theory and practice, **32**, 44-50 (2010)
10. M.S. Yakubov, U.R. Kubaev, Science and the World, **3 (31)**, 101-103 (2016)
11. N.I. Matuzov Law culture, **1**, 8-24 (2012)
12. A.S. Kiselev *Proceedings from Revolution and modernity* (Vladimir, Russia: Vladimir branch of Ranepa, 2017)
13. D.S. Mikheev, Mari legal Bulletin, **1 (12)**, 50-52 (2015)
14. V.P. Kirilenko, G.V. Alekseev, Management consulting, **9(93)**, 21-29 (2016)

Overcoming psychological barriers in discourse interaction in the era of information

Ekaterina Bobyreva [1*], *Marina* Zheltukhina [2], *Mariana* Busygina[3] and *Olga* Khudobina [4]

[1] Volgograd State Socio-Pedagogical University, Volgograd, Russia, new_life@mail.ru
[2] Volgograd State Socio-Pedagogical University, Volgograd, Russia, zzmr@mail.ru
[3] Volgograd State Socio-Pedagogical University, Volgograd, Russia, busygina_maryana@mail.ru
[4] Ugra State University, Humanitarian Institute of Nothern Studies, olga_hdb@mail.ru

Abstract. The article deals with the types of communicative competence in different types of discourse. Communicative competence refers to the ability to predict a communicative situation in a specific area of communication which becomes important in the era of information, to manage the process of communication in a certain situational environment. Levels of language personality formation are distinguished. The concept of a psychological barrier arising in the process of communication is given. Psychological barrier refers to a condition that impedes full development and manifestation of a personality. Multilevel character of communicative competence is shown, levels of its formation are distinguished; such types as lexical, textual and discourse competence are considered. As a person enters discourse space and develops communicative competence, his emotional state changes – from negative or neutral emotions to positive emotions. The question is raised about the possibility of communicative failures caused by communication barriers arising during communication. Types of communicative barriers, their characteristics and ways to overcome them are considered; besides, barriers related to insufficient proficiency in a particular level of language (phonetic, lexical, stylistic) are considered. Among the reasons for communicative barriers are: features of communicants' intelligence; difference in interpreting concepts; difference in vocabulary; misunderstanding of communicative situation; psychological characteristics of partners; social, political, professional, religious differences.Many of these barriers appear in modern society which overcomes specific changes in the era of information. Significance of psychological factor as well as level of formation of emotional competence to overcome emerging communicative barriers is shown.

1 Introduction

In the process of communication in different types of discourse, there "may arise states that impede full manifestation of the personality, block its activity, such states may be realized both in the intellectual and practical sphere, impede creative expression, lead to general dissatisfaction with the learning process and reduce the effectiveness of professional training" [1]. Such conditions can be called forth by the psychological barriers – obstacles and difficulties that arise in the process of communication in almost any type of discourse. Problem of perception, understanding and interpretation of the received information is one of the main problems of modern society in general [2].

2 Problem Statement

In many respects psychological barriers are the result of the contradiction between new communicative tasks and opportunities that a discourse participant has, which imply a good level of knowledge of the subject sphere of communication, general level of knowledge and motivated emotional communicative activity in the process of social interaction [3]. Problem of overcoming communicative barriers appears to be relevant because such barriers make process of understanding during communication significantly complicated [4].

3 Research Questions

The question is raised about the possibility of communicative failures caused by communication barriers arising in the process of communication. Types of communicative barriers (avoidance, authority, misunderstanding), their characteristics and ways to overcome them are considered; besides, barriers related to difficulties or insufficient proficiency in a particular level of language (phonetic, lexical, stylistic) are considered. Among the reasons arising communicative barriers are: features of communicants' intelligence; difference in interpreting concepts; difference in vocabulary; lack of correct understanding of the situation of communication; psychological characteristics of partners; social, political, professional, religious differences.

* Corresponding author: new_life@mail.ru

4 Purpose of the Study

The article deals with the types of communicative competence in different types of discourse. Communicative competence refers to the ability to predict a communicative situation in a specific area of communication, to participate and manage the process of communication in a certain situational environment.

5 Research Methods

The following methods have been used to conduct research: semantic analysis, discourse-analysis, method of cultural meanings' analysis. Levels of language personality formation are distinguished. The concept of a psychological barrier arising in the process of communication and impeding normal communication is given. Psychological barrier refers to a condition that impedes full development and manifestation of a personality. Multilevel character of communicative competence is revealed, levels of its formation are distinguished; such types as lexical, textual (genre) and discourse competence are considered. The dependence of the correct expression of emotions on the level of formation of communicative competence is distinguished.

6 Findings

We consider it possible to divide all communicative barriers into three groups: 1) language barriers – insufficient knowledge of linguistic means (in particular, terminology), which are used in a certain sphere of discourse (medical, scientific, religious); 2) discourse barriers – lack of knowledge of norms and rules of communication in a certain communicative space within a certain discourse; 3) psychological barriers – barriers related to the will or emotional sphere of a person which can be explained by the insufficient development of a person's psychological readiness to enter and participate in the process of communication.

In general, five levels of linguistic personality formation can be distinguished: 1) level of correctness (assuming sufficient knowledge of lexical and structural units of the language which allow to build statements and texts in accordance with the rules of the language); 2) speed level (ability to perceive and produce utterances with natural speed); 3) level of saturation of vocabulary and language forms (in terms of the variety of linguistic expressive means use); 4) level of selection of the required language unit (from the point of view of the language means used, tonality which suits this sphere of communication and communicative situation, as well as necessary communicative role); 5) level of adequate synthesis (correspondence of the text produced by a person to the complex of informative and communicative tasks) [5].

Speaking about the peculiarities of discourse communication, it is necessary to consider the question of communicative competence. By communicative competence we mean the ability of a participant of communication to predict a situation in a particular sphere of communication as well as to participate in and manage the process of communication (both at a rational and emotional levels) in a certain situational environment.

The concept of communicative competence has a multi-level character, the following types of competence are distinguished:

1. Lexical competence: knowledge of lexical units and the main terms characteristic of this sphere of communication. Note, that in any type of discourse communication attention of the researcher is primarily concentrated on the meaningful side of the utterance. And only after that the researcher can turn to the form into which this content is enfolded. The shape, therefore, plays only the role of the outer shell, a "storage" for information. Form, therefore, doesn't matter on its own. Though, this statement is not relevant, for example, for the religious discourse. In religious discourse form is significant in itself, it has priority and in some cases it can be more important than the content [6]

2. Text (or genre) competence: the ability to navigate in the genre space of a particular discourse; understand and distinguish the main genre samples, as well as build certain genre fragments according to certain rules. Text competence presents the level of competence that follows lexical competence. After a communicant has elementary lexical base formed and begins to navigate in the main concepts and terms in a particular sphere of communication, he begins to form text or genre competence.

Text (genre) competence can be implemented in two types – passive and active. Formation of text competence begins at a passive level. Passive stage involves understanding of the main genres and the ability to distinguish (without any doubt, only at the level of an amateur, according to certain external characteristics) one genre from another. Following the passive level of text competence active level is formed, and it involves not only recognition and distinguishing the key genres but also the ability to build some of them. Active level of genre competence is a rather deep immersion in the communicative space of discourse. After the participant has entered this discourse rather deeply and acquired the necessary knowledge the text competence which was formed at the passive level passes to the active level, where the participant is able not only to distinguish the most important genres but, if necessary, to build certain genre fragments.

3. Discourse competence: represents the highest level of communicative competence; knowledge of language and ability to communicate in various spheres of discourse. Discourse, in a broad sense, is a text immersed in a situation of communication. Thus, discourse competence involves, besides language knowledge (this level of competence implies presence of all the above types of competence – lexical, text\genre) the ability to navigate within the framework of this social institution.

Discourse competence is based on lexical competence, as well as text (genre) competence, which being built one on another like blocks create some basis for the further formation of a more complex and complicated discourse competence. Although some components that form this type of competence (for example, knowledge of some elements of ritual) are already formed in the early stages of a person's "entry" into a communicative space of a certain discourse.

To certain extent, it will be fair to say that discourse competence, having been formed in its full scale, really follows lexical and textual (genre) competence being the phenomenon of a higher level. However, it will not be an exaggeration to claim that some components of this competence are formed by a participant of communication at the very beginning of his entry into the communicative space of a certain discourse. However, it is possible to talk about complete and not fragmented formation of discourse competence only after a communicant has rather deeply immersed in the specific communicative field of this discourse. The development of discourse competence is a more complex, compared to the competence of the lexical and textual (genre) continuum.

In any type of discourse, a certain emotional message is hidden behind practically each word and each text. Emotions colour the text, give it additional meaning and sometimes change meaning of a message. In particular, even understanding itself begins at the level of emotions – at the beginning of any process of perception there is a corresponding reaction to the situation and emotional state of the partner, and only then meaning is decoded at the cognitive level [7].

Speaking about the process of communication, it should be noted that more often there are not words that are understood but what stands behind them – a communicative personality with its emotions [8]. Moreover, different emotional mood of communicants can lead to different accuracy of understanding, and in some cases to misunderstanding and impossibility to continue communication – the recipient perceives information passing it through his emotional field, his "emotional echo" may or may not meet the expectations of the addressee [9].

In the process of communication (as linguistics of emotions states), there is a change in the emotional field of the speech partner, "infection" with his emotions, a certain emotional interaction. At the same time, asymmetric emotions can be a subjective factor of the communicative barrier. E.V. Rudenskiy points out that there are some emotions that are negatively stable and thus create a communicative barrier, among them – suffering, anger, aversion, fear, bad mood [10]. These emotions violate the level of communicative contact since negative emotional state of one communicant can be transmitted to his partner. Emotional interaction of partners is inevitable in the process of any communication.

Conducted analysis made it possible to conclude that as a person enters the discourse space and his communicative competence (active or passive) is being developed, there is inevitably a change in his emotional state – from negative or neutral emotions at the beginning of communication to positive emotions as the communicative competence of a person develops. As we move up the ladder of communicative competence the emotional background of the communicative personality radically changes. Even though the level of communicative competence and the emotional background, the mood of the communicative personality do not determine each other, their development and change go in parallel and it seems possible to establish several regularities in their co-relation. Each position, each level of communicative competence corresponds (to a greater extent) to a certain emotional mood of the participant in communication, which, on the one hand, is determined by general communicative practice and, accordingly, communicative competence of a person, and, on the other, depends to a certain degree on the level of formation of communicative and general discourse competence.

From a psychological point of view, three types (forms) of communicative barriers are distinguished: a) avoidance, b) authority, c) misunderstanding. This classification of barriers clearly shows that communication barrier is in many ways a mechanism for protecting against unwanted information.

Psychological barrier that the recipient puts on the way of unwanted, tedious, or dangerous information may have various degrees of transparency [11]. An opaque barrier is avoidance. Unwanted information can be avoided both physically (avoiding communication) and psychologically (ignoring information). Psychological barriers are among the most frequent and common ones. Such barriers are characteristic of communication in the framework of scientific, religious, and medical discourse [12].

The second barrier is authority, action of which can be interpreted as follows: information enters consciousness, but on its way, it is significantly depreciated through a subjective decrease in the authority of its source – that is, in the end, it becomes unreliable and insignificant. Such barriers arise in everyday communication when communicants have equal status [13]. In this case, the addressee may decide that this information is not interesting for him or not relevant, and unilaterally he may "refuse" it and leave communication without any "sanctions".

The third barrier is misunderstanding. The reasons for communication barriers can be found in the content and formal characteristics of the message itself, as well as in the logic of its construction. Logical barrier occurs when partners cannot find a common language. Each person sees the world, the situation, the problem that is discussed, from his point of view, which may not coincide with the position of the partner. In addition, the same words in a particular situation may have a completely different meaning, which is always individual-personal: it arises in the consciousness of the one who speaks, but is not necessarily understandable to the one who listens.

From a linguistic point of view, barriers can be phonetic, semantic (lexical), structural (incorrect division of the heard phrase).

Phonetic barriers (obstacles that are created by the peculiarities of the speaker's language) usually arise when participants in the process of communication speak different languages, dialects, have defects of speech and diction, distorted grammatical construction of statements.

Psychologists have also proved that when something is reported to a person in elevated tones, understanding is often blocked. Psychologically, the reason lies in the fact that all attention of the individual in this case is aimed at the flow of offensive words and the intonation pattern of the phrase. As a result, a protective reaction occurs and we can observe a shift of attention.

Semantic barrier often arises due to the absence of matches in the value systems of communication partners (thesaurus). Semantic barrier is, firstly, a problem in jargon and slang; secondly, in the limited vocabulary of one of the interlocutors; thirdly, social, cultural, psychological, national, religious, professional, group and other features of communication.

T. Dridze considers communicative situations in which this effect arises: a. clear discrepancy between the language means used by the communicator and the language resources of the recipient; b. discrepancy arises even earlier – at the stage of conversion thought into words; c. mutual understanding can be hampered by certain individual characteristics of the recipient – ability to operate with language as a means of thinking [14].

Stylistic barrier arises when style of speech of the communicant, situation of communication and actual psychological state of the recipient do not correspond. For example, a communicant may not accept critical remark of the interlocutor, because it is said in a friendly manner. A stylistic barrier can also arise provided that the form and content of communication do not correspond. A stylistic barrier also occurs when information is transmitted in a style that is not characteristic of this area of communication.

Let us list the most general reasons for communication barriers: a. peculiarities of communicant's intelligence; b. different understanding (interpretation) of the subject of the conversation; c. different vocabulary (lexicon, thesaurus); d. absence of a common understanding of the situation of communication; e. certain psychological features of partners; f. social, political, professional, religious differences [15].

Barriers associated with the communicative characteristics of participants are social or psychological in nature.

In the process of communication, it is important not just to transmit information, but to be able to manage the "direction" of thinking of the partner of communication. The effectiveness of communication depends significantly on how deeply partners are involved in the process of communication.

7 Conclusion

To overcome communication barriers, the addressee must possess the art of "managing the partner's attention". Research in the field of psychology has shown that attention can be attracted by external and internal factors. External include novelty (unpredictability) of the message, intensity, and physical characteristics of the signal; to internal – relevance, significance, importance of information for a person.

To increase the effectiveness of communication, it is important to be able to direct thinking of communicative partner to the "right" direction. To be understood by the interlocutor, you need to take into account logic of the partner. To do this, you need to take a partner's position, to understand his individual and socio-role characteristics. Understanding the partner, adequate interpretation of his point of view, goals, individual features are the main conditions for overcoming psychological barriers arising in the process of discourse communication.

References

1. V.I. Karasik, *Language manifestation of the personality*. (Volgograd: Paradigma, 2014)
2. E.V. Bobyreva, M.R. Zheltukhina, O.A. Dmitrieva, & M.V. Busygina, Advances in Social Science Education and Humanities Research **97,** 52-56 (2017)
3. A. A. Ivin, *Axiology*. (M.: Yurait, 2018).
4. G.I. Goddess, *Model of linguistic personality in its relation to varieties of the text*: diss.... *doct. philol. sciences* (Leningrad, 1984).
5. E.K. Chernichkina, *Artificial bilingualism: linguistic status and characteristics: autoref. dis. ... doct. philol. sciences* (Volgograd: Peremena, 2007).
6. O.F. Khudobina, *Pedagogical conditions for overcoming psychological barriers by students in the process of bilingual education at a university: autoref. diss... kand. pedagogical. sciences* (Volgograd: Peremena, 2007).
7. S.E. Krivykh, *Axiology of world religions: a comparative typology of value structures: dis. ...cand. phil. Sciences.* (Armavir, 2009)
8. E.V. Bobyreva, M.R., Zheltukhina, K.G Korovina,. & M.V. Busygina, *SHS Web Conf.* **69**, 00021 (2019).
9. V. I. Shakhovskiy, *Linguistic theory of emotions* (Moscow: Gnozis, 2008)
10. E.V. Rudenskiy *Conceptual foundations of psychology of an adapting personality: an introduction to the psychological theory of the social functioning of the personality* (Novosibirsk: Institute of Personality Psychology, 1997)
11. E. I. Melibruda, *I – You – We: psychological opportunities for improving communication* (M.: Progress, 1986)
12. D.Z. Akhmetova, T.S. Artyukhina, M.R. Bikbaeva, I. A. Sakhnova, M.A. Suchkov, & E.A. Zaitseva, Higher education in Russia, **29(2),** 141-150 (2019)
13. E. A. Akhverdiev, & A. V. Ponomarev, *International Scientific and Practical Conference "Current Issues of Linguistics and Didactics: The Interdisciplinary Approach in Humanities and Social Sciences": SHS Web Conf., 50.* (2018)
14. T.M. Dridze, *Next activity in the structure of social communication* (Moscow: Nauka, 1984).
15. E.P. Panova, & N.R Saenko, Service plus, 94-102 (2020)

Suffering in world religions within paradigm of modern information

Ekaterina Bobyreva [1*], *Olga* Dmitrieva [2], *Tatyana* Gonnova[3] and *Anna* Oganesyan [4]

[1] Volgograd State Socio-Pedagogical University, Volgograd, Russia, new_life@mail.ru
[2] Volgograd State Socio-Pedagogical University, Volgograd, Russia, dmoa@yandex.ru
[3] Volgograd State Socio-Pedagogical University, Volgograd, Russia, concept-t@mail.ru
[4] Volgograd State Social Pedagogical University, oganesyananna1994@mail.ru

Abstract. Any type of discourse, along with its characteristic concepts, operates with its own values. The basis of religious discourse form universal values providing moral guidelines that represent a standard for people of different cultures and eras, which are associated with the ideals of justice and are timeless. Among the universal are cultural, social and moral values. The bulk of religious values form cultural ones in any modern society. Religious beliefs form inner culture of a person. Some religious values can be referred as social and include: meaningful (meaning of life, happiness), universal (life, health), values of interpersonal communication (benevolence), values of public recognition (hard work), democratic values (freedom of speech). In modern society among religious, social and moral (mercy, compassion) values can be distinguished. Religious values can be found within each subgroup of universal human values. The article interprets phenomenon of suffering in modern society, which is an integral component of any world religion and forms the category of value in Christianity. The analysis of suffering in the article was carried out along with the analysis of religious discourse - a special type of institutional communication that combines features of institutional-oriented and personal-oriented phenomenon. The article shows reasons for the occurrence and existence of suffering in any modern society, as well as approaches that exist in various religious systems (Christianity, Islam and Buddhism) to interpreting suffering, attitude to it, need and possibility to overcome it as well as linguistic means to express phenomenon of suffering in world religions.

1 Introduction

Understanding and interpretation of various phenomena and objects of the surrounding life is the basis of their acceptance by a person [1]. Religious discourse, within the framework of which phenomenon of suffering will be considered, distinguishes from all other existing types of institutional communication the open postulation of values [2]. According to E.A. Kozhemyakin, one of the features of religious discourse is the "presumption of sincerity of the subject" [3], whose statements can be assessed as "false" or "true", "correct" or "wrong", but they can never be "deliberately false" or "manipulative". One of the main argumentative tactics in religious discourse is appeal to authority. The authorities of the assessment are canonical figures, as well as "subjects whose public prestige increases the significance of their actions" [4]. In its turn, evaluation in religious discourse is based on moral and moral postulates that form the basis of this religious concept. The basic opposition "good::evil", which determines the key provisions of a particular religious doctrine, is expressed in specific categories of assessment "good::bad", "acceptable::unacceptable", "permissible::unacceptable".

2 Problem Statement

The basis of communication is information environment or information paradigm, in a modern interpretation. In the development of the information paradigm it is advisable to distinguish three stages: classical, non-classical and post-classical (modern); these stages are related to cultural and historical codes of civilization. All existing information can be presented in the ontological, gnoseological, axiological aspects. Social structure is heterogeneous, analysis of given heterogeneity of society showed coexistence in any historical period of time of different value systems [5].

3 Research Questions

* Corresponding author: new_life@mail.ru

Religious values present timeless phenomenon, they can be defined as "principles", "laws", "attitudes", "commandments", "canons", "spiritual axioms", etc.
Universal human values that form the basis of religious discourse: 1) provide fundamental moral guidelines and represent absolute standard for people of all cultures and eras; 2) connected with the ideals of humanism, justice and dignity of a person; 3) do not correspond to a specific historical period of society's development or a specific ethnic tradition.

4 Purpose of the Study

In the category of universal values: a) cultural values, b) social values, c) moral values can be distinguished. The bulk of religious values refer to cultural values. Religious beliefs and values are the main components of a person's daily life, forming his internal culture. At the same time, a number of religious values can be referred to social values, which include: meaningful values (ideas about happiness, purpose and meaning of life); universal values (life, health, personal security, family, law and order); values of interpersonal communication (honesty, disinterest, benevolence); values of social recognition (hard work, social status); democratic values (freedom of speech, conscience). Such values as life, family, honesty, unselfishness, etc., are central in a number of religious teachings. Therefore, among religious, a number of social values can be found. Moral values are certainly an integral part of religious values - mercy, compassion are significant values of Christian creed. Thus, religious values can be found inside each subgroup of universal human values.

5 Research Methods

The methodological framework of this research is the system approach. The purpose of the research is achieved by the application of the methods of semantic analysis, discourse-analysis, method of cultural meanings' analysis.

6 Findings

Human values are the focus of attention of various approaches and are considered by different researchers. The biogenetic concept of universal human values considers the latter as universal cultural phenomena common in their function (they satisfy the vital needs of man and society - survival, adaptation to the environment, etc.)[6].

Speaking about the values of religious discourse it seems appropriate to present the system of values in the form of peculiar oppositions. Almost all concepts that are interpreted as values have their absolute opposite – "anti-value". "A person would never think about the value of common things if he did not constantly run into" forks". The choice requires motivation. And it is, as a rule, sought in practical expediency, convenience, ease, reliability, desirability and other private assessments" [7].

The peculiarity of religious values is their open statement. These values are based not on rational, but on emotional base. Higher religious values are values through which a person comes closer to God. Acquiring higher values an individual gets the opportunity to transcendence, to go beyond his private existence, to rise above it [8].

The expression "Christian values" came into at the beginning of the 20th century. But the concept *'moral value'* first appeared in the ethics of E. Kant [9].

Christianity proceeds from the understanding of value as an absolute good that is relevant in any respect and for any person. According to Yu.S. Garanov, religious (Christian) values are "absolute spiritual and moral sources of the person gaining connection with the transcendent, achieving the highest improvement, grace, salvation (liberation), spiritual immortality, mystical knowledge". The desire for these values characterizes the level of spirituality, morality of a person, orientation towards the highest ideal. Classifying Christian values, Yu.S. Garanov identifies: 1) religious-ontological values (God, existence, nature); 2) religious and anthropological values (man, soul, freedom of will, faith, hope, love, grace, salvation, etc.); 3) religious and ethical values (good, love for both neighbor and enemy, non-resistance to evil, forgiveness, suffering, martyrdom, holiness, virtue, etc.); 4) religious and aesthetic values (beauty as a reflection of God, beautiful as a form of good, etc.) - endowed with sacred, mystical meaning the highest qualities of a person's sensual-figurative attitude to the world; 5) religious and cognitive values (Holy Bible - Scripture and Tradition, practices of praying); 6) religious and symbolic values (icons, rites and rituals); 7) religious and social values (church, Christian community) [10].

In the most general form suffering is "a combination of unpleasant and painful sensations of a living creature, in which it experiences physical and emotional discomfort, pain, stress, torment; thus, in the most general meaning, suffering is physical or moral pain, torment" [11].

The primary basis of human suffering is inherent negative emotion of both animals and humans, "signaling about difficulties in implementing basic instinctive programs of behavior (experiencing unfulfilled or suppressed desire)" [12]. However, due to the rational nature of a person's consciousness, his suffering moves from the level of strong emotions to the level of feelings. Human suffering, therefore, is an extremely strong negative feeling in which a negative situation is experienced as a 'fragment of a dysfunctional world" which makes it impossible for a person to continue a comfortable life.

Suffering is an integral part of human existence that is why this phenomenon has a special place in each world religion. Any religion develops its own attitude to suffering and offers its own way to overcome it [13].

Christianity makes suffering absolute. An example of the highest suffering that ultimately led to salvation are sufferings of Jesus Christ himself and his death on the cross: "And whoever does not take his cross, does not follow me, is not worthy of me" [14].

Suffering in Christianity is considered not as a cause for sorrow but as a cause for joy because suffering endured in person's present life promises (and, to some extent, even guarantees) happiness in another life: "But as you participate in Christ's suffering, rejoice, and in the manifestation of his glory you will rejoice and triumph" [14].

Christianity to some extent convinces a person that his sufferings are only a "small fee" for the well-being and happiness that a person will get after death. Perhaps that is why, having true faith, it is easy for a person to endure physical and moral deprivation: "I think that the present temporary suffering is worthless compared to the glory that will be revealed in us" [14]. Current life, and at the same time current sufferings are defined as "temporary" ones, while the "Kingdom of God" awaiting for a suffering person is presented as eternal - such opposition "temporary::eternal", having some psychological effect on a persons who believes in God, helps him to endure current deprivations.

The idea of inevitability of suffering is also reflected in the language: "The Lord tolerated and commanded us to tolarate!". Many monks' orders consciously try to bring their lives closer to the life of Jesus. The very belief that suffering is welcomed by God led to the idea of asceticism; Christian ascesis (starvation, self-destruction, asceticism, wearing verigs, etc.) is considered one of the ways to improve spiritually.

Suffering has a special pace in Christianity. According to John Domaskin, suffering - is a special state of man, followed by "pleasure or displeasure". "Fruits of sufferings", therefore, depend on the choice of a person himself. Christian history says that two "criminals" were crucified, but for one of them sufferings meant, while another one became even more cruel.

Christianity views present sufferings of man as a temporary stage which will necessarily be followed by "reward", "retribution" for all hardships suffered during earthly life: "As the sufferings of Christ multiply in us, Christ multiplies and our comfort is increased" [14]. Moreover, Christianity considers suffering itself as a reward sent to a person from above, and not without a reason the Bible uses the expression "you are given" but not "you have to" or "you are forced", but namely "given" - a person as a creature chosen by God is given a reward to suffer in the name of a future happy life: "You are given for the sake of Christ not only to believe in him but also to suffer for him" (Bible, 1973). This is the reason for the Christian concept that the more a person suffers the more Lord loves him.

From the point of view of Christianity suffering is not an absolute evil. Having, in fact, a negative vector of evaluation, suffering leads to good consequences. Suffering plays a purifying role: "Before my suffering I was misled; and now I keep your word" [14]. Speaking about sufferings in the Christian conception, it is worth pointing out that, first of all, they mean not body sufferings, but mental sufferings arising from inability to have what is necessary for a person in his life - food, home, family, etc. So, what it sometimes means is that very often it is not possible to meet person's basic physical or mental needs. However, according to the Apostle Paul: "The kingdom of God is not food and drink, but righteousness and peace and joy in the 'Holy Spirit'" [14], so the righteous, awaiting for future bliss, must and can enjoy the kingdom of God, which is within them.

Christianity almost always sees the cause of suffering in sins, in violation of morality and law by a person: "We suffer for our sins" [14]. In its turn, suffering has a beneficial effect on a person; Christianity teaches that suffering almost always has a certain cause and convinces that in the life there is not moral chaos but a slender order based on the truth. And it is through suffering that a man is improved. Therefore, Christianity includes suffering in the category of values: "Life is a huge workshop in which the souls of people become cleaner and prepare for a transition to a different, better world".

Christianity delimits just suffering and suffering for "good deeds", "suffering for the truth". It is suffering "for the truth" that brings man closer to God: "But if you suffer for the truth, you are blessed" [14]. Suffering is considered as something sent to a man by God, and if so, any suffering is inevitable and must be endured: "Bear suffering, therefore, as the good warrior of Jesus Christ" [14].

Besides, sufferings endured by a person teaches him to be condescending to other people. So sufferings improve a person. "It is unworthy to taste only good. Through sufferings the soul improves" - this quotation of one of the firm's character clearly conveys Christian attitude to the idea of suffering. It is true that believers view sufferings as a way to improve, an opportunity to approach God, that is, sufferings in Christian conception is perceived as God's mercy, his benevolence to a man, designation of a man as one of the chosen ones. That is why suffering itself can be regarded as a Christian value.

Unlike Christianity, where suffering is elevated to a cult and is an integral part of human life, "guarantee" of salvation and a happy existence in another world, the philosophy of Buddhism focuses on the idea to make a person free from suffering.

Like any other religion, Buddhism promises that people can get rid of sufferings, passions, fear of death, etc. However, denying the immortality of soul, Buddhism sees no sense to pursue eternal life (in heavens) since eternal life from the point of view of Buddhism is only endless series of reincarnations, a change of bodily shells. Unlike other religions, Buddhism does not recognize providence and emphasizes that the fate of a person depends only on his own efforts: "*Canal builders let water, archers obey an arrow, carpenters subjugate a tree, sages humble themselves*".

The most important concept for Buddhists is the concept of dharma (law, truth, path), it represents the teachings of Buddha, the highest truth that he revealed to all beings. Dharma literally means "support", "something that supports". In Buddhism the word "dharma" means moral virtue, primarily moral and spiritual qualities of Buddha, which all believers must imitate.

The doctrine of Buddhism rests on four noble truths: about suffering; about the cause of any suffering; about eliminating the cause of suffering and about the way to end suffering: "The good truth is that there is suffering. Birth – is suffering, old age – is suffering, illness – is suffering, death – is suffering; connection with what is unpleasant is suffering; leaving what is pleasant - is suffering; lack of possibility to achieve the desired - is also suffering".

The main goal of Buddhism is to achieve enlightenment or nirvana - a special state of consciousness in which understanding of the true meaning of things is achieved and occurs deliverance from passions and ignorance that tie everyone to sansara - an endless chain of earthly incarnations.

Buddhism knows the 8-stages way (path), the way to get rid of sufferings of present life: "This 8-stages way (path) is: the right views, the right intentions, the right speech, the right actions, the right lifestyle, the right efforts, the right awareness and the right concentration". Such 8-stages way (path) contains three main groups of exercises: exercises in morality, contemplation and wisdom: culture of behavior (correct thoughts, correct words and actions), culture of meditation (correct awareness and concentration) and culture of wisdom (correct views - knowledge of noble truths). Buddhism does not only testify to the possibility of liberation from suffering, but also states the way in which each person is able to achieve freedom and even become a Buddhist (enlightened). All these points differ from other well-known religions – not any other religious doctrine recognizes that a person can make himself a God-like creature by his efforts, having achieved the highest goal - to stop sufferings, to achieve liberation from all that belongs to the earth (nirvana).

Islam, to certain extent, interprets sufferings of this life as trials sent to a person; and in this it echoes Christianity: "We will certainly test you with insignificant fear, hunger, loss of property, people and fruits. But please the patient ones who, when they are in distress, say: *"Truly, we belong to Allah, and we will return to him"; "They receive blessing of their Lord and his mercy. They follow a direct path"* (Quran, 1999). Suffering is a kind of test that a person must endure to enter paradise in the future: "Or did you believe that you would enter Paradise without experiencing what befell your predecessors? They were afflicted by poverty and disease. They experienced such upheavals that the Messenger and those who also believed said to him: "When will the help of Allah come?" Truly, Allah's help is close" [15].

Trials and sufferings are given to everyone. Islam emphasizes that even the prophet Muhammad was given difficulties and trials: "Most of all people are tested by the prophet, then those who were closer to them (by their faith), then those who were closer to these. And a person is tested according to the degree of his religion (faith). If he was steadfast in religion, then his trials also increased. If there was weakness in his faith, then he was tested according to the degree of his religion. And the slave will suffer deprivation and distress until they make him walk on the earth free of his sins". In our opinion, this position of Islam, to some extent, echoes the Christian message – "To whom much has been given, from that one much will be asked", recorded in the paremiological fund of the Russian language. Along with suffering, Islam pays special attention to patience, which is considered as grace and blessing of the Most High, which is given to people and allows them to overcome all difficulties of the life.

7 Conclusion

Thus, sufferings occur in person's life in order to teach him to follow moral laws established by the Almighty. Sufferings and misfortunes are a test to a person's faith and morality. Islam teaches that whenever a person faces some misfortune, he must ask himself: "Didn't he violate any law of Allah?". And if it is seen as a kind of punishment, a person must ask for forgiveness and try to change his life; if it is a kind of test a person should make some efforts and pass this test.

We should notice that since human existence cannot be completely devoid of difficulties, troubles, experiences that inevitably entail the emergence of negative emotions, none of the world religions can circumvent the comprehension of such a phenomenon as suffering. Each religion offers slightly different ways to overcome suffering (for example, Christianity and Islam, on the one hand, and Buddhism, on the other), however, all world religions agree on one thing – it pays maximum attention to the phenomenon of human suffering and states the necessity to help a person if not to get free of the sufferings completely (which is impossible) at least to make the existing state of sufferings easier.

References

1. E.V. Bobyreva, O.A Dmitrieva, M.R Zheltukhina. & M.V. Busygina, Advances in Social Science Education and Humanities Research, **97**, 52-56 (2017).
2. E.V. Bobyreva, *Religious discourse: values, genres, linguistic characteristics (based on Orthodox creed)* (Volgograd: Peremena, 2007).
3. E.A. Kozhemyakin, *Discourse approach to the study of institutional culture* (Belgorod: Publishing House of BelSU, 2008)
4. A.A. Ivin *Axiology: book for academic undergraduate studies (* M.: Yurait, 2018)
5. E.V. Bobyreva, M.R. Zheltukhina, K.G. Korovina, & M.V. Busygina, *SHS Web Conf,* **69**, 00021 (2019).
6. B. Malinovskiy, *Selected: Dynamics of culture* (M.: ROSSPEN, 2004).
7. N.D. Arutyunova, *Language and the world of a man* (M.: Languages of Russian culture, 1999).
8. I.E. Sedankina, *Axiology of religion* (Kazan: Publishing House of Kazan University, 2013).

9. I. Kant, *Study of clarity of the principles of natural theology and morality* (M.: Nauka, 1985)
10. Yu.S. Garanov, *Christian values in modern domestic cinema industry: philosophical analysis: autoref. diss. .... cand. philol, sciences* (Astrakhan, 2014).
11. S.I Ozhegov, *Interpretive Dictionary of the Russian Language* (M.: Azbukovnik, 2000)
12. Yu.V. Gritskov, *Phenomenon of suffering and the ways of its development in culture:.... diss... doctor of philol.. sciences (*Omsk, 2006)
13. E.V. Bobyreva, *Linguistic and cultural characteristics of religion as the basis for the formation of values' guidelines of society: monograph* (Saint Louis, Missouri, USA: Science and Innovation Center Publishing House, 2019)
14. Bible *Books of the Holy Scriptures of the Old and New Testaments* (Brussels, 1973)
15. Quran. *Translation of meanings and comments by V. Prokhorova* (Moscow-Damascus: Al-Furkan, 1999).

New terms in the information society: nominative characteristics

Yuliya Chemeteva*[1] and *Marina* Davydova[2]

[1] Volgograd State University, Prosp. Universitetsky, 100, 400062 Volgograd, Russia, chemeteva@volsu.ru
[2] Volgograd State University, Prosp. Universitetsky, 100, 400062 Volgograd, Russia, davidovaml@volsu.ru

Abstract. The paper deals with the issue of new terms in the information society. The authors consider the concepts of neologism, legal term and terminoid giving examples. The paper states that currently the process of neologization in the Russian-language legal discourse is characterized by the fact that one of the active sources of terminology are terminoids which initially operate in new fields of law and function for nominating innovative technologies in law. The authors mark two ways of appearing new terms in legal terminology with regard to the language of civil contracts. Focusing on borrowing as one of the most productive sources of replenishing legal terminology, the authors consider new lexemes connected with innovative concepts and digitalization. The paper shows the analysis of terminological units borrowed from the English language. The authors conclude that the current term system of both legal discourse in general and the sphere of civil law in particular is flexible, mobile, susceptible to previously unproductive ways of forming or borrowing terms, and even the use of terms with the preservation of English spelling, as well as metaphor terms.

1 Introduction

Due to natural processes of development of the Russian language, the vocabulary of various fields of activity is constantly updated. Neologisms regularly appear in legal discourse which is actively involved in social processes and interacts with other discourses. High rates of digitalization influence this process significantly contributing to the appearance of terminological units connected with information technologies and innovations. The branch of civil law that regulates a wide range of legal relations and is subject to continuous changes affecting the language of law in general and the legal term system in particular, is one of the most active "suppliers" of new terms in legal discourse.

2 Problem Statement

Neologization is a traditional direction in the development of legal discourse associated with the use of words that came to law from different fields of knowledge. In the modern Russian legal language, the specificity of this process is largely associated with the processes of borrowing terms from the English language. Analyzing specific examples of such terms should show what the typical ways and stages of turning a new word into a legal term are.

3 Research Questions

- What are the terms-neologisms and terminoids used in the language of civil law?
- How is the process of neologization effected in legal discourse?
- What is the specificity of borrowed terms-neologisms?
- In which areas is the borrowing process most active?

4 Purpose of the Study

The purpose of the study is to review new terms that have appeared in the Russian-language legal discourse in recent years and to assess the ways in which they are included in the legal terminology system.

5 Research Methods

The research methodology is based on philosophical, general scientific, as well as special methods: induction, generalization, explanation, synthesis, continuous sampling method, contextual analysis method.

In the process of studying the research material the selection of terms-neologisms and terminoids was undertaken using the continuous sampling method. The method of contextual analysis allowed explicating the concepts denoted by new lexemes. The method of induction, synthesis, as well as generalization and explanation techniques were used to extrapolate the

* Corresponding author: chemeteva@volsu.ru

results obtained from the analysis of the material to the process of neologization in legal discourse as a whole.

The empirical base of the research consists of legal discourse texts, including civil contracts, as well as materials of the law enforcement practice.

6 Findings

6.1Terms-neologisms

There are a lot of new lexemes, including terms, in legal discourse. Legal terms-neologisms are a particularly interesting subject of research, as they illustrate the existence and functioning of changes taking place in the legal system. The improvement of legislation, the introduction of innovations in various spheres of life and the need for their legal regulation, the development of the international cooperation, the emergence of new social relations that require legal regulation, new legal institutions, etc. are among the factors that determine the processes of neologization in legal discourse. Civil contracts, regulating the relations of citizens on various issues in almost all spheres of life, are saturated with terms-neologisms. Neologisms are often reflected in civil contracts even before new legal relations are legally established.

To proceed to the consideration of the characteristics of terms-neologisms in legal discourse, a definition of the concept of "neologism" should be given. A neologism is "a word or phrase created to denote a new subject or to express a new concept" [1]. These include, for example, *self-employment*, *smart contract*, *e-government*, *legal tech*, etc. The question arises as to how long the concept considered as a neologism remains new.

Lexemes remain neologisms as long as they are perceived as new by native speakers [2]. With a wider spread of a term and users getting used to it, the lexeme becomes common. For example, the term "материнский капитал" (maternity capital), which was once a term-neologism, has passed into the common vocabulary layer, and is no longer perceived as a new lexeme. It is worth noting, however, that the border between neologisms and words that have passed into the layer of commonly used vocabulary is rather conditional.

When talking about terms-neologisms, we should also indicate our understanding of the concept of "legal term". After V. Yu. Turanin, under the legal term we mean a term that meets the following criteria significant for the legal language:

1) accuracy of the designation of the corresponding legal concept;

2) unambiguity of the perception within the legal language;

3) presence of a definition [3].

It should be noted that currently not all lexical units that function as terms in legal discourse meet the listed criteria. For example, there is a lack of consistency in the definition of terms in legal discourse currently. Some documents provide an extensive list of definitions, while others define concepts contextually or do not define them at all. In addition, definitions of the same concept can be found in two legal acts, and such duplications are often characterized by declarative nature [4].

6.2 Terminoids

The above mentioned criteria allow distinguishing terms from so-called terminoids. Terminoids – lexemes that denote new concepts but do not yet have a clear definition – are actively involved in the process of neologization in legal discourse. These lexical units do not have such terminological properties as accuracy of meaning, contextual independence, and permanent stability. The set of features of the concept denoted by a terminoid has not yet been finalized. Without clear semantic boundaries, they can have several definitions and be defined differently depending on the context [5]. Many terminoids that are included in the texts of documents (for example, civil contracts) and are widely used in legal discourse or are established in legislative texts and thus acquire an official status (often an official definition), become neologisms.

Most terminoids function in new branches of law and are used for nominating innovative technologies in jurisprudence. Here are examples of such lexemes that function in legal discourse and civil law: *фьючерсный контракт* (futures contract), *франчайзинг* (franchising), *web-сайт* (website), *провайдер* (provider), *облачное хранение* (cloud storage), *SaaS*, *SLA*, *удаленный доступ* (remote access), *CRM-система* (CRM system), etc. Most of the above terminological units are borrowed from the English language. As already mentioned, terminoids are of a fragile nature. So, the spelling of a lexeme-terminoid can be variable. For example: *web-сайт*, *веб-сайт*, *вебсайт* (website). One contract shows the use of the first option: "Подрядчик обязуется выполнить следующую работу: разработку и обслуживание WEB-сайта Заказчика, именуемую в дальнейшем «Работа»" (The Contractor shall be obliged to perform the following work: development and maintenance of the Customer's website, hereinafter referred to as "Work") [hereinafter examples of using lexemes are translated by Yu.V. Chemeteva]. Another similar document contains the second option: "Исполнитель обязуется по заданию Заказчика доработать для него программное средство «веб-сайт», а Заказчик обязуется принять результаты работы Исполнителя и оплатить их" (The Contractor shall be obliged to modify the "website" software for the Customer on the Customer's instructions, and the Customer shall be obliged to accept the results of the Contractor's work and pay for them).

Legal terminology is a continuously changing system that depends on various processes occurring both within legal discourse and in its interaction with other types of institutional discourses. Terminoids that entered legal discourse from other areas and have become frequent are included in the terminology system of law over time. With regard to the language of civil contracts, there can be marked two ways of appearing new terms in the legal terminology system:

- establishment of new legal institutions in legislation, from which they are transferred to the texts of contracts;
- contractual regulation of relations that have not yet been properly reflected in the legislation. This is where, as a rule, the use of terminoids takes place, since there is no official normative term yet and the new word is only being tested in the contractual practice. Subsequently, the legislation can accept it, establishing the spelling and meaning of the term, and thus the lexical unit will pass into the category of terms-neologisms.

6.3 The process of neologization legal discourse

To form a visual representation of the processes of neologization occurring in legal discourse in general and in the field of civil law in particular, examples should be given. This study considers terms as well as terminoids related to neologisms. The word combination "terminological unit" combines the concepts of "term" and "terminoid".

Here is an example of a neologism related to the current issue about the establishment of a special tax regime in Russia "Tax on professional income": "самозанятый" (self-employed). The concept of self-employment is relatively new in Russia. For the United States, the terms "self-employment" and "self-employed" seem quite common, since they have long occupied the status of "official" in the system of legal terminology. The legislation of the Russian Federation in this issue currently uses only the term for the tax paid by officially registered self-employed people: "налог на профессиональный доход" (professional income tax). At the same time, the following variants of terminology for this concept are found in the media: *налог для самозанятых* (tax for the self-employed), *налог на самозанятость* (tax on self-employment), *налог на самозанятых* (tax on the self-employed). Probably, these unofficial equivalents of the term "налог на профессиональный доход" (professional income tax) were formed by analogy with the name of taxes in Russian law (*налог на прибыль* (income tax), *налог на добавленную стоимость* (value added tax), *налог на доходы физических лиц* (personal income tax), etc.). Persons to whom this tax is applied are designated in Federal law No. 422-FZ dated November 27, 2018 "On conducting an experiment to establish a special tax regime 'Tax on professional income'" [6] as *налогоплательщики* (taxpayers); *физические лица, применяющие специальный налоговый режим* (individuals applying a special tax regime). This Federal law does not contain any succinct and at the same time short term denoting persons who pay professional income tax (for example, "самозанятый" (self-employed)).

Despite the fact that the term "самозанятый" (self-employed) is not established by law, it is used in legal discourse, including in civil contracts, for example: "Исполнитель является самозанятым лицом и по доходам, получаемым в рамках данного договора, применяет в своей деятельности специальный налоговый режим «Налог на профессиональный доход»"(The Contractor is a self-employed person and applies a special tax regime "Tax on professional income" for the income received under this agreement). This terminological unit can also be found in information materials of the authorities: "Налог на профессиональный доход — это новый специальный налоговый режим для самозанятых граждан, который можно применять с 2019 года" (The professional income tax is a new special tax regime for self-employed citizens, which can be applied from 2019) [7].

The establishment of this tax regime in Russia is still experimental. Probably, with the widespread introduction and adoption of additional legislation regarding the regulation of self-employment in the Russian Federation, a need for a concise, simple term that refers to individuals that fall under this tax regime will appear. With a high probability, such term may be the lexeme "самозанятый" (self-employed).

It should be noted that the conceptual content of the term "self-employment", from which, probably, the terminological unit "самозанятость" (self-employment) was formed by means of calque, does not completely coincide with the Russian-language conceptual content of this term-neologism. The concept of self-employment in the United States is not identical with the emerging concept of self-employment in the Russian Federation. In the United States, it also includes home businesses which are classified as microenterprises employing maximum 25 employees. In the Russian Federation, the special tax regime "Professional income tax" does not cover persons who have employees with whom they are in an employment relationship [7].

As can be noted, a large number of so-called terminoids used in legal discourse are borrowed from various areas of special knowledge. Many of the terminological units mentioned above have entered legal discourse and are used in civil law texts due to the convergence of law and IT technologies. Terms-innovations in the field of law, functioning at the intersection with other fields of knowledge, are of particular interest. As a rule, such terms become widely used quickly. Here are examples of such lexemes: *блокчейн* (blockchain), *legal tech*, *смарт-контракт* (smart contract), *криптовалюта* (cryptocurrency), *биткоин* (bitcoin), *электронное правительство* (e-government), *краудфандинг* (crowdfunding), etc.

The terminological unit "криптовалюта" (cryptocurrency) is currently quite common in the Russian-language legal discourse and is familiar to the Russian-speaking recipient, which is probably due to its frequent use in the media. The concept of cryptocurrency in the Russian Federation is not yet established by law, but the lexeme is quite often used in legal discourse. It is worth noting the presence of a synonymous variant of the considered terminoid "виртуальные валюты" (virtual currency / cryptocurrency). However, the calqued version "криптовалюта" occupies a dominant position in legal discourse and is likely to replace the option "виртуальные валюты", which is less similar to the English-language term "cryptocurrency".

The term "блокчейн" borrowed from the English language (blockchain) by means of transcription currently has no official, legally established definition in Russian. Its spelling in Latin (blockchain) is also found in texts of legal discourse. It is likely that the development and introduction of blockchain technologies in the sphere of law will not only legally establish the concept under consideration, but also contribute to its more frequent use in legal discourse in general and in texts of civil contracts in particular.

One of the characteristic features of neologization in legal discourse at the moment is using metaphors, which is not peculiar to the language of law. The status of a term is often given to words and word combinations used in a figurative sense: *искусственный интеллект* (artificial intelligence), *экосистема цифровой экономики* (digital economy ecosystem), *экосистема умной городской среды* (smart urban environment ecosystem), *облачные платформы* (cloud platforms), *облачные и туманные вычисления* (cloud and fog computing), *регуляторные песочницы* (regulatory sandboxes), *сквозные цифровые технологии* (end-to-end digital technologies), *виртуальная лаборатория* (virtual laboratory), *дорожная карта* (road map), etc. Despite the fact that metaphorization is a widespread method of term formation [8], most legal scholars are critical of the use of this method of neologization in legal discourse emphasizing that the accuracy and adequacy of scientific thought are not achieved using metaphor terms [9]. One of the reasons for metaphorization in legal terminology is, in our opinion, the active borrowing of terms from the English-language legal discourse, which is more characteristic of imagery than the Russian-language legal discourse. The issue of borrowing terminological units should be given special attention.

6.4 Borrowings as sources of neologization in legal discourse

Borrowing (as a rule, from English) is currently one of the most productive ways of replenishing terminology [10]. The influence of global English on the nominative process is indisputable [11]. A borrowed term is most easily adapted in the Russian language when it is used in a transcribed or transliterated form [12]. We should add that often terms are borrowed by means of calque: *самозанятость* (self-employment), *криптовалюта* (cryptocurrency). Many of the metaphor terms mentioned above are borrowed from English by means of calque: *умные города* (smart cities), *облачные вычисления* (cloud computing), *облачные услуги* (cloud services), *регуляторные песочницы* (regulatory sandboxes), etc.

An example of the term-neologism that appeared in the text of a civil contract due to the adaptation of a borrowing is the term "инсайдерская информация" (insider information): "В случае получения доступа к инсайдерской информации Заказчика Исполнитель обязуется..." (In case of obtaining access to the Customer's insider information, the Contractor undertakes...). Initially, the lexeme "инсайдерская информация" was borrowed from the English-language legal discourse (insider information) by means of the transcription of the element *insider* and the use of the Russian-language equivalent of the English lexeme "information". The term was also grammatically adapted so as to sound more familiar for native speakers of Russian (suffix *–ск* and the inflexion matching the adjective with the noun were added).

The component "инсайдерский" has quickly adapted to the Russian-language legal discourse, and currently the terminological units containing it can be found in legal texts: *инсайдерская оговорка* (insider clause), *инсайдерская торговля* (insider trading), *инсайдерская атака* (insider attack), *инсайдерская угроза* (insider threat*), инсайдерская активность* (insider activity), *инсайдерская сделка* (insider transaction). Some of the mentioned phrases are already found in civil law and in civil contracts in particular, for example: "6. Инсайдерская оговорка" (Insider clause – name of the contract clause).

Probably, in the near future other terms containing the component "инсайдерский" will appear in civil law texts. In English-language contracts, the functioning of the term "insider trading" can be observed: "Insider Trading. Sellers and Buyers hereby certify that they have not themselves, nor through any third parties, purchased nor caused to be purchased in the public marketplace any publicly traded shares of the Company". The concept has not yet been clearly defined in Russian and is not included in the texts of legal documents. The Russian-language legal discourse uses a similar term to the English-language term "инсайдерская торговля" (insider trading), however it is not currently fixed in legislative texts and is not used in civil contracts. The text of the Federal law "On countering the misuse of insider information and market manipulation and on amendments to certain legislative acts of the Russian Federation" [13] contains the following word combinations similar to the term "insider trading": *неправомерное использование инсайдерской информации* (misuse of insider information), *раскрытие инсайдерской информации* (disclosure of insider information), *доступ к инсайдерской информации* (access to insider information), *передача инсайдерской информации* (transfer of insider information), *предоставление инсайдерской информации* (provision of insider information), etc.

The considered terminoid is used in the Russian-language legal discourse outside the framework of legislative texts: "…незаконной инсайдерская торговля становится только в случае использования важной непубличной информации в качестве основы для принятия решения о совершении той или иной транзакции" (...insider trading becomes illegal only if important non-public information is used as the basis for making a decision about making a particular transaction) [14]. The unsettled nature of the terminoid and the lack of a clear definition contribute to its ambiguity, which leads to the variability of its use: *инсайдерская торговля*, *инсайдерский трейдинг*, *трейдинг инсайдеров*, *инсайд-трейдинг*. It is possible to assume that with the improvement of legislation in the field of securities and the growing interest of citizens in investing, the term "инсайдерская торговля" will be used in the texts of civil contracts and regulatory legal acts in the Russian-language legal discourse. The most likely variant of the adaptation of the terminoid with the subsequent transition to the category of legal terms-neologisms is the word combination "инсайдерская торговля", since it is built on a similar model with the word combination "инсайдерская информация", which is quite common in the Russian-language legal discourse.

Another relevant example is the process of borrowing the English-language lexeme "smart contract" which denotes one of the most discussed and innovative subjects of law. In the Russian legal science the terminoid is used in the borrowed variant formed by means of calque: "смарт-контракт". However, there are different variants of the terminoid: *смарт контракт*, *умный контракт*, *смарт-контракт*. Given the innovative nature of this concept and the fact that the smart contract technology is used in the most advanced industries involved in international interaction, the calqued version of the lexeme is probably the most appropriate for the use. In addition, Russian scholars use this variant in their scientific articles, and the spelling with a hyphen is more common. Here is an example from a scientific article on jurisprudence: "Сказанное означает, что смарт-контракты могут быть применены далеко не во всех сферах имущественных отношений, что как раз и подтверждается складывающейся практикой заключения указанных соглашений" (This means that smart contracts cannot be applied in all areas of property relations, which is confirmed by the current practice of concluding these agreements) [15].

Another example of a borrowed neologism is the terminological unit "электронное правительство" (electronic government / e-government). While in the United States the term "E-Government" is officially established, in the Russian-language legal discourse the term is currently not reflected in legislation, though it is used both in scientific literature and in the media in a unified way (электронное правительство), which gives reason to consider it a term-neologism. When borrowing the lexeme "E-Government" from the English-language legal discourse, an explication technique was used: the component *e–* (electronic) was explicated.

A special case is the borrowing of the lexeme "legal tech" from the English-language legal discourse. Currently, the terminoid "legal tech" is used in the Russian-language legal discourse in the completely borrowed version, i.e. in Latin characters. There can be found the spelling of the lexeme in quotation marks, which indicates that it does not yet have a specific place in the system of legal terminology in Russian.

The examples given in this paragraph demonstrate the activity of different ways of borrowing terms (transcription and transliteration, calque). What is more, we give examples of unproductive methods and techniques (explication, full borrowing). It is worth noting that a large number of new terminological units are entering legal discourse from the field of IT technologies, which indicates the development and improvement of law in the information sphere.

7 Conclusion

The issue of terms-neologisms, acting as indicators of changes occurring in the law, is of great relevance. The process of neologization in legal discourse, against the backdrop of the legislation improvement, the innovations implementation, the spreading of digitalization, the development of international cooperation, the emergence of new social relations requiring legal regulation, is characterized by the fact that one of the active sources of terminology are so-called terminoids which initially operate in new fields of law and function for nominating innovative technologies in law.

A large number of new lexemes in legal discourse are borrowed from various fields of special knowledge. This is due to the fact that the law currently makes maximum use of rapidly developing information technologies and adapts them to its activities. In addition, the interaction of individuals and communities in the information sphere is reaching a new level, and there is a need for legal regulation of relations in this area, taking into account all their features and ongoing transformations. Besides, the international interaction is increasing, which leads to the intensity of borrowing terminological units from the English-language legal discourse. Under these conditions, the term system of both legal discourse in general and the sphere of civil law in particular is flexible, mobile, susceptible to previously unproductive ways of forming or borrowing terms and even the use of terms with the preservation of English spelling, as well as metaphor terms.

Acknowledgment

The reported study was funded by RFBR, project number 20-312-90026 "Legal analytics in the media discourse space".

References

1. D.E. Rosenthal, & M.A. Telenkova,. *Dictionary-reference of linguistic terms* (Moscow: Prosveshchenie Publ. 1985).
2. Yu.N. Nesvetaylo Vestnik Stavropolskogo Gosudarstvennogo Universiteta **2**, 144-148 (2008)
3. V.Yu Turanin,. *The phenomenon of legal terminology*. (Moscow: Yurlitinform Publ. 2018)
4. M. L Davydova Science Journal of Volgograd State University. Linguistics **19 (4)**, 52-63 (2020).
5. S.V. Grinev-Grinevic *Terminology Studies*. (Moscow: Academiya Publ. 2008)
6. *Federal Law* (November 27, 2018 № 422-FZ, Moscow, Russia, 2018)
7. Federal tax service of Russia URL: https://npd.nalog.ru/ (date of access: 16.03.2021)

8. A. Assal Terminologie et traduction, **2**, 235-242 (1994)
9. A.F. Cherdantsev, *Logical and linguistic phenomena in law* (Moscow: Norma Publ.: INFRA-M Publ. 2012)
10. G.F. Musina, Theory & Practice, **12(3)**, 149-151 (2017)
11. E.Yu. Novikova, V.A. Mityagina, & E. Charfaoui, XLinguae Journal, **10** (3), 285-299 (2017).
12. E.V. Moskaleva, Vestnik LGU im. A.S. Pushkina, **1**, 61-71 (2012).
13. *Federal law* (July 27, 2010 №. 224-FZ, Moscow, Russia, 2010).
14. ITI_CapitalURL:https://habr.com/ru/company/iticapital/blog/273337.(date of access: 16.03.20)
15. O.M. Rodionova, Gaps in Russian legislation, **6,** 183-185 (2017)

https://doi.org/10.1051/shsconf/202110901012

The legal mechanism of bringing to account in the epoch of digitalization

Tatiana Deryugina*[1], *Lyudmila* Chegovadze[2], *Albert* Tumakov[3] and *Felix* Vertlib[4]

[1] Kikot Moscow University of the Ministry of Internal Affairs of Russia, st. Academician Volgina, 12, Moscow, Russia, sofija96@mail.ru
[2] Kikot Moscow University of the Ministry of Internal Affairs of Russia, st. Academician Volgina, 12, Moscow, Russia, chegov@mail.ru
[3] Kikot Moscow University of the Ministry of Internal Affairs of Russia, st. Academician Volgina, 12, Moscow, Russia, albert-tumakov@rambler.ru
[4] Kikot Moscow University of the Ministry of Internal Affairs of Russia, st. Academician Volgina, 12, Moscow, Russia, vertlib.f@gmail.com

Abstract. An analysis of problems occurring during the application of civil liability in legal relationships using information and communication technologies is being conducted in this scientific research. Apart from the absence of legal norms, there are conflict-of-laws rules, contradictions in which are to be resolved. The research of the turnover using digital technologies takes a big part in the legal doctrine. It must be noted that several fundamental works have appeared that make it possible to create a harmonious and non-contradictory system of legal regulation. However, there are still some unresolved questions left; for instance, ones concerning the mechanism of legal regulation of civil liability of subjects that have entered relations using digital technologies. The analysis of the civil law doctrine, the propositions of current Russian and international legislations, and legal practice have revealed a complex of problems connected to the civil liability of subjects entering legal relations using information and communication technologies. It became possible to systemize them depending on the subject and their role in the initiation or termination of the relation. A range of methods (general, general scientific and legal) has made it possible to make conclusions about the reasons why legal regulation is imperfect and why conflicts exist (to prevent them from happening in the future), formulate a legal regime of liability of subjects of the main or supporting group, suggest alterations to the current Russian legislation, aimed at bringing the owner of the information system to account.

1 Introduction

The development of digital technologies positively influences various spheres of society, including the ones connected to entering legal relations. Fast and expense-wise economical deal conclusion using digital information systems, document exchange, submission of claims, the possibility to access big information volumes – all this makes communication of significantly easier and finally increases the efficiency of economic activity. However, the negative sides of these processes cannot be ignored.

After the introduction of information and communication technologies the condition of protection of the individual and the society from informational threats requires particular attention of the state. This in turn requires the development of new means of legal guarding and protection of subjective rights and the interests of the subjects of legal relations. Despite the numerous researches aimed at addressing the questions of legal regulation of relations connected to using the digital environment, bridging the gaps and solving conflicts, which have appeared in the law field in the last five years, a lot of problems remain unresolved. The problem of improving the legal mechanism of bringing to account is among them.

The problems listed can be divided into two groups: ones caused by flaws in the legal regulation of relations appearing with the help of digital technologies and ones connected to the low awareness level of subjects on the peculiarities of using information and communication technologies in entering legal relations, and peculiarities of legal regulation of such relations and the protection of violated rights and rules. In order to achieve a positive result and the required effect, the problems of both groups should be solved. Forming new legal mechanism and perfecting the existing ones including those from the field of bringing to account is one of the priority areas of development of institutes and branches of law, while

* Corresponding author: sofija96@mail.ru

https://doi.org/10.1051/shsconf/202110901012

creating a fundamental basis for effective regulation of relation is a primary objective of science.

Thus, the absence or divergence of legal rules regulating civil liability upon entering relations using information and communication technologies, the imperfection of mechanisms and procedures of their application that take the peculiarities of digital relations into account, predetermines the necessity of conducting the present research.

2 Problem Statement

The problems of civil liability that occur due to the use of digital technologies are of a multidimensional nature, which was mentioned on several occasions both in the Russian legal [1] science, as well as in civics [2-4] and the foreign one [5-7]. The questions of liability in the digital technology development epoch go beyond one branch of law, a set of problems appears at the intersection of different branches, which requires a multidisciplinary approach.

For instance, along with problems of civil character connected to the identification of a contract's subject, determining the place of an agreement's conclusion, determining the liability subject in entering relations using an aggregator, etc., the questions of protection of personal data of a contract's subjects, the protection of confidential information, etc. are of great importance. Therefore, it is necessary to address adjacent sciences of informational, labour, constitutional, criminal law, as well as the legal theory.

Solving the problems of civil liability of subjects entering relations using digital resources is complicated by the multiplicity of subjects. Apart from the traditional contract sides, there are following subjects that can participate or influence the sides: the owner of the information resource used to enter the relations (generally aggregators); the right holder of exclusive rights on the result of intellectual activity – software used by the sides to enter the relations. What is more, some authors think that the owners of technical means of information networks (technological platforms, Internet services, etc.) are subjects of the relationships as well [8]. However, the former can directly influence not only the possibility to conclude contracts, but their execution as well.

For instance, in early March 2020 a hosting provider "Masterhost" has stopped working; among its clients there are SPA "Energomash" and others [9]. The reason for the failure was the owner blocking the technical equipment after a conflict with the provider. As a result, the users of the hosting provider could not access their electronic resources for a week, which had a direct impact on their economic activity.

Undoubtedly, such events should be qualified as force majeure that exempt the sides from liability, when deciding on bringing the sides to account using the resource. However, this example illustrates the possibility of influence by a third party and their actions on realizing the rights and fulfilling the duties in legal relations that use information and communication technologies and the Internet.

Two groups of subjects should be distinguished upon reviewing the subjects of civil legal relations that appear using digital resources. The first group (let us call them main subjects) are the persons who directly conclude contracts, address claims to each other, etc. Relations between the subjects are sufficiently regulated by civil law.

However, the peculiarities of relations initiated using information and communication technologies cause the presence of another group of subjects (let us call them the supporting subjects). They are the subjects that provide mediation (generally not of legal character) connected to providing an Internet resource for the sides to communicate; for instance, the aggregators of Internet trade, information aggregators [10], authors of software, persons in possession of information, equipment owners. Even though they do not directly participate in the relations, they cannot be called a passive side. Moreover, they are given authority and are able to influence the possibility for the main participants to enter the relations. For instance, an electronic commerce aggregator has a right to restrict access to electronic resources [11], information aggregator can store and transfer information that can influence the conclusion of contracts, or their alteration and annulation, allow or restrict access to it, etc.

It follows from all the above that the existence of diverse legal subjects, their ability to influence the initiation and termination of legal relations using digital technologies is in direct correlation with the necessity to solve questions of civil liability of the subjects for breach of obligations. The crucial question is the correlation of liability of one or other of the subjects of legal relations with their role in the relation, which will allow distributing liability depending on the influence of the subject on adverse effect.

3 Research Questions

The emergence of digital technologies and their introduction to the economic trade creates new interdisciplinary tasks for Russian (and world) science. They can be solved by conducting research of various legal phenomena, including the legal mechanism of bringing to account and its transformation during the epoch of formation of the digital ecosystem.

In order to achieve the purpose of the study, it is necessary to consistently and systematically solve a number of theoretical and practical questions; for instance, to analyse the present Russian and foreign doctrines; to study the legal framework of the question determining the set of problems connected to mechanisms of bringing to account the subjects entering the trade using information systems, to review the practice of application of the existing rules of law in order to identify questions that have to be regulated by the rules of law, to solve existing conflicts, to systemize the problems identified and suggest ways to solve them based on the conducted research.

 https://doi.org/10.1051/shsconf/202110901012

4 Purpose of the Study

The present study is aimed at the fundamental development and scientific reasoning of solving problems connected to legal mechanisms of bringing to account in the epoch of formation of the digital ecosystem, subjects entering legal relations using information and communication technologies. The scientific positions developed will become a base for the creation of adequate modern legal constructions meeting the requirements of the digital economy and for further improvement of the actual legislation.

5 Research Methods

While conducting the research the authors relied upon (1) common scientific methods (dialectical, system approach), which allowed to examine the mechanism of liability of persons entering relations using information and communication technologies as a united system of legal rules, establish a connection and intersectionality of the processes; (2) general scientific methods (analysis and synthesis, systemic functional method, induction, deduction, analogy, modelling) that made it possible to establish particular elements of the content of legal categories, discover general patterns of liability in the digital age, establish the influence of separate factors on the development of a relation; (3) specifically legal methods (dogmatic, comparative legal, legal prediction) that made it possible to identify formal logical connections, formulate perspectives of legal regulation of relations connected to civil liability of subjects in the digitalization epoch.

6 Findings

Various sources in the scientific literature are dedicated to the question of civil liability, the liability in the age of digitalization included, both Russian [12, 13] and foreign [14-16]. The problems specified in the scientific literature can be divided into two groups: ones concerning main subjects of legal relations (subjects of contract, transaction, etc.) and ones concerning subjects performing supporting functions (non-legal mediators: owners of digital networks, websites and other kinds of information infrastructure).

Among the problems of main subjects that require immediate solution there are problems of identification of the subject entering legal relations using information and communication technologies; the problem of the national treatment (the law that will be applied to the relations); the problem of determining the scene of a legally important action (the transfer of goods, the payment, etc.)

As for the problems of subjects of the supporting group, the main place is taken by the problem of their influence on the initiation, alteration and termination of relations between main subjects and, consequently, the possibility of bringing them to account in cases when they are not among the sides of the principal obligation. And here, perhaps, the question of the legal nature of relations between the owners of Internet services, websites and information systems becomes the most important.

For instance, considering a case on causing death as a result of a car accident by a taxi service, the first and the second instances have rejected the claim to the information aggregator ("Taxi Prestige" LLC) through the services of which the order was made, pointing at the circumstance that the sides (the carrier and the passenger) have concluded the contract of transportation, of which the "Taxi Prestige" LLC is not a side and, consequently, is not responsible for it. Only the Supreme court of the Russian Federation has qualified the actions of the aggregator as these of an agent, pointing at the responsibility of the agent according to article 1005 of the Civil code of the Russian Federation [17]. In another case, where the question of the possibility of bringing an aggregator into account on the contract of transportation was discussed, the court has qualified the relations of the aggregator and the carrier as organizational, pointing that since the aggregator creates the conditions for the transportation, they take part in the activity that creates high risks. Consequently, they can be brought to account [18].

The question of bringing to account the owners of infrastructure (subjects of supporting group) is directly connected to the recent discussion of whether it is possible for artificial intelligence to be a subject of law. In this case the opinions differ from positions stating the necessity to bring its developer to account or its user to bringing to account the intelligence itself [19-22].

Unfortunately, it is to be stated that despite the rapid development of economy using information and communication technologies, civil regulation of such relations falls behind significantly. The most advanced one is the consumer protection legislation that implies mutual rights and obligations of the aggregator's owner, the executor, the seller, as well as the liability of the owner for improper information on the goods [23].

Among recent alterations of the legislation aimed at introducing digital technologies there are the emergence of digital law as a subject of civil rights (article 128 of the Civil code of the Russian Federation), the right to vote at a meeting using electronic or technical means (article 181.2 of the Civil code of the Russian Federation), the possibility for the sides of an obligation to execute it using information technologies determined by the terms of the contract (article 309 of the Civil code of the Russia Federation), the possibility to compose a contract in electronic form and the possibility to exchange electronic documents (article 434 of the Civil code of the Russian Federation), the possibility of purchase of digital rights (article 454 of the Civil code), formulating the features of the information provision contract (article 783.1 of the Civil code), the introduction of liability for the information mediator exclusively for intellectual rights (article 1253.1 of the Civil code of the Russian Federation) and others [24]. Apart from this, the concept of utilitarian digital rights was introduced to the legislation, as well as the legal regulation of the utilitarian digital rights turnover on investment platforms that are information systems [25].

Among positive aspects of civil liability regulation there are: (1) the appearance of the liability of investment platforms operators, news aggregators, search systems operators, owners of audio-visual services, information mediators (for intellectual rights), which demonstrates the focus of the legislator on the possibility of applying sanctions to the subjects of the supporting group and (2) determining the law nature of the contract with an aggregator as a services contract. In this case there is a claim to the legislator on the imperfection of major contract terms, which also negatively influences the possibility to bring the actor to account.

As for the negative aspects, they are: (1) the possibility of establishing the content of a digital right and the rules of its realization by the information system operator without determining the system of restrictions and limits in the legislation, which, due to a faulty formulation of the article 141.1 of the Civil code of the Russian Federation that does not list property rights included in the digital right, can make the body applying the law think that the content of property right can be established not only by the law but by the information network operator as well; (2) the presence of gaps in the law, including questions of civil liability. The legislator has not established the content of digital law (and has not determined its borders), which makes it impossible to develop general dispositions of bringing law subjects to account; (3) the absence of unified terminology used without consideration of the dispositions of the Federal law (2006) on information, information technologies and the protection of information [26]; (4) the unsystematic attempts to regulate particular kinds of liability of information mediators (news aggregators, search system operators, owners of audio-visual services (Federal law (2006) on information, information technologies and the protection of information), investment platform operators (Federal Law (2019) on attracting investments using investment platforms and on altering the legislation of the Russian Federation). It is necessary to solve general questions of civil liability of the information system operators, and then regulate particular kinds of liability.

Addressing the international statutes, one can note their focus on bringing to account (in case of offense) all subjects of both main and supporting groups. For instance, in the European Parliament resolution of 16 February 2017 with recommendations to the Commission on Civil Law Rules on Robotics[27] the necessity to bring to account the manufacturers, users, operators and owners of networks is noted. The liability must accrue without guilt. What is more, it is specified in the act that it is impossible to dismiss bringing to account based on the fact that the damage was caused not by a human (which seems to suggest that the European parliament is in favour of the possibility of bringing artificial intelligence). Undoubtedly, the rules of liability mentioned mostly concern artificial intelligence, though the main idea is that all involved must be brought to account.

It must be noted that in the developed draft of the Federal law "On altering the Civil code of the Russian Federation in perfecting the legal regulation of relations in the field of robotics" the necessity to bring to account both the owner and the owner of the robot-agent was noted. In our opinion, the great advantage of the draft law was the definition of the nature of arising relations.

The analysis conducted demonstrates the presence of systemic problems of legal regulation that can be determined as (1) problems connected to flaws in legal regulations; (2) the absence of systemic approach; (3) the inconsistency of the legislator, also in conceptual framework; (4) no solution to the principal question of the composition of the subjects of civil liability.

7 Conclusion

The research conducted has made it possible to formulate several scientific inferences and to develop recommendations on improving the existing legislation.

Presently the improvement of legal regulation follows the path of establishing the specificity of civil liability for some subjects of legal relations (news aggregators, information mediators, search system owners and others) that arise using information and communication technologies, which is not entirely right considering the diversity of subjects. The institute of civil liability is filled with contradicting rules with such an approach, which is unacceptable. What is more, some of the questions of liability are out of legal regulation. It is necessary to regulate general questions of liability of subjects participating in relations using information and communication technologies and determine the specificity of liability for particular participants of the market afterwards, based on this.

The subjects participating in civil legal relations using information and communication technologies are to be divided into two groups depending on the rights and obligations that influence the rise or alteration of the relation. They are the first group (main subjects), persons that directly conclude contracts, address claims, etc.; the second group (supporting subjects), the subjects that act as mediators (generally not of legal character) connected to providing resources on the Internet for the sides to communicate (for instance, Internet trade aggregators, information aggregators), the authors of software, the holders of information, search system owners, equipment owners.

The civil liability of the subjects of the supporting groups must directly correspond to their role in the legal relation risen, which will make it possible to spread the liability depending on the influence of the subject on the occurrence of negative consequences.

The creditor and the debtor (the main group of subjects) and the information system operator (the supporting group) must be the subjects of civil liability in rising relations using information and communication technologies.

It must be secured at the legislative level that the information system operators are prohibited from restricting the kinds and the volume of reparation, restrict the forms of compensation in cases when digital rights serve as civil law objects.

When determining the legal regulation of civil liability of subjects entering legal relations using information and communication technologies, a legal construction consisting of two contracts must be used: the services contract concluded by the consumer and the information system operator and a services contract with the information system operator and the executor as sides.

The condition of the obligatory execution of the contract in favour of an undefined group of persons must be a significant term of services contract concluded between the information system operator and the executor.

The subsidiary liability of the information system operator and the executor (the contactor, the seller) must be determined at the legislative level.

References

1. V.N. Lopatin, Information law, **2,** 14 - 19. (2018)
2. E.A. Abrosimova, *The organizers of trade: the legal status and functional purpose* [Dissertation of Doctor of Law]. (Moscow, Russia: Lomonosov Moscow State University, 2014).
3. A., Inshakova, E. Inshakova, A. Ryzhenkov & M. Sevostyanov *Proceedings of the International Scientific Conference in Memory of Oleg Inshakov (1952-2018)*), 421-429 (2020).
4. A.E. Molotnikov & E.V. Arkhipov, Entrepreneurial law, **4**, 38-47, (2017)
5. C. Bublitz, A. Wolkenstein, R.J Jox & O. Friedrich, International Journal of Law and Psychiatry, **65**, 101-123 (2019)
6. P. Čerka, J. Grigienė, & G. Sirbikytė, Computer Law & Security Review, **31,** 376-389 (2015).
7. W. Huang & X. Li, Computer Law & Security Review, **35,** 105-112 (2019)
8. V.N. Lopatin, Information law, **2(33),** 10 - 13 (2013).
9. D. Shestoperov, H. Aminov, N. Schurenkov, & E. Volkova, Kommersant, 39, 10-13 (2020)
10. State Duma of the Russian Federation. (2006a). *The Civil code of the Russian Federation (part four).* (December 18, 2006 No 230-FL; edition of 26 July 2019)
11. The government of the Russian Federation. The order *"On creating a united trade aggregator"* (April 28, 2018 No 824-p.)
12. V.V. Arkhipov, & V.B. Naumov, Zakon, **5**, 159-170 (2017)
13. V.B. Nagrodskaya, *New technologies (blockchain / artificial intelligence) at the service of law: a methodological teaching aid* (Moscow: Prospect, 2019).
14. L. Darrell, & Ph. D Ross, Civil Liability in Criminal Justice, **5,** 1-22 (2009)
15. J. Klick, & J. MacDonald, International Review of Law and Economics, **63**, 105-112 (2020)
16. B. Turvey, Forensic Victimology, **2**, 429-455(2014)
17. The Supreme court of the Russian Federation. *The determination of 9 January 2018 No 5-CC 17-220.* Moscow, the Russian Federation (2018)
18. The Supreme court of the Tatarstan Republic. *The ruling of appeal of 5 May 2016 on case No 33-8017/2016.* Kazan, the Russian Federation (2016)
19. V.V., Arkhipov, & V.B. Naumov, *SPIIRAS Proceedings*, **6**, 46-62 (2019).
20. Daar, J. The Legal Liability Landscape and the Person/Property Divid. URL: https://reader.elsevier.com/reader/sd/pii/S2666334120300519?token=16A3DD4D2D1796C496AB8AD5A76213184B4AB105B7046BA58DDBFA46AD76F8A7C2CBA0398C8413ADAC2045EDC6ECE787 (date of access: 17.03.2021)
21. V.A. Laptev, Pravo. Zhurnal Vysshey Shkoly Ekonomiki, **2,** 79-102 (2020).
22. P.M. Morhart, *Artificial Intelligence: the legal perspective.* (Moscow, Buki Vedi. 2017).
23. State Duma of the Russian Federation. *The law of the Russian Federation "On the protection of consumers' rights"* (February 07, 1992 No 2300-1, edition of 24 April 2020)
24. State Duma of the Russian Federation. *The Federal law "On the alteration to parts one, two and the article 1124 of part three of the Civil code of the Russian Federation"* (March 03, 2019 No 34-FL)
25. State Duma of the Russian Federation*The Federal law "On attracting investments using investment platforms and on altering some rules of the legislation of the Russian Federation"* (August 02, 2019 No 259-FL, edition of July 20, 2020)
26. State Duma of the Russian Federation. The Federal law *"On information, information technologies and information protection"* (July 27, 2006 No 149-FL, edition of April 03, 2020)
27. The European parliament. *"Civil Law Rules on Robotics" resolutions* (Strasbourg, Switzerland. February 16, 2017)

https://doi.org/10.1051/shsconf/202110901013

Network crowdsourcing and youth extremism prevention: law enforcement agencies and civil society

Pavel Fantrov[1*], *Vladimir* Shinkaruk[2], *Wilhelm* Guhl[3], *Natalya* Solovyeva[4]

[1] Universitskiy Prospekt, 100, Volgograd, Russia, pavelfantrov@volsu.ru
[2] Universitskiy Prospekt, 100, Volgograd, Russia, shinkaruk@volsu.ru
[3] Augsburg, Germany, gulyaich@yandex.ru
[4] Universitskiy Prospekt, 100, Volgograd, Russia, solovieva_na@volsu.ru

Abstract. The fight against the manifestation of extremism among young people is an urgent problem that goes beyond law enforcement and requires a comprehensive socio-political solution. At the same time, preventive measures and technologies for the prevention of this negative phenomenon of public life play a significant role in countering extremism. The effective implementation of crowdsourcing network technologies unites national security entities for a coordinated solution of an urgent problem and helps to level the extremist threat. Their effectiveness should be measured based on the number of identified facts of an extremist orientation, and also the quality of leveling the negative consequences caused by extremist activity. To achieve the goal and objectives of the study, a system of scientific methods based on the systemic principle, the institutional approach, and the sociological method was used. All these methods made it possible to identify the importance of network crowdsourcing technologies for the prevention of extremism among the youth in the context of interaction between the state and citizens. When determining the effectiveness of crowdsourcing network technologies, the study evaluated the projects "Cibersquad", "MediaGuard" and "STOPTERROR", which participated in the prevention of extremism among young people. Largely due to their promotion, the level of extremist crime in 2019-2020 was reduced (compared to 2018). The article substantiates the effectiveness of crowdsourcing network technologies and concludes that their widespread adoption can "socialize" the system of preventing extremism among the youth, provided that not only government agencies, but also constructively-minded citizens' associations are its most important actors.

1 Introduction

Involving citizens in preventing extremism among the youth should become an important trend in the development of partnerships between society and the state [1]. However, at present, there is a kind of "communicative failure" between society and the government, reinforcing the controversies and antagonisms of youth and power. So, as one of the manifestations of this is not only the growing disappointment of youth in the possibility of realizing their potential provided by the state but also the deepening split between various groups of Russian society, for example, the old federal elite and young regional-level leaders [2]. The growing aggravation of social contradictions becomes a significant factor in the radicalization of public sentiment. This is also reflected in the results of a sociological survey conducted by the Levada Center on September 26 - October 2, 2019. So, 53% of respondents between the ages of 18 and 24 want to leave Russia. In this regard, the issues of technological support for the systemic interaction of citizens' associations and state structures to level extremist sentiments among Russian youth are of particular relevance. Currently, there is a need to establish a productive interaction between the active forces of the community to develop young generation and its humanistic spiritual and moral values as a systemic factor in opposing extremist ideologies. The implementation of network crowdsourcing technologies can play a significant role in solving these pressing problems.

2 Problem Statement

To combat the manifestations of extremism, complex systematic work of citizens' associations and state structures is needed. Scientifically based methods are required to carry out the task. The relevance of the problem is determined: firstly, by the increased risks as a result of the development of some negative phenomena (social contradictions, nationalism, extremism, religious radicalism, legal nihilism, xenophobia, etc.) among the youth; secondly, by a change under the influence of globalization of the social space caused by the transition to the information society and the need to use modern

* Corresponding author: pavelfantrov@volsu.ru

technologies for solving social problems; thirdly, by the needs of the society for sound scientific recommendations on the developing of technologies to solve acute social issues; fourthly, by the need to clarify and update the content of the concept of "extremism", which may be used incorrectly in the process of law enforcement.

3 Research Questions

- Identify controversies arising from the interaction of law enforcement agencies and civil society structures.
- To assess network crowdsourcing as a modern technology for the prevention of network extremism.
- Describe crowdsourcing projects implemented by the state and society in the context of the prevention of youth extremism in modern Russia.
- Justify the effectiveness of network crowdsourcing technologies for the prevention of youth extremism.

4 Purpose of the Study

To characterize the peculiarities of interaction between law enforcement agencies and civil society structures in the context of their implementation of network crowdsourcing technologies to prevent extremism in the youth environment.

5 Research Methods

The methodological basis of the study is a system of principles and methods of scientific knowledge that were used to identify the role of network crowdsourcing technologies implemented in the context of the prevention of extremism among young people. In the research process, the following scientific methods were used:

- general scientific (ascent from the abstract to the concrete, analysis, synthesis, induction, deduction, modeling, comparison, etc.);
- systemic - in assessing the interaction of state structures and citizens' associations, which are the main actors in the prevention of extremism;
- institutional, focused on the study of institutions that are subjects and objects of the implementation of crowdsourcing network technologies;
- sociological, allowing on the basis of quantitative and qualitative data to evaluate the effectiveness of crowdsourcing network technologies used to integrate the activities of constructive citizens and law enforcement agencies.

6 Findings

Due to their age-specific psychophysical and sociocultural characteristics, youth is objectively a necessary public resource for modernization changes in Russian society [3]. After all, it always strives for the new and unknown, often decisively breaking the prevailing foundations and conservative stereotypes of thinking. It is often cramped in the strict framework of social conventions and restrictions, which often leads to nihilism and breach of the peace. Manipulators of various kinds can easily drag young people into extremist activities due to its lack of great life experience [4]. In Russian society, there are several adverse social factors that contribute to the inclination of young people to deviant, socially dangerous activities. These, first of all, include excessive polarization of higher and lower social groups; cognitive dissonance caused by globalization; socio-economic problems determining the decline in the demand for the creative potential of youth; ideological pressure from radical public organizations; the slowness of the state apparatus in solving acute social problems and its use of outdated and ineffective technologies [5].

Unfortunately, in the 21st century, such phenomena as xenophobia, nihilism, nationalist sentiments and extremism continued to develop. Their carriers began to widely use modern digital technologies to promote their views [6]. Social networks are increasingly used by supporters of extremism for the brainwashing of the public and its incline towards a deviant activity. The rapid development of information technology allows supporters of extremism to actively use social networks for disseminating illegal information and involving young people in ultra-radical and terrorist organizations. So, most extremist crimes in Russia are committed on the Internet, in 2018 they amounted to about 80% of the total number of such crimes. Of the 1,265 extremist crimes, more than 75% of those that were registered in the Russian Federation in 2018, were identified on VKontakte social network.

Modern electronic technologies that provide financing for criminal activities have become widely used by the criminal community. Thus, cryptocurrencies are gaining popularity, and allow the sender and receiver to interact in a qualitatively different way. Cryptocurrencies create great difficulties in identifying channels for financing extremist activities. Features of cryptocurrency encryption will allow achieving a high degree of anonymity of participants in financial transactions. As a result, this "innovative" nature eliminates all available measures to combat the financing of extremist organizations. According to experts, the current situation requires the introduction of appropriate amendments to international criminal law [7]. Given the current lack of any state-legal regulation of cryptocurrencies at the national level, it seems appropriate to unify international and national legislation governing payment services, electronic payment systems, and virtual currencies. One way to overcome these obstacles is to empower law enforcement agencies to obtain information on the legality of the movement of funds, their origin, on the property and property status of close relatives of persons who committed an extremist act, if there is reason to believe that this property was obtained as a result of criminal activity [8]. The use of modern information technologies in the consolidation of state structures and associations of citizens aimed at protecting national interests would help to solve these problems effectively.

Network crowdsourcing should be highlighted as one of the innovative and effective network technologies in the field of extremism prevention among the youth. According to A.I. Kurylenko, this means the transfer of some of the functions that the state previously performed to a certain circle of entities - representatives of civil society [9]. E.G. Smolina defines crowdsourcing in a slightly different way as "... the modern practice of solving professional problems by laypeople on the principle of "the whole world" based on an online platform" [10]. The attribution of only laypeople to the crowdsourcing entities seems to us not quite correct since professionals can also act in their capacity. Ya. A. Tsukanov even demonstrates a slight disregard for crowdsourcing, recognizing it as a "crowd": "... the very interpretation of crowdsourcing "leads" to the main character - the crowd, which exists within the framework of information and communication technologies with a high level of cooperation" [11]. If citizens demonstrate a "high level of cooperation", then this is no longer a crowd, but an informal association of citizens.

To implement network crowdsourcing in modern Russia, a system of horizontal public relations is gradually being created, and their importance is also recognized by the authorities. So, the President of the Russian Federation V.V. Putin expressed the view that it is necessary to make greater use of the so-called crowdsourcing, modern technology of public examination. Every Internet user can offer proposals and solutions within its framework. The introduction of crowdsourcing into law enforcement practices enhances national security [12].

In terms of efficiency of using resource and technological support for the joint activities of law enforcement bodies and civil society structures, network crowdsourcing technologies are the optimal mechanism for the prevention of extremism among the youth. They "... can predict opportunities important for the state and build strategic steps for their implementation" [13], and they can also involve a significant number of citizens in the process of implementing national security policies through "new media" and crowdsourcing projects. The development of network technologies as a resource for crowdsourcing technologies such as "Cibersquad", "MediaGuard", and "STOPTERROR" has shown their effectiveness in solving the problems of extremism among young people.

In recent years, Russia has seen a downward trend in the number of recorded crimes of an extremist nature. So, in 2019, 56.9% less extremist crimes were recorded compared to 2018. These indicators are undoubtedly associated with the effective implementation of crowdsourcing network technologies. Law enforcement practice shows that the prevention of extremist crimes is the simplest and cheapest in terms of cost for eliminating the consequences of extremist crime, but at the same time, it is the most effective way to eradicate extremist sentiments from the minds of the younger generation.

As in 1960-70, the streets of Soviet cities, along with police officers, were patrolled by members of public order squad, and now, cyber squads are patrolling the great interwebs. The United Russia faction has prepared a bill on the creation of cyber forces in Russia, which, together with law enforcement agencies, will fight extremism on the network. The CyberSquad network crowdsourcing project, created by the Safe Internet League, has already been operating in the Russian Federation since 2011 and unites more than 20 thousand volunteers in 36 constituent entities of the Russian Federation; however, the activities of cyber-workers are not regulated at the federal legislative level. Citizens' associations can arbitrarily search for extremism on the net, but law enforcement officials are not required to take note of information received from cybercriminals. The activities of cyber-workers can be legalized in the future [14]; the bill is currently being finalized. Its significance is as follows: it is proposed to create cyber-forces on the initiative of citizens who have expressed a desire to oppose the spread of illegal information on the Internet, in the form of a public organization, with notification to the relevant executive authority. It will be possible to join the ranks of cyber-volunteers freely at the age of 18. Undoubtedly, the approval of this bill will increase the effectiveness of the prevention of extremist activities carried out in the implementation of network crowdsourcing technologies by cyber squads and law enforcement agencies. Confirmation of this is the fact that over the past 5 years more than 10 thousand sites and pages on social networks have been blocked and hundreds of criminal cases have been initiated with the help of cyber soldiers.

A successful example of the fight against extremism among young people is the implementation by the Young Guard of United Russia (MGER) of the MediaGuard network crowdsourcing project, which has been operating since March 13, 2013. The goal of this project is to consolidate the efforts of MGER activists to identify online resources specializing in the distribution of illegal content. For example, in 2018, "MediaGuard" activists checked more than 50,000 materials. More than 5,000 Internet resources with illegal content were identified and blocked, and methodological recommendations "How to Avoid Getting Into the Hands of ISIS Recruiters" made on the Internet and educational institutions RF.

For a number of economic, political, cultural and historical reasons, the south of Russia is most prone to extremism. Therefore, for the subjects of the Southern Federal District (SFD), the priority area of ensuring national security is the fight against extremism, including the youth. According to the Secretary of the Security Council of Russia N.P. Patrushev, in the southern regions of Russia in January-August 2019, about 88 thousand websites with extremist and terrorist information were blocked. With more than 100 thousand network resources, illegal information has been deleted. In addition, about 460 thousand requests for the detection of unlawful information of an extremist and terrorist nature were processed, more than 50 thousand court decisions on the ban on its dissemination were executed. These facts indicate that the system of civic associations should be further developed as an important component of the fight against extremism among young people. And some progress has been achieved.

Since the fall of 2015, the Commission of the Public Chamber of the Russian Federation for the harmonization of interethnic and interreligious relations has been implementing the STOPTERROR network crowdsourcing project, in particular in the southern regions of the Russian Federation. It is organized to monitor the Internet space, including social networks, for the availability of information propagandizing extremism and inter-religious hatred. During the first 2 years of its work, the project coordinators received over 7000 citizens' applications with links to specific resources of illegal content, and more than a thousand of them were blocked. As rightly notes I.A. Bronnikov, "... network civic activism has significant potential for increasing the effectiveness of systemic dialogue practices between the state and civil society institutions" [15].

7 Conclusion

The authors conclude that the development of modern technologies should have a scientific basis, which allows analyzing extremist crimes, emerging law enforcement practices, proposals to improve legal regulation in this area, and also developing measures to eliminate social factors that are catalysts for extremist behavior.

The authors also conclude that with the active involvement of constructive citizens' associations in the process of preventing youth extremism, the potential of modern technologies as systemically important tools for maintaining the sustainable development of the Russian superethnos increases significantly. At present, for the effective interaction of the state and civil society in the context of the prevention of extremism among young people, the following conditions must be achieved: a powerful resource and technological base for the functioning of civil society institutions and mechanisms to ensure their "entry" into the national security system; coordination of priorities and consolidation of the parties based on patriotic values; raising the level of legal awareness and legal culture of the young generation.

The effectiveness of network crowdsourcing technologies lies in solving problems of information security (identifying and blocking extremist websites, disseminating information on the Internet to combat illegal content, etc.) by constructive associations of citizens. Crowdsourcing projects aimed at preventing and combating network extremism among young people ("Cyberdruzhina", "Media Guard" and "STOPTERROR") are being successfully implemented in Russia. The use of crowdsourcing technologies in public practice implies obtaining a relatively quick and expected result in the process of solving acute social problems. In modern realities, it is necessary to develop an algorithm and criteria for the use of these network technologies in the activities of government structures and associations of citizens, while not compromising the interests of the latter. Therefore, for their successful implementation, it is important to achieve partnerships between citizens' associations and state institutions, since the prevention of youth extremism is in the interests of both sides.

Thus, the introduction of network crowdsourcing technologies for the prevention of youth extremism can help mitigate risks and threats among the youth and is a positive factor in improving national security, creating healthy relationships in society and preventing the threat of the transformation of certain youth groups into the social base of terrorism. Their wider implementation can initiate the long-term trend of "socialization" of the national security system, when not only state structures, but also public organizations of patriotically inclined young citizens act as a system-forming component. For this, it is necessary to create a real system of cooperation between authorities and public organizations, when officials are not suspicious of a popular initiative. This is quite possible with the openness of state authorities and the wide informing of citizens about decisions being made to solve social problems.

Acknowledgment

The reported study was funded by RFBR, project number 20-011-00688 "Model for the prevention of extremism among the youth: integration of the activities of civic associations and law enforcement agencies in the context of legal socialization".

References

1. I.L. Morozov, PolitBook **1**, 34-47 (2018)
2. S.Yu. Biryukov, M.V. Bobovkin, M.A. Shmatov, Legal Concept **2**, 13-20 (2020)
3. O.A. Yakovleva, Legal Concept **2**, 84-88 (2020)
4. V.N. Gulyaikhin, O.E. Andryushchenko, P.P. Fantrov, E.V. Galkina, Bulletin of Volgograd State University. Series 4: History. Regional studies. International relationships **2**, 77-85 (2018)
5. I.L. Morozov, Questions of political science **5**, 910-917 (2019)
6. A.O. Inshakova, Legal Concept **2**, 6-12 (2020)
7. A.Yu. Tutukov, A.G. Bondar, Problems of Economics and Legal Practice **5**, 277-280 (2019)
8. P.P. Muraev, D.N. Yakovlev, V.V. Shalimova, Legal Concept **2**, 43-49 (2020)
9. A.I. Kurylenko, Theory and Practice of Social Development **2**, 15-17 (2016)
10. E.G. Smolina, Islamic Studies **4**, 63-77 (2015)
11. Y.A. Tsukanov, Bulletin of Volga Institute of Management **3**, 130-137 (2018)
12. P.P. Fantrov, V.M. Shinkaruk, Legal Concept **2**, 33-38 (2018)
13. S.A. Pankratov, S.V. Shakarbieva, Bulletin of Saratov University. New series. Series: Sociology. Political science **3**, 329-333 (2016)

14. D.G. Kakhhorov, V.M. Abdrashitov, Sh.Kh. Machidzoda, Legal Concept **2**, 43-49 (2020)
15. I.A. Bronnikov, PolitBook **2**, 6-24 (2019)

SHS Web of Conferences **109**, 01014 (2021) https://doi.org/10.1051/shsconf/202110901014
LISID-2020

Current security issues in the information society

Anastasia Filippova[1*]

[1] Kikot Moscow University of the Ministry of Internal Affairs of Russia, Russian Federation, 117437, Moscow, st. Academician Volgina, 12., e-mail fipovaa@mail.ru

Abstract. That scientific research highlights specific items of modernization in the field of public society information security. Taking into account professional and public orientation, government support is aimed at promotion of human recourse, provision of information support. Currently human mind is one the most valuable item. In this regard, jurists are responsible for explanation of political, economic and legal points. Moreover, it is rather important to reveal an experienced specialist in it. Among the new challenges and threats, the problem of information security is the main one. Information technologies cause considerable impact on management of humans being. Scientific analysis of problems in the field of information security has shown that there is an insufficient level of public awareness. More precisely the problem is in ability to use password with high degree of protection, spread of unnecessary information in social networks. In that research are touched upon various concepts of information society. Also, main theories and elements are analyzed, which are included into meaning of information society. The difference between main information and other types of social values are covered. Specifics of the above problem, its reasons and impact on development of society are outlined. The conclusion encompasses problems of legal regulation of information security (fragmentation, absence of system). In cognitive activity, scientific complex, including legal, economic, scientific and practical approaches was used.

1 Introduction

Information exists in all spheres and aspects of life. Information is something that occurs in mind, as a reaction (to hear/see), that generates consciousness. Information is unalive process. Security issues in the information society are multilateral, which has been many times drawn attention both in Russian legal science [1], civil science [2], [3], [4], [5], [6], [7], [8] as well as in foreign science [9], [10], [11]. Moreover, the article analyzes Russia's conceptual and doctrinal approaches of international information security ensuring, as well as Russia's initiatives in international organizations (United Nations, BRICS, and Organization for security Co-operation to Europe). For the first time, the term "information" appeared in the 16th and 17th centuries and was translated as an explanation. In the 20-30s, the term was interpreted as a tool for consciousness. Now, there are different approaches to understanding "information": humanitarian, cybernetic, mathematical, philosophical, and attributive. The most significant types are: reactive (to respond adequately), resource (accumulation on resources), background (the most subtle adaptation mechanisms, on properties of which hidden coding technologies are built). Creation of cyber viruses, cyber terrorism, and cybercriminals can destroy the structure of any state.

Society sticks with a massive informational impact on digital sovereignty. In other words, there are interventions in national and internal affairs, provoking inter-ethnic and inter-religious strife.

We must protect ourselves. We must produce immunity. Currently information technology allows to occupy foreign country peacefully. Therefore, in the Russian Federation has been developed a national security strategy that considers these factors. In that, strategy is set the following task: to counteract, thus not to be affected by impacts of information flows that exist in TV, social networks, business cannot work without a computer.

Crime is both destructive and constructive. Law enforcement agencies considered that it deserves to make "prevention" of healthy state, healthy relationships. Thus, law enforcement agencies should train a specialist for practical activities using a competence-based approach (reactive vector). Accordingly, there should be a clear model of a specialist with skills and functions, for example, if there are programs that violate someone's copyright, then you can close the domain itself and the server.

The absence of necessary information is caused by the unsystematic data security, inadequate coordination of state measures to protect information, departmental disunity in sphere of preservation of information, and lack of control over the export of Russian scientific technologies, weapons, and management. Measures for the protection of state secrets, commercial and official secrets in state authorities and management and in defense enterprises have been seriously weakened. Also,

* Corresponding author: fipovaa@mail.ru

protection of personal data, tax, customs, and property information is poorly organized.

The lag of Russian information technologies is caused by purchases of import equipment that cannot resist cyber threat, which increases the possibility of unauthorized access to databases and data banks, and also increases the dependence of the Russian Federation on foreign manufacturers of computer and telecommunications equipment and information products.

2 Problem Statement

The situation with information security in the Russian Federation is such that it does not allow to join the world information system on an equal basis and requires urgent solutions from law enforcement agencies to solve the following key problems:

1. to improve the legal and regulatory framework to ensure information security, including development of registry of information resources, the regulation on information exchange for law enforcement authorities, regulatory, consolidation responsibility of officials and citizens for observing requirements of information security;
2. develop mechanisms and devices for the implementation of citizens' rights for information;
3. arrange information security systems that are an integral part of the country's overall national security system;
4. develop advanced methods and technical tools that provide a comprehensive solution to information security problems;
5. develop criteria and methods for evaluating the effectiveness of information security systems and tools and their certification;
6. to study the forms and methods of civilized influence of the state on the formation of public consciousness;
7. implement a comprehensive approach to the activities of information systems personnel, including methods to increase motivation, moral and psychological stability and social security of people working with secret and confidential information.

Based on the provisions, law enforcement agencies to ensure information security must carry out all measures to protect information in the political, economic, defense and other areas of state activity. Each sphere of state activity requires a special organization of law enforcement agencies. It has its characteristics, this is due to the nature of solving the tasks set, the presence of weak elements and vulnerable links related to in each area of information security.

3 Research Questions

An implementation of information technologies into its modern circulation sets new interdisciplinary tasks for national and world science, which can be solved by conducting research on various legal phenomena, including identifying current problems of ensuring security in the information society.

To achieve the research goal, it is necessary to consistently and systematically resolve a number of issues of a theoretical and practical nature, namely: to analyze the existing domestic and foreign doctrine; to study the legal framework of the issue, identifying a set of problems related to the mechanisms for bringing to justice entities that involved into circulation by means of information systems; to consider the practice of applying existing legal norms in order to identify the range of issues which should be regulated by law, as well as systematize the identified problems and identify ways to solve them.

4 Purpose of the Study

This research is conducted in order to identify issues and their scientific justification related to information protection in society, namely, subjects involved into legal relations using information and communication technologies. The developed scientific positions will become the basis for creating adequate modern legal structures that meet the requirements of the digital economy, and further improvement of legislation.

5 Research Methods

In research the author was based on the following methods: (1) general cognition methods (dialectic, systemic approach), which allowed to consider actual problems of security of the information society as a unified system of legal norms; (2) general scientific methods (analysis and synthesis, system-functional, analogies, modeling), which allowed to establish individual elements of information security, to identify General patterns of information protection in society, (3) special legal (comparative legal, legal forecasting), which allowed to identify formal and logical connections, to formulate prospects for ensuring security in the information society.

6 Findings

In scientific literature, numerous sources are devoted to security issues in the information society, in domestic literature: L. Karsavin, A. F. Losev, V. N. Muravyov, in foreign: K. Knorr-Cetin, J. URR, E. Giddens, and A. Toffler.

Problems identified in particular scientific literature can be divided into several groups: problems related to general methodological support of information security; problems related to ensuring technological independence; problems related to scientific and educational support for continuous training in the field of information security. Thus, there is a necessity to study the theory of object-centered sociality, the sociology of mobility, and the theory of postmodernity.

 https://doi.org/10.1051/shsconf/202110901014

According to regulatory legal acts, to ensure information security, law enforcement agencies must take all measures to protect information in the political, economic, defense and other areas of the state. Each sphere of state activity requires a special organization of law enforcement agencies, they have their own characteristics, this follows from: the nature of solving the tasks set, the presence of weak elements and sensitive links involved in each area of information security.

According to research by the analytical, center [12], in Russia, in 2020 year the number of informational leaks increased by 46% in comparison with 2018 year. Another analytical, center [13] predicts that by 2021, the volume of (informational) - leaks will increase by 66%. In 2019 year, informational leaks from Russian companies and government agencies amounted to 395 cases, which is 15.7 % of the global number of leaks. The most common cases are: 76.7% - leakage of payment information and personal data, among them – 72.1%, when employees of organizations are found guilty and 4.6% - when management is responsible for. Russia, for the seventh year, ranks second after the United States in the list of countries most severely affected by informational leaks.

According to the Internet Security Alliance, problems, which the information society is collide with, are caused by leakage of personal data, cyber warfare, terrorism and commercial espionage. Thus, there is a task for an expert to improve artificial intelligence and related products, in addition to adequately and effectively use machine learning to prevent, detect and predict attacks.

Each problem has a solution in its area, so the recommendations by Information technology Alliance on Internet security suggest modernize research in the field of information technology and digital programs, adapting cooperation between the public and private sectors, and introducing new methods of investigation in the field of computer power [14].

For example, the interactive Contour BI platform of Contour Components companies [15] provides access to corporate data, the advantages of which are expressed in highly productive Analytics, publication of statistics and data security (for example, customers of the software industry in the United States are – Decision technology, Inc.Software Spectrum; in Germany – Focus DV-Technologies' GmdH.; in Italy – FinWin Srl.). These interactive reports and data analysis are used by thousands of organizations in more than 70 countries in the following industries: public sector, healthcare, consulting, mass media, education, software, industry, statistics, etc.

7 Conclusion

The review of current problems of ensuring security of information society leads to the conclusion that content needs to be filtered considerably, and monitoring of information resources in particular should be conducted. Of course, the work on ensuring security in the information society includes the implementation of goals and objectives that are indicated in the document "Economic security Strategy of the Russian Federation for the period up to 2030", as well as in documents of international organizations.

To achieve efficiency in the legal regulation of information security issues, it is necessary to:

- develop a register of information resources necessary for the exchange of information for law enforcement agencies,
- establish the responsibility of officials and citizens for information security requirements compliance;
- structure the information security system, which is an integral part of the overall national security system of the country;
- develop methods for evaluating the effectiveness of information security systems and tools and their certification;
- develop a system of regulatory support in order to train personnel in the field of information security;
- perform new social mobility as a measure of security;
- improve methods, special and educational literature in the field of creation of effective mechanisms for the use of information technologies in the educational sphere, including technological support for training personnel in the field of information security;
- analyze the problems of information security related to the influence of the Internet on the consciousness of the subject (public).

Currently, the potential of negotiations on security issues in the information society is not exhausted, this is due to objective factors that create necessary basis for the subsequent development of a global document on information security.

References

1. V. N. Lopatin, Information law, 14 – 19 (2018)
2. A. A. Vnukov, *Fundamentals of information security: information protection: textbook for secondary professional education,* 161 (Yurayt, Moscow, 2020)
3. V. V. Gafner, *Information security: a textbook*, 324 (Phoenix, Moscow, 2017)
4. S. V. Zapechnikov, N. G. Miloslavskaya, A. I. Tolstoy, D.V. Ushakov, *Information security of open systems. Means of protection in networks*, 558 (Hotline – Telecom, Moscow, 2018)
5. A.V. Ivanov, Evaluation of information Security against leakage through channels of side electromagnetic radiation and interference, 64 (Liters, Novosibirsk, 2019)

6. A.V. Krutskikh, *International information security: theory and practice*, 384 (Aspect Press, Moscow, 2019)
7. O.V. Koshko, E.M. Romanova, *Economics and management: problems and solutions,* 37 – 47 (Moscow, 2019)
8. R.V. Meshcheryakov, A.A. Shelupanov, *SPIIRAS Proceedings* **3(34)**,136 – 159 (2014)
9. M. Walker, *CEH Certified Ethical Hacker All-in-One Exam Guide*, 1092 (McGraw-Hill Education, 2016)
10. S. Winterfeld, J. Andress, *The Basics of Cyber Warfare: Understanding the Fundamentals of Cyber Warfare in Theory and Practice* (Syngress, 2012)
11. A. Toffler, *The third wave*, 544 (Morrow, New York, 1980)
12. Information security software products and solutions, URL: https://www.infowatch.ru/analytics/reports (date of access: 01.12.2020)
13. Accenture, URL: https://www.accenture.com/ru-ru (date of access: 01.12.2020)
14. T. Stonier, *Information wealth: a profile of the post-industrial economy*, 393, 395 – 396 (Thames Methuen, London, 1983)
15. Contour BI, URL: https://www.contourcomponents.com/ru (date of access: 01.12.2020)

https://doi.org/10.1051/shsconf/202110901015

Model of registration of road accidents: problems of law enforcement

Marina Fokina [1], *Lilia* Voitovich [2,*], and *Olga* Egorova [3]

[1]Russian State University of Justice, Moscow, Russia, universitet.pravosudiya@mail.ru
[2]North-West Branch of the Russian State University of Justice, St. Petersburg, Russia, voitovich@ya.ru
[3]Russian State University of Justice, Moscow, Russia, olga_0_00@mail.ru

Abstract. This publication focuses on the theoretical investigation of issues arising in the Russian Federation in law enforcement related to the recently introduced digital procedure for the registration of road accidents through the "DTP.Europrotokol" electronic application running on the Android or Apple iOS operating systems. The purpose of the study is to examine the above method of independent registration of a road accident without authorized police officers, developed by the Russian Association of Motor Insurers, with a substantiation of the possibility of subsequent attaching evidentiary value to the information contained in such an application in legal proceedings for the recovery of insurance compensation in connection with the occurrence of this road accident. The authors assessed the effectiveness and feasibility of this digital traffic accident model, including the lack of sufficient regulatory guarantees for the safety of the information uploaded by the user to the mobile application, the existence of virtually unlimited powers of the application developer to control its functioning. They drew attention to certain enforcement problems associated with the assessment of the information available from the "DTP.Europrotokol" application, which is not possible to use in all cases, data the accuracy of which depends on many technical factors.

1 Introduction

The doctrine notes the progressive movement of technological progress, with qualitative shifts clearly visible as early as the middle of the 20th century [1]. Digital technology has permeated all areas of human activity. There is already discussion about creating a new reality that has no parallel in the old world. The digitalization of society and its political and economic components suggests the urgent need to fill national legal systems with standards that will acquire a different degree of compatibility within forming the macro-environment of legal regulation [2]. The proliferation of digital technologies led to the creation of new ways and means of regulation [3]. However, there is no doubt about the need for regulatory models to be consistent with international human rights standards [4].

The society with digital relationships is called the digital society [5]. The digital paradigm of activity in the Russian Federation, becoming more widespread year by year, has had a significant impact on motor insurance, affecting not only the possibility of an electronic insurance policy (E-CMTPL) confirming that the insured has concluded an insurance contract with a particular insurance company, but also the possibility of applying a digital model to the procedure for handling road traffic accidents. Such a digital model of processing road accidents through the use of the "DTP.Europrotokol" and "CMTPL Assistant" electronic applications running on the Android or Apple iOS operating systems is certainly not possible in all cases. The law regulates the circumstances and conditions where it can be implemented independently by parties involved in a road traffic accident without the help and presence of authorized police officers. The statutory legal provision for the absence of authorized police officers at the scene of the accident itself confirms that the digital procedure can only be used when there are no obvious and directly dangerous consequences for the parties involved. This study will focus on the effectiveness of the electronic (mobile) application "DTP.Europrotokol" because there is no universal solution to the problem of legal effectiveness to date [6]. Although this electronic application, by filling in all the data and conditions contained therein, significantly increases the statutory limit of the insurer's liability, the statutory electronic model of processing a road traffic accident contains some questionable legal provisions. It is important to note that the execution of a road traffic accident in a simplified form by using the electronic application "DTP.Europrotokol" does not negate the possibility to further initiate legal protection of their rights by the injured party to the accident, if the insurance company fails to perform (not properly perform) the civil law

* Corresponding author: voitovich@ya.ru

https://doi.org/10.1051/shsconf/202110901015

obligations imposed on it by the car insurance contract. In this case, a proper judicial legal procedure to protect the victim's rights is not possible without a thorough judicial evaluation of the information and data contained in the "DTP.Europrotokol" electronic application, the accuracy and integrity of which are dependent on many technical factors.

2 Problem Statement

There is the need to revise universal human rights and to extend them in view of the era of information technology [7]. The issue of balancing interests in the sphere of insurance relations, including in the digital transformation of public activity, in this regard, annually attracts the indispensable interest of the public authorities. There are various reasons for this: legislative actualization of an alternative extrajudicial (quasi-judicial) procedure for settling motor insurance disputes by appealing to a competent public person; regulatory adjustment of the contractual relationship between the insured and the insurer to expand the permissible legal framework for the use of electronic means in this case; seeking the legal introduction of a simplified procedure for the registration of a road accident (insurance event) which may subsequently be recognized as an insured event. Certainly, the latter must have a legal formulation, interpreted in such a way as to prevent uncertainty whether the means envisaged by the Act are necessary to bring the parties to the insurance relationship into a balance. It is justified to note the relevance of the task of society in view of the full realization of the political and legal potential of the Constitution through law-making and its direct application, improving the activities of the authorities, increasing their constitutional and legal responsibility [8]. Policy-making should not exclude a focus on information security and the protection of human rights [9].

The initial prerequisites of the simplified procedure of registration of road accidents introduced by the Russian legislator and justified, inter alia, by the accelerated procedure of such registration and recording of the event, have shown their practical importance in a positive aspect and led to its further legislative development and implementation, but with the increasing digitalization of society. This has resulted in the adoption of legal regulation concerning the digital ability of road accident participants to independently complete a road accident without the presence of authorized police officers at the scene of the accident by using the special electronic (mobile) application "DTP.Europrotokol" which operates on the Android or Apple iOS operating systems. The introduced legal regulation involving a displacement of the accident registration procedure by authorized police officers in its classical sense to a digital transformation of the recording of the accident has a limited scope of the circumstances of the accident.

We should note that the simplified procedure for drawing up a road traffic accident was originally introduced into the Russian public domain by the legislator under Federal Law No. 40-FZ of 25 April 2002 "On Compulsory Civil Liability Insurance for Owners of Motor Vehicles" (hereinafter, the CMTPL Law). It does not presuppose a digital format, currently provides for several conditions for its implementation. Thus, there should be no more than two vehicles involved in a road traffic accident, the civil liability risk of both drivers must be insured under a valid compulsory motor third party liability insurance contract (hereinafter the CMTPL contract); the damage caused must be exclusively material in nature and be caused only to vehicles whose drivers do not have any disagreement about the circumstances of the accident and the list of vehicle damages. The right arising from the compliance of the participants in a road traffic accident with these conditions entitles them to claim insurance compensation from the insurance company in an amount not exceeding 100,000 roubles.

The use of the electronic (mobile) application "DTP.Europrotokol" is possible not only when the participants of a traffic road accident subjectively wish to increase the possible amount of the insurance payout, with a maximum amount of 400,000 roubles (in the territory of the constituent entities of the Russian Federation - the cities of the federal significance of Moscow and Saint Petersburg, and by using digital technologies of the global navigation satellite system of the Russian Federation (GLONASS) and GPS coordinate data), but also if the participants disagree about the circumstances of the accident (including the fault of a particular participant and the list of vehicle damages). "DTP.Europrotokol" is optional without one of these conditions in relation to the circumstances of a specific crash, expressed in appropriate legal definitions to be assessed in a combined manner. Artificial intelligence embedded in the application software has some attributes, such as substantiality, autonomy, indicated in the doctrine [10].

The legislative incorporation of this simplified digital model of registering a road accident has not precluded the initiation of legal proceedings to protect the rights of insured persons, whose procedural features have also changed as a result of this digital modification of the classic procedure for registering accidents. There is a strong case for the non-specificity of Russian law enforcement in resolving issues related to the regulation of electronic processes [11].

3 Research Questions

The practical implementation of a remote electronic digital procedure for the registration of a road traffic accident, similar to the traditional registration of an accident without authorized police officers, has its legal regulation set out in the provisions of Article 11.1 of the CMTPL Act providing for the previously outlined conditions where it can be used by technical means to record violations, the circumstances. Moreover, Paragraph 2.5 of the Russian Federation Road Traffic Rules also establishes the obligation for drivers to self-record the accident on the legally prescribed accident notification form.

"DTP. Europrotokol" is a mobile application developed by the Russian Association of Motor Insurers, with software for the main mobile operating systems (Android, Apple iOS), which allows for photographic documentation of vehicle damage resulting from a road accident, to a portable (mobile) device (smartphone, tablet) of the user and transfer the received photo-images with additional information about the road accident via GSM communication (mobile Internet) to ST-GLONASS AIS RAMI (automated information system of the Russian Association of Motor Insurers).

But there is the lack of an objective possibility to reproduce the circumstances of the road accident, the location of the vehicles at the territorial police body or a stationary post after the drivers had performed their duties to document the scene of the road accident without traffic police officers, the lack of the possibility, when relevant questions arise, to comprehensively, fully and objectively examine evidence of fixing the event that occurred [12].

3.1 Guaranteeing the integrity of user data: unlimited possibilities for the mobile app developer

As the developer of the "DTP.Europrotocol" software, the Russian Association of Motor Insurers has, accordingly, certain powers to ensure the functioning of this mobile application, establishing, among other things, the right to modify or delete any information posted by the user on the mobile application at any time. This right also extends to the possibility of modifying or deleting photo content (photo images) uploaded to the app by the user at any time for documenting damage to the vehicle as a result of a road traffic accident. Thus, it appears that the objective security of the user's account and all the information uploaded to the mobile application is not only legally established, but actually made dependent on the actions of the application's developer. It is an open question whether it is justifiable and reasonable for a developer to have the power to intrude into the original data of a user's account, who may initiate legal proceedings to vindicate his right, which would require information initially recorded in the mobile app, but during its realization such information may disappear.

3.2 Lack of responsibility of the software developer: is it justified?

As noted, the information about damage to vehicles recorded by photographic images with accompanying information about the accident is transmitted via GSM communication (mobile internet) to the ST-GLONASS AIS RAMI, whose principle of transmission is its invariability. Immutability, in this case, implies that the information transmitted is consistent with its original recording, without giving drivers involved in the crash the right to change it, including by excluding any photographic data.

However, users accept the terms and conditions of the "DTP.Europrotocol" mobile application as set out in the relevant agreement for its use, stipulating that the right to use the mobile application is granted on an "as is" basis. The Russian Association of Motor Insurers makes no warranties or representations regarding the error-free operation (functioning) of the mobile application or its individual parts (materials), independent information elements, or the compliance of the mobile application with the specific purposes and expectations of the user.

Furthermore, the Russian Association of Motor Insurers does not guarantee the authenticity, accuracy or completeness of the information (materials) contained in the mobile application, or provide any guarantees or representations. It is not liable for any consequences associated with the use of the mobile application, including due to errors or malfunctions in its operation (functioning). The app is not guaranteed to function continuously, quickly and without technical failures, reliably and error-free or to provide error-free, complete, accurate and correct results from the mobile app.

4 Purpose of the Study

The purpose of the study was to investigate issues arising from the use of a digital procedure to process road traffic accidents through the use of the "DTP.Europrotocol" electronic application running on the Android or Apple iOS. The study aims to identify legislative gaps in the regulatory framework for the use of this digital method of recording accidents.

5 Research Methods

The study relies on the application of comparative legal analysis, system-structural analysis and the method of synthesis.

6 Findings

6.1 Guaranteeing the integrity of user data: unlimited possibilities for the mobile app developer

It seems that the possibility provided by law for the software developer to modify or delete, at his subjective discretion, the initial information contained therein by the user, the road traffic accident participant, including information about the latter, is an unjustified "extension" of the permissive framework of control over the functioning of this mobile application. The existence of this power, which is outside the responsibility and control of the insured (app user), can complicate the latter's evidentiary activities in litigation against the insurance company, de facto creating a situation where it can exclude the submission of evidence in support of its position (including when challenging the guilt of a particular road traffic accident participant). Lacking sufficient safeguards to protect the user's information, effectively creating a strong protective "bastion" in his possible subsequent evidential activity, there is no reason to argue for compliance with the principle of procedural fairness in litigation against an insurance company and based on the examination of such data. There is no instrumental nature to the legal procedure noted in the doctrine [13], which implies the possibility of providing subjects with the most favourable regime for exercising their rights. The abstractiveness of textual characterization of legal statehood can be compensated by doctrine and jurisprudence [14].

6.2 Lack of responsibility of the software developer: is it justified?

We believe that a proper legal argument about the liability of the developer of the mobile application "DTP.Europrotocol", legally prescribed for mandatory use by road accident participants in certain cases (circumstances), on the one hand, actually absent and, on the other hand, in principle, enshrining the absence of any guarantees on its part about the reliability of fixation and transfer of information about the road accident to ST-GLONASS AIS RAMI, necessary for investigation when later initiating a court dispute, appears to be an unreasonable legislative decision. The data and information contained in the mobile app are certainly of evidentiary value in a judicial review of an auto insurance dispute, both in terms of determining who was at fault in the accident and for determining the list of damages to the vehicle, the total value of which, together with other circumstances, makes it possible to determine the cost of the repair. The latter affects the procedural possibility of appointing a forensic examination in a court case to determine the cost of repairing the damaged vehicle, which can only proceed on the basis of available photo content (for example, if the vehicle has been taken out of the possession of the user of the mobile app). The lack of provisions on the truthfulness, quality, accuracy, reliability and correctness of the information and data makes it clear that the mobile app user cannot use all possible procedural ways of protecting his subjective right to receive insurance compensation in the relevant proceedings. At the same time, the idea of compulsory protection of the weaker party is increasingly becoming a feature of the law [15].

7 Conclusion

The study confirms the growing importance of digital transformation of insurance legal relations, while at the same time demonstrating the need for legislative adjustments to certain provisions of the digital procedure for the functioning of the "DTP. Europrotocol" mobile application already in use, including from the perspective of legally establishing the principle of invariability and integrity of user information, while limiting the rights of the software developer to subjective control and interference with it and its content. The introduction of this digital traffic accident model, except an accelerated procedure for recording the accident, not only establishes other guarantees of increased protection for the user (consumer of financial services), but also puts into question the possible exclusion by the developer at his discretion of information uploaded by the user which could affect the outcome of the legal case. The use of a digital procedure for processing road traffic accidents will only be more practical if the results of its implementation, used in court proof of the position of the disputing parties who are parties to the insurance legal relationship, are aimed at increasing the level of protection for the insured (mobile app user) as a weak party by excluding possible procedural defects in the functioning of the software. We have achieved the purpose of this study by identifying gaps in the enforcement of the digital crash model and making proposals to address them.

References

1. A.E. Evstratov, I.Y. Guchenkov, Law Enforcement **4(2)**, 13-19 (2020)
2. O.A. Stepanov, Law. Journal of the Higher School of Economics **3**, 4-23 (2018)
3. E. Gromova, T. Ivanc, BRICS Law Journal **7(2)**, 10-36 (2020)
4. A. Lushnikov, M. Lushnikova, Russian Law Journal **4(4)**, 95-111 (2016)
5. V. Laptev, V. Fedin, Russian Law Journal **8(1)**, 138-157 (2020)
6. I.P. Kozhokar Bulletin of Perm University. Legal Sciences **48**, 196-225 (2020)
7. Y.A. Chernysheva, Psychology and Law 9(4), 90-102 (2019)
8. I.A. Kuzmin, A.V. Yurkovsky, Comparative Politics **10** (1), 58-73 (2019)
9. C. Anderson, R.L. Baskerville, M. Kaul, Journal of Management Information Systems **34(4)**, 1082 – 1112 (2017)
10. I.V. Ponkin, Bulletin of the Peoples' Friendship University of Russia. Series: Legal Sciences **22(1)**, 91-109 (2018)
11. V.N. Grigoryev, A.P. Sukhodolov, S.S. Ovanesyan, M.G. Spasennikova, V.V. Tyunkov, All-Russian Journal of Criminology **13(6)**, 873-883(2019)
12. F.N. Zeynalov, Russian Justice **9**, 65 - 67 (2018)
13. A.G. Repyev, Bulletin of Tomsk State University. Law **34**, 21-28 (2019)
14. A.V. Dolzhikov Bulletin of Tomsk State University. Law **6**, 56- 69 (2020)
15. A.V. Kuzmina The formation of the doctrine of weak party protection in contract law // Bulletin of Perm University. Legal Sciences **46**, 698-727 (2019)

Digital society, artificial intelligence and modern civil law: challenges and perspectives

Vladimir Kamyshanskiy[1,*], *Dmitry* Stepanov [2], *Irina* Mukhina[3], and *Dina* Kripakova [4]

[1] Kuban State Agrarian University named after I.T. Trubilin, st. Kalinina, 13, Krasnodar, Russia, vpkam@rambler.ru
[2] Kuban State Agrarian University named after I.T. Trubilin, st. Kalinina, 13, Krasnodar, Russia, o02323@icloud.com
[3] North Caucasus Branch of the Russian State University of Justice, st. Levanevsky, 187/1, Krasnodar, Russia, rapkrassnodar@mail.ru
[4] Astrakhan State University, st. Tatishcheva, 20 a, Astrakhan, Russia, kripakova2013@mail.ru

Abstract. The article analyses the problems of civil law regulation of relations arising in the digital society. The property relations, the civil turnover sphere is increasingly shifting to various kinds of electronic platforms. The process was further accelerated by the COVID-19 coronavirus pandemic. The development and application of artificial intelligence technologies bring forward the problem of ensuring safety for humans from the negative impact of such technologies, minimizing threats to their life and health. Due to the virtuality of artificial intelligence as a social phenomenon, a product, a good created by man, the application of traditional legal means to regulate social relations connected with the application of artificial intelligence technologies at present does not provide a full solution to a number of uncertainties arising on the market of goods, works and services. To this end, it is proposed to form a new legal fiction in civil law at the level of scientific doctrine to resolve legal uncertainties arising in civil turnover due to the use of artificial intelligence technologies. It is about formalizing artificial intelligence technology as a subject of law as an "electronic person" or "electronic legal entity" bringing together several legal entities with a special legal personality.

1 Introduction

The concept of "digital society" and digital rights is a relatively new object of civil rights. The digital economy is an economic activity based on digital technology related to e-business and e-commerce and the electronic goods and services produced and marketed by them. Payments for services and goods in the e-economy are increasingly made in electronic money. The social relations arising from the digitalization of society and the introduction of artificial intelligence technologies have specific characteristics to be considered when choosing the legal means for their legislative regulation. Legal science and the science of civil law face the very difficult task of developing new doctrinal approaches to the legal regulation of social relations associated with the emergence and use of new technologies in the economy and other spheres of life in modern society.

2 Problem Statement

Practical problems for lawmakers arise from the lack of new doctrinal approaches in the science of civil law in addressing the problems of civil law regulation of relations arising in society involving artificial intelligence technologies as one of the stages of digitalization of society and the creation of digital space in a post-industrial society. The lack of norms regulating peculiarities of the legal regime for carriers of artificial intelligence, the results of intellectual labour created by them, introduces uncertainty into the regulation of arising relations and comes into conflict with the current legislation in the intellectual law sphere, as well as norms defining the legal status of civil law subjects and the grounds for legal liability.

3 Research Questions

The article deals with the legal regulation of relations arising in society as a result of the introduction of digital technology and artificial intelligence. It substantiates the necessity of forming a new legal fiction in civil law that would give the status of a subject of law to a carrier of artificial intelligence. The proposed scientific positions can form the basis for the formation of a new institution of civil law and the further improvement of the current legislation.

4 Purpose of the Study

The purposes of the study are to shape the notion of artificial intelligence in civil law scholarship, analyse the characteristics of social relations as an object of legal regulation arising in the digital society, justify the

* Corresponding author: vpkam@rambler.ru

https://doi.org/10.1051/shsconf/202110901016

feasibility of forming a general classification for carriers of artificial intelligence technology, depending on their degree of autonomy and separateness. Based on the study, it is necessary to develop proposals for further improvement of the current legislation in the digital economy and digital society with the use of artificial intelligence technologies.

5 Research Methods

The methodological basis of the study consists of dialectical and general scientific methods of knowledge, including system analysis, synthesis, induction and deduction, as well as special legal methods (formal-logical, which allowed to develop a number of conclusions regarding the legal status of the carrier of artificial intelligence in the system of key categories of civil law. The study of the problems of civil law regulation of social relations in the context of the digitalization of society also used abstraction and modelling methods. The comparative-legal and legal foresight method allowed formulating the prospects of legal regulation of relations in a digital society with the use of artificial intelligence technologies.

6 Findings

Russia is on the threshold of a new stage in the development of modern civilization, associated with the formation of information space and the development of digital technologies in all spheres of public life. According to V.P. Filimonov, the digital society is a global project that aims to build a new slave society governed by the information and communication technologies based on the use of microelectronics, local and global computer networks that collect, process, generate and distribute information through global telecommunication network systems [1]. It is rather difficult to agree unequivocally with such a categorical definition of the digital society by Valery Pavlovich Filimonov, a publicist, specialist in cybernetics and control systems, academician of the Petrovsky Academy of Sciences and Arts. It is clear that we can avoid a new slave-holding society by placing the advances of modern technology at the service of humanity, by subordinating their use exclusively to human interests through legislation. The continuous growth of the global population and the need to mobilise new resources to sustain it leads to the fact that the e-economy is no longer limited to business and e-commerce and services, but penetrates all areas of human life, including health care, education, science, culture, etc. [2].

The objective reality is that the sphere of property relations, civil turnover, is increasingly shifting to various kinds of electronic platforms. It is such platforms that allow active scientific collaboration at various international and national scientific conferences, maintaining a communication environment even in the forced isolation associated with the COVID-19 pandemic coronavirus infection.

Any digital platform is a collection of interacting hardware and operating systems running various applications and development tools. The Russian market for goods, works and services is increasingly moving to such platforms. There is a transformation of the traditional and tangible property relations familiar to us into a virtual reality where activities take place autonomously, not entirely subject to the human will.

In this context, we should look in more detail at artificial intelligence technologies as objects of civil law regulation. The rapid development of artificial intelligence technology is pushing legal scholars to explore this new phenomenon as an object of study in its own right. The development of artificial intelligence technologies and their introduction into various spheres of social life, production and consumption is a powerful stimulus for the improvement of national and international legislation. The creation of artificial intelligence, the digitalization of the economy and a range of social processes is creating a new social reality that needs to be scientifically understood and legally mediated.

It is equally evident that the legal regulation of relations using artificial intelligence technologies lags far behind the needs of the civil sphere. In this situation, it is only natural that the science of civil law should form its own position and provide answers to the challenges of the times.

In this context, it is worth noting a number of authors who are actively engaged in developing civil law issues of artificial intelligence. Among them, it is necessary to highlight the doctoral dissertation of Peter Mechislavovich Morhat "The legal personality of artificial intelligence in intellectual property law: civil law problems" and the candidate thesis of Evgeny Polikarpovich Sesitsky "Problems of the legal protection of results created by artificial intelligence systems" defended in 2019 and a number of other works [3-6].

The problem is that the social relations arising from the use of artificial intelligence technologies have already formed a number of areas that are now impossible to regulate using the traditional legal means and modes of legal regulation enshrined in the current civil legislation.

In this connection, it is necessary to initially define the essence of the category "artificial intelligence" and then try to apply such traditional institutions of civil law as object and subject of law, subjective right and legal relations, civil liability and tortious capacity to it.

Artificial intelligence is the property of intelligent systems to perform creative functions autonomously and independently of their creators that are traditionally considered the prerogative of humans. An intelligent system can be a technical or software system capable of solving creative problems in a specific subject area, with knowledge stored in its memory. An intelligent system comprises three main units: a knowledge base, a problem solver and an intelligent interface [7].

Today, artificial intelligence has become a global trend. The leading economies are in a race to be the first in this field. One of the most important problems in the development and application of artificial intelligence technologies, especially in such sensitive areas as national defence and state security, human life and health, is the problem of excluding any error in its activities, as well as working out ways to compensate for the costs of artificial intelligence errors and to personalize the subjects of such responsibility.

The development and application of artificial intelligence technologies bring forward the problem of ensuring safety for humans from the negative impact of such technologies, minimizing threats to their life and health. Artificial intelligence, for all its power and breadth of possibilities, is just as prone to error as humans. The cost of such mistakes can be very high. For example, errors in the provision of fee-for-service medical diagnoses or consultations to patients by means of artificial intelligence can lead to tragic and sometimes irreversible consequences in the case of a single individual. Artificial intelligence error in national defence and state security could lead to the unauthorized use of nuclear weapons, other weapons of mass destruction and the destruction of humanity.

In this regard, experts in mathematical modelling, concerned with the problems of developing artificial intelligence, see the way out as developing a new generation of artificial intelligence capable of correcting its errors and inaccuracies on its own without human intervention [8]. In itself, the idea expressed by the programmers commands respect and deserves support. It meets the needs of human society. Solving this problem makes it much easier for programmers to work around the bugs of the artificial intelligence built by them. There is no need to retrain such an AI technology carrier for a full "retraining" course based on the identified error.

The emergence of failures in artificial intelligence technologies brings to the fore the problem of personifying the subject of civil liability for harm or loss arising from the use of artificial intelligence technologies. This means that we must be aware that between the malfunctioning of the artificial intelligence and the removal of the malfunctioning in the artificial intelligence, there may be consequences that result in the violation of someone's subjective rights, inability to fulfil obligations assumed, causing harm or damage to citizens or legal entities.

For the structure of the intelligent system, failure could occur in all three of its blocks: the knowledge base, the problem solver and the intelligent interface. These software products have numerous authors developers. Errors in the operation of artificial intelligence could be the fault of individual developers, or of a group of individuals involved in the development of artificial intelligence technology. The errors could also arise irrespective of the fault of the authors of the artificial intelligence technology. For example, it may result from applying an open knowledge base, which is later substantially updated from the original knowledge base generated during the development phase of the intelligent system.

This logically raises the question of who should be held legally responsible for the harms or losses caused by artificial intelligence? Some problems of legal liability associated with the use of artificial intelligence were investigated in the literature by both Russian [9-10] and foreign authors [11]. Legal regulation of integration of artificial intelligence into human society in the context of legal relations that arise in this case requires the solution of such an issue as the distribution of legal liability for harm (damage), losses caused to participants in civil turnover by activities related to the use of artificial intelligence technologies [12].

This problem requires identifying the place of the artificial intelligence carrier among such basic concepts of civil law as "object", "subject" and "legal relationship". In civil law, it is these basic concepts that allow civil law to give certainty to the regulation of property and personal non-property relations. The further assertion that a carrier of artificial intelligence can only be an object of civil rights and not a subject in any way does not reduce the number of problems in the science and practice of civil law.

These problems may well be solved by the formation of new legal fiction in civil law to resolve the legal uncertainties arising in civil turnover due to the use of artificial intelligence technologies. It is about formalizing artificial intelligence technology as a legal fiction, as a legal technique. Legal fiction is known to be a legal device that suggests the existence of a certain fact contrary to existing reality. This recognizes a known non-existent fact as existing. Fiction was already present in the laws of Hammurabi. According to experts, there is no legal order without fictions as legal techniques [13].

At the same time, the literature attempts to make sense of the possibility of giving artificial intelligence the status of an entity as an electronic person. For example, Professor O.A. Yastrebov suggests interpreting the e-person "as a personified unity of rules of law that bind and empower an artificial intelligence (e-person) that possesses the criteria of 'reasonableness'" [14]. Sharing O.A. Yastrebov's position on the admissibility, at least on a doctrinal level, of recognizing the carrier of artificial intelligence technologies as a legal entity, the proposed definition of an electronic legal entity needs critical reflection. It is difficult to treat the e-person as a subject of law as a "personified unity of law" that binds and empowers them. A personified unity of law has a greater right to exist as an independent civil law institution regulating social relations involving a carrier of artificial intelligence.

However, the science and practice of civil law already face a number of challenges that need to be addressed. This refers to the authorship of the results of intellectual activity, paintings and musical works created by an artificial intelligence carrier. As I.V. Ponkin and A.I. Redkina correctly point out, artificial intelligence is capable of creating unique results of an intellectual activity without human involvement [15]. The right of authorship of an AI carrier challenges all the underlying concepts of intellectual property law in relation to the creative element in copyright [16]. There is also a need for legal certainty regarding the fate of the income generated by the artificial intelligence carrier, the subjects of liability for damage caused by the activities of the artificial intelligence carrier due to a malfunction in its operation, and so on. Their solution would eliminate the legal uncertainty associated with the use of artificial intelligence technologies. Therefore, we must seek answers to these questions today.

The peculiarity of legal regulation of relations with artificial intelligence technologies relates that human activity involves the use of artificial intelligent systems capable of autonomously exercising certain creative functions apart from the will of a human being. The strategy for the development of legal regulation of relations connected to the use of artificial intelligence technologies requires defining the following directions.

First, it requires determining a list of those technologies that may pose a threat to individuals and society because of their autonomous nature and their inability to fully control human use.

Second, it needs to define the boundary between the carriers of artificial intelligence, which may well qualify as objects, including those relating to sources of increased danger, by classifying the features of their legal regime.

Third, it is necessary to form a scientific doctrine of the carrier of artificial intelligence as a subject of civil legal relations, endowing it with a certain special legal personality. This can be a single artificial intelligence carrier with the conventional name of "e-person" or a combination of them with a specific legal status of "e-legal entity" operating in a specific environment on various electronic digital platforms.

In this context, it is time to elaborate the content of a special civil law institute with the title "Participation of Artificial Intelligence carriers in relations governed by civil law" at the doctrinal level as an integral part of a subsection of the civil code with the title defining the legal status of persons acting as subjects of civil law.

This will resolve a number of issues currently in need of resolution. In particular, who owns the copyrights to works created by the robot; who will be held liable for the activities of an artificial intelligence carrier that cause harm (losses) to third parties; what restrictions should be placed on the use of military robots; what should be the legal restrictions and forms of enforcing control over the development of a strong artificial intelligence comparable in intelligence to the human brain.

The use of a robot with artificial intelligence in communication with humans suggests the need to legislate against and limit machine activity of so-called anthropomorphic behaviour (transferring the human image and its properties onto inanimate objects). In this case, it is not enough to have a "red button" to disconnect an artificial intelligence carrier from its power source and prevent it from causing harm or loss to humans, public interest harm or other negative consequences caused by its actions.

For example, we will consider a situation where a robot with artificial intelligence is tasked to represent the state on the board of directors of a joint-stock company with state participation or the interests of a customer when forming the terms of a state contract for the supply of goods for state and municipal needs, or to act as an expert on legislative acts for their anti-corruption nature or limiting competition in markets for goods, works and services. On the contrary, in these cases, the so-called red button in the hands of the person concerned, as well as the manual management of the Russian economy that we have periodically observed, is capable of dashing the long-held expectations of our society.

The successful development of robotics and artificial intelligence is possible if the state has the will and consistency to implement investment and legal policies in the relevant area. In any country, the state is the biggest investor in the national economy. The mass introduction of robots with artificial intelligence into various spheres of social life will lead to significant job losses in government, healthcare, education, banking and the real economy.

Sberbank of Russia, for example, intended to free up around three thousand jobs in 2017 by introducing robot lawyers to handle claims for retail customers. It has plans to move all typical lawsuits to an automated format in the future, which will free lawyers from routine work and focus on resolving complex legal issues [17]. Consequently, we cannot rule out strong opposition to robotization processes and the development of artificial intelligence for a number of social groups in society, including lawyers.

7 Conclusion

The adoption of regulations containing peremptory norms seems best in the context of law-making in the development and application of robots with artificial intelligence. Any regulations that contain dispositive principles that open up the possibility of any discretion will serve as a brake on the development of this process and prevent it from becoming irreversible. In the first place, this should apply to the spheres of public life related to the activities of state and local government, courts, the prosecutor's office, health care and education. To alleviate the negative effects of such opposition, it is now necessary to develop state programmes for retraining and retraining of those subject to redundancies due to robotization and to create new jobs for them.

Thus, the rapid development of artificial intelligence and robotics technologies requires appropriate legal regulation of the new social relations that are emerging. Numerous issues currently need to be resolved. In particular, it should be thought about solving the issues of who owns the copyright of the works created by the robot. Equally important is the personalization of responsibility for performance using artificial intelligence technologies. There is a need to develop a system of legal restrictions on the military use of robots, as well as an international system to control the development of strong artificial intelligence technologies comparable in intelligence to the human brain.

References

1. V.P. Filimonov The digital society and the end of history
https://zen.yandex.ru/media/id/5e274bc843863f00acd7e

d97/cifrovoe-obscestvo-i-konec-istorii-5f3403d4f8858d5978533028
2. V.P. Kamyshansky, Power of the Law **1,** 14-18 (2019)
3. P.M. Morhat, Power of Law **1**, 82-95. (2019)
4. P.M. Morhat, *Law and Artificial Intelligence: a Monograph. Russian Academy of Intellectual Property.* (Moscow, Russia. UNITY-DANA, 2018)
5. M.E. Cheremisinova, Legal status of subjects in virtual space: a monograph (Moscow. Russia. Publishing house "Jurisprudence", 2020)
6. Yu. Tikhomirov, S.B. *Nanba The legal concept of robotization: a monograph* (Moscow. Russia. Brochure, 2019)
7. A.N. Averkin, M.G. Gaase-Rapoport, D.A. Pospelov, Radio and communications (1992) //http://www.raai.org/library/tolk/aivoc.html#L208
8. Y. Medvedev, Rossiyskaya Gazeta **207**, 9 (2020)
9. T.V. Deryugina., O.A. Inshakova, E.Yu. Malikov, *Advances in Intelligent Systems and Computing. Switzerland. Springer International Publishing,* **1100**, 693-705 (2020)
10. V.P. Kamyshansky, E.Y. Rudenko, E.A. Kolominetz, E.O. Osadchenko, Studies in Computational Intelligence **826,** 299-306 (2019)
11. Bublitz Christoph, Wolkenstein Andreas, Jox Ralf J., Friedrich Orsolya, International Journal of Law and Psychiatry **65**, 101399 (2019)
12. J. Daar, The Legal Liability Landscape and the Person/Property Divide F&S Reports URL: https://reader.elsevier.com/reader/sd/pii/S2666334120300519?token=16A3DD4D2D1796C496AB8AD5A76213184B4AB105B7046BA58DDBFA46AD76F8A7C2CBA0398C8413ADAC2045EDC6ECE787 (date of access: 27.02.2019).
13. A.M. Shirvindt, *The Significance of Fiction in Roman Law* (Moscow, Russia Statut, 2013)
14. O.A. Yastrebov, *Proceedings of the Institute of State and Law of the Russian Academy of Sciences* **13 (2)** (2018)
15. I.V Ponkin, A.I. Redkina, Intellectual Property. copyright and related rights **2**, 35-44 (2018)
16. A. BridyStanford Technology Law Review **5,** 1–28 (2012)
17. Sberbank Robot lawyers will replace three thousand Sberbank employees (2019), URL: https://tass.ru/ekonomika/3936662 (date of access: 27.02.2019).

Application of LMS Moodle for distance learning foreign languages at university

Oksana Kandrashkina[1], and *Elena* Revina[2]

[1] Samara State Technical University, Molodogvardeyskaya Str., 244, Samara, Russia, petergrifon@mail.ru
[2] Samara State Technical University, Molodogvardeyskaya Str., 244, Samara, Russia, rev.e.v@mail.ru

Abstract. The article deals with the application of the Moodle system for teaching the subject of foreign languages (English language) at the distance learning department of Petroleum engineering at Samara State Technical University (Russia). The paper provides an outline of the course structure, tools for practical classes, independent work and the forms of assessment procedure. The course content is presented and divided into modules in a logical way according to the main education programme. The tool chosen for instruction for practical classes has been the Page tool. Tasks for independent work are provided via the Assignment tool. The efficiency of the course has been analyzed according to the data provided by the system: students` online activity in the course, fulfilled assignments and assessment reports based on the quiz results. The research has proved that the tools chosen for the course to be relevant to the subject. Analysis has shown that the learning materials and files with exercises are easily accessible. Moodle system has been considered to be appropriate for distance learning of students with a low level of language proficiency.

1 Introduction

The technologies for distance learning have been gaining more importance now than ever before. The wide use of personal computers and access to the internet has changed the attitude to studying and learning. Besides, the internet has undoubtedly become one of the major sources of information. It can provide a student-centered learning environment if the materials and methods are designed to take advantage of the interactivity and resources the Internet provides [1].

The quick development of information technologies nowadays has also added to the increase in various forms and tools for learning, e.g. online courses, video conferencing, etc.

It is getting more and more difficult to deal with information pressure being online, so the main task for students as well as for teachers is to orientate themselves in the great amounts of information that quickly becomes outdated and irrelevant and is constantly updating. Students must learn how to search for and analyze the necessary sources of information. These skills are important for future specialists and their professional development.

1.1. The outline of Moodle as a distant learning system

The key elements of any distant learning technology are interactive online courses on different websites or hubs that providing support for the educational process and assessment techniques.

There are a lot of distant learning systems, e.g. Claroline, ATutor, Ilias, Open ACS and others. Moodle (Modular Object-Oriented Dynamic Learning Environment) is considered to be the most popular and relevant e-learning platform according to recent research [2]. It has an organized interface for virtual learning and allows teachers to create online courses that students can access as a virtual classroom. It is open and free of charge platform that can be downloaded from the official website.

Educational software environment was initially oriented for teachers who did not have any special computer skills in administration of databases, websites, etc. The system is easy to use due to the accessible and clear interface. Educational management system allows designing, creating and hereafter controlling and monitoring the resources of the educational environment.

One of the advantages of Moodle is a great number of libraries and functions in PHP programming language that makes it a convenient tool for developing various information internet systems.

Another important feature is the powerful monitoring system that allows the administrator or teacher to monitor participation of students in the course and their progress. It also provides detailed information about any course module, unit and element. So students` online activity and participation in the course can be analyzed [3], e.g. how many tasks have been done by a student and how often he or she has accessed the course or was online.

In comparison with other distant learning systems Moodle is based on principles proposed by such scientists as John Dewey, Jean William Fritz Piaget, Ernst von Glasersfeld. The developer of Moodle Martin Dougiamas defined five principles of the learning system and united them under the name of "social constructivism" [4].

While most virtual learning systems are instructor-oriented and largely concerned with how course content is delivered, Moodle is based on a learner-oriented philosophy, in which students are involved in constructing their own knowledge" [5].

Nowadays the largest universities worldwide as well as Russian educational establishments make use of the Moodle system.

2 Problem Statement

Distance education presupposes the application of the information technologies aiming at lessening distance between a student and an instructor physically and psychologically, and to increase interactivity and interaction among students, learning sources and facilitators [6]. It also suggests using multimedia resources as well as the Internet open access resources relevant to the course of study. Distance learning technologies have become one of the common ways of learning. The first reason for it is the Internet and information and communication technologies that influence every sphere of life. The other reason is a global pandemic situation which has imposed strict regulations on the face to face interaction. Distance learning technologies for language learning have been examined by Russian and foreign scholars, especially in the UK [7]. Russian teachers focus mainly on different multimedia technologies and virtual learning systems as ways to get students more motivated and interested and publish various articles about the implementation of them in the learning process [8], and papers dedicated to advantages and disadvantages of using LMS Moodle [9].

3 Research Questions

It is suggested that quite simple tools and activities developed for the online course in Moodle can provide an effective learning process for students of the distance learning department. It is expected that Moodle system is easy to use so students will not face any difficulties accessing and navigating the course content and assignments provided.

4 Purpose of the Study

The main aim of the research is to analyse the efficiency of the Moodle system for teaching a foreign language to students from the distance department of Petroleum engineering. It examines the relevance of the tools chosen for the course content.

5 Research Methods

A content analysis was applied. The data provided by the system serves as a basis for the study. It includes reports that reflect participation of students in the course (time spent online in the system), total number of fulfilled assignments uploaded and assessment reports on tests taken during the term. The use of the Moodle was analysed through the frequency of access, the purpose of the accesses and the formats of the information accessed, posted or uploaded.

6 Findings

6.1 Outline of distant form of learning in Samara State Technical University

Higher education is changing very rapidly and trying to meet the needs of the very diverse stakeholders including instructors, learners and administration staff [10]. Russian higher educational institutions being aware of the global educational changes have also modified the forms of learning according to the needs of the students. Russian universities have begun using virtual learning systems primarily for distance learning and blended learning.

Samara State Technical University (SSTU) has been applying online systems for distance learning because of the increasing demand for computerization of educational process. The technological platform chosen for distance learning has been the Moodle system. Application of LMS Moodle forms the information-education environment that compensates for lack of personal communication between a teacher and a student in distance learning and supplements full-time form of education with the technologies used in online courses development [10].

At first Moodle was used primarily for the distance learning department. Distance learning is not considered to be an innovative way of studying as it has existed for decades. Several years ago, learning at the distance department was carried out in the form of short-term face to face sessions including lectures and seminars. Students had to study for 5 or 6 years. It was obligatory for students to attend classes one or two weeks during the semester and to get credits and take exams at the end of the term at the university. The educational process was conducted without any application of information technologies or e-learning systems. It presented some problems for students who lived in other cities or worked in shifts and had to travel the long way to the university to attend classes and take exams.

Since then the form of learning has changed completely. Traditional form of learning has been replaced by distance learning that is organized with the application of the Moodle system. The education process can be characterized as blended because some subjects in the curriculum require students' attendance. As far as the foreign languages are concerned, the education process is carried out via LMS Moodle.

To work in LMS effectively, the university provides training courses for teaching staff where everyone can try and create his/her own online course. Different subjects require the use of different Moodle tools, e.g. lectures must be accompanied by video recordings or presentations.

It should be noted that teachers having basic computer skills and Internet skills connected have no difficulties in learning how to work in the Moodle system. Only with the help of a reference system a teacher is able to create an online course in any discipline that will include:

- information activity-oriented content (forum, wiki-projects, online meetings, lectures, quizzes, multimedia resources, additional materials);
- assessment content (the analysis of the test/quizzes results in every group, a report of students' activity during the course);
- administrative content (user registration, role confirmation, assigning roles) [11].

6.2 Course Structure

Most students entering the department have elementary or beginner level of English. It is explained by the fact that there are mostly adult students who finished schools or colleges many years ago and have not had any foreign language experience since then. It specifies the choice of teaching methods and the content of the course. The course has been developed according to the main educational programme on the subject of the foreign language, approved by the Federal Educational Standard of the Russian Federation and Samara State Technical University.

The course structure is important and must correspond with the education program. All language courses must include materials for so-called "practical studies" and individual training activities, so the course structure must be clear and well-organized. Moodle provides the tools needed to design and implement appropriate language learning courses that use online activities. Within Moodle three types of tools are distinguished: for structuring and design; for content and activities; for collaborative learning. Each tool contains a variety of modules that can also be connected to other platforms.

As for the subject of foreign language itself, there are some components essential for the quality of language learning programmes:

- general didactic (individualization of instruction, active role of a student)
- linguistic (language accuracy and language norm)
- methodical (communicative and professional relevance of content)
- digital (interactive forms of learning, module content structure, multimedia and online resources) [12].

The above-mentioned aspects should be taken into consideration for online course development.

At the distance department of Petroleum engineering of SSTU the subject of Foreign Language is taught in the 3rd and 4th terms. Each term has academic hours for practical classes, for independent work and for assessment procedure. Practical classes take up from 8 to 4 hours depending on the semester. Independent work accounts for a very substantial amount of time - 124 hours each term. Assessment form depends on the semester; students get credit after the 3rd term in the form of a test and take an exam in the end of the 4th term. The education programme is divided into the following modules: Personal Profile, Education, My Specialty and Engineering, Professional Background and Skills. Each term covers two modules from the programme.

Course content in Moodle is also divided into modules according to the education programme. Each module consists of several sections where learning material is presented according to the syllabus. It includes instruction materials for practical classes, tasks and exercises for independent work and quizzes in the form of tests. Each course element has a short description for students. Taking into consideration the low level of language proficiency of the students the descriptions for the course elements and grammar explanations are written in the native language.

Each activity or tool within one module has a date and time restriction. Each of the following pages or assignments can be seen and studied according to the date and time set by the teacher. So course materials can be reached over the internet and can be saved and looked through later. All lesson related activities are shown in the weekly format and a student can easily see and follow all activities that will take place in a semester [2]. So in general a student has one particular activity or task every week in such a way that practical studies are followed by assignments for practicing new skills and abilities [13]. In the foreign language course a new activity is accessible every other Monday. In such a way of organizing a course a student does not have access to the whole course content from the beginning of the term.

6.3 Practical studies

Course content for practical classes provides students only with the basic educational material. Students first must get familiar with the material and then apply the knowledge of new material in order to do exercises for individual work (e.g. practice grammar skills, exercises on new vocabulary or reading comprehension).

Practical classes are created in the form of a Page tool in Moodle. A Text page is a simple page created in the form of a plain text and looks like a page of a Word document. Text pages are simple, but they are appropriate for delivering information or instructions. Page tool is convenient for practical classes as the information is displayed right on the screen in the system and students do not need to download any files. Different media like pictures or graphics can also be added on the page.

Each page has a short description of the material provided. The description is written in the native language due to the elementary level of language of the students.

 https://doi.org/10.1051/shsconf/202110901017

Practical classes include such language aspects as grammar, vocabulary and reading. The course content is designed in such a way the practical studies alternates with independent tasks and tests. First there is a practical class available for studying and then tasks for practicing new language material presented earlier.

Grammar is represented by grammar rule explanation and examples. The example of the grammar section is shown in Fig. 1. It is a part of a grammar rule – a table with different forms of pronouns.

The Vocabulary section has a page with a list of words and phrases connected with the studied material. New lexical terms are given with the translation. Students have to read and learn the words that later will be used and practiced in different exercises.

The Reading section usually consists of a simple text and tasks for general understanding and understanding of the main idea. The tasks are of the following types: matching, True/False statements, questions and gap-filling exercises.

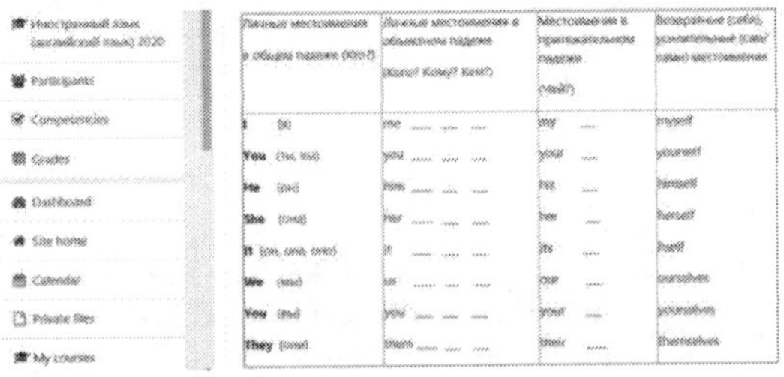

Fig. 1. The Page representing a grammar rule

The analysis of students` frequency of access to these sections has shown that no one has had any difficulty in getting familiar with new material from the technical point of view. All students managed to access the content in time after it has become open for them in the system. We can conclude that the Page tool allows presenting the instruction material in a simple and appropriate way.

6.4 Independent work

Any distance learning technologies involve a great amount of independent work that students have to do. So various types of assignments form the most course content. Besides, there are links to open access Internet resources to facilitate the understanding of new material, especially connected with grammar points. The online course includes mainly independent work of the reproductive type and cognitive-search-type.

The tasks in our course have been created with an Assignment tool. Each assignment is provided with a file that a student has to download and do exercises in it. As soon as fulfilled tasks have been submitted and uploaded, a teacher can see how many students have done the task and then check it.

The types of exercises in the Assignment sections for practicing various language aspects are the following:

Exercises for practicing grammar:

fill in the gaps with the correct forms/words given;

match questions with responses;

put words into correct order to make a sentence

change the form of a word according to the grammar rule

Exercises for practicing vocabulary:

match words and phrases and their translation;

fill in the gaps with words.

Exercises for reading comp:

answer the questions to the text;

define True/False statements ;

choose the correct answer.

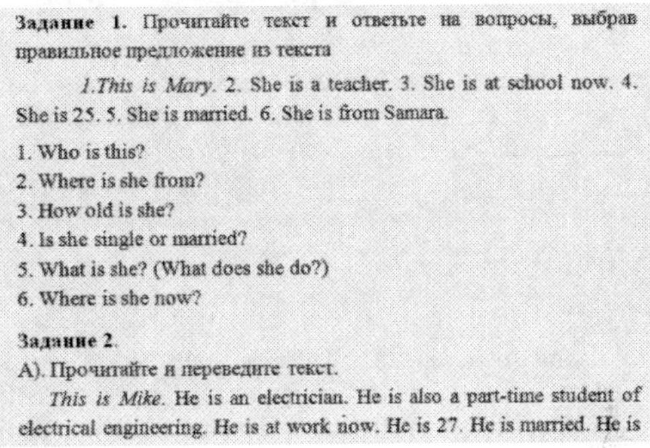

Задание 1. Прочитайте текст и ответьте на вопросы, выбрав правильное предложение из текста

1.This is Mary. 2. She is a teacher. 3. She is at school now. 4. She is 25. 5. She is married. 6. She is from Samara.

1. Who is this?
2. Where is she from?
3. How old is she?
4. Is she single or married?
5. What is she? (What does she do?)
6. Where is she now?

Задание 2.

A). Прочитайте и переведите текст.

This is Mike. He is an electrician. He is also a part-time student of electrical engineering. He is at work now. He is 27. He is married. He is

Fig.2. Example of the exercise to practice reading comprehension

Fig. 2 shows an elementary reading comprehension task – answering the questions to the text that presupposes matching questions and answers. It is a very simple exercise created with the Assignment tool and uploaded in the course in the form of word-office document. The Assignment tool is used in the course to upload the files with the tasks that students must first download, then complete and then upload as a submission to the assignment. The teacher then check the files. The assignments are taken into consideration and give additional grades to a student at the end of the term.

Having analyzed students` files with accomplished tasks, it should be noted that this kind of the form of organizing independent work appeared to be relevant and easily accessible for students.

6.5 Assessment

One of the most difficult methodical problems of distance learning is the assessment. In order to choose an assessment form it is necessary to match the chosen form of assessment with the education content and with the teaching techniques. If one needs to check the understanding of facts (what is to be done), the knowledge of processes (how it is to be done) or strategic vision (why it should be done in such a way), then a quiz seems to be the best option. Quizzes in the forms of tests allow getting the results quickly and analyze them.

The online delivery of teaching, perhaps on mobile devices, enables smooth progression from teaching resources to assessment and back again. Questions can be asked at the most appropriate point in the teaching,

https://doi.org/10.1051/shsconf/202110901017

and attempted wherever and whenever the students are doing their studying.

Even the simplest of multiple-choice quizzes can enable students to check their understanding of a wide range of topics, whenever and wherever they choose to do so [14].

In our course we make multiple-choice questions test. First we create a test in Microsoft Excel and then convert it into the Moodle quiz form that can be uploaded to the system. There are several progress short module tests for one section consisting of 8-12 questions and one final test at the end of the term.

It is convenient to save all the questions in a question bank in Moodle that is an overall stock of all the questions and tasks you have uploaded on the platform.

Each module includes 2 tests so there are 4 tests per term and one final test. The final test consists of the questions from the module tests and has additional multiple-choice questions. The scores from the module tests are taken into account for the final grade (credit or exam).

All the questions are uploaded and saved in the Question bank, so the teacher can change and modify the test if required and choose questions relevant to the particular module or language skill.

There is a time limit set to each attempt and restricted access (by date) to every test. Moreover the number of attempts is also limited to 3 times. All the quiz settings allow obtaining credible results and decreasing the level of cheating.

Quiz reports show the time of the beginning and completing a test, overall grade and correct/incorrect answers to the questions. This powerful monitoring system helps the teacher to understand what part of the course content has not been acquired on full scale and what topics are more difficult to study for some students.

Assessment reports are valuable and convenient because a teacher doesn`t have to follow every student separately – you can easily get the grades of a whole group from the report provided by the Moodle system. As it is shown in Figure 3, the most difficult questions of the test can be identified and then the weak points might be revised or studied again, or the teacher can replace more difficult questions with the easier ones.

Surname	First name	Email address	State	Started on	Completed	Time taken	Grade/10.00	Q. 1 /1.00	Q. 2 /1.00	Q. 3 /1.00	Q. 4 /1.00	Q. 5 /1.00	Q. 6 /1.00
Vapilov	Dmitriy	[illegible]@mail.ru	Finished	1 March 2020 6:54 PM	1 March 2020 7:22 PM	27 mins 40 secs	8.00	1.00	1.00	1.00	1.00	1.00	1.00
Nikitin	Nikita	nd.nikitin99@mail.ru	Finished	1 March 2020 7:04 PM	1 March 2020 7:30 PM	26 mins 27 secs	8.00	1.00	0.00	1.00	1.00	1.00	0.00
Tishakov	Dmitriy	tishakov.tolya@mail.ru	Finished	6 March 2020 9:34 PM	6 March 2020 9:44 PM	10 mins 4 secs	8.00	1.00	1.00	1.00	1.00	0.00	1.00
Pavlov	Denis	Pawlow.d2018@yandex.ru	Finished	14 March 2020 2:01	14 March 2020 2:27	26 mins 15 secs	8.00	1.00	1.00	1.00	1.00	1.00	0.00
Krasnov	Peter	krasnov.[illegible]@[illegible]m.ru	Finished	15 March 2020 4:08 PM	15 March 2020 4:13 PM	5 mins 8 secs	9.00	1.00	1.00	1.00	1.00	1.00	1.00

Fig.3. Quiz report with a question number (Q.) answered or failed

We have analysed several data provided by the Moodle system concerning students` learning and completion of one semester. Table 1 provides the data from the system.

Table 1. Data provided by the Moodle system

Total number of students enrolled	Students completed the assignments	Students passed final tests	Students not accessing the course
180	170	167	2
100 %	94,4 %	92,7 %	1,1%

Table 1 shows that about 94% of students have completed the course in the 3rd term within the term`s deadlines. About 7 % of students fulfilled the tasks and took tests after the term`s deadline. The percentage of students who passed the final test proves the statement that the Moodle system is easily accessed and has a clear interface.

7 Conclusion

The paper analyzed the efficiency of the online course in the Moodle system for teaching the foreign language at the distance learning department. The online course content structure and Moodle tools chosen for the course have been examined. To conclude, the online course has proved to be effective due to the simple and clear structure and a user-friendly interface. The study has shown that the students didn`t face any difficulties accessing the course content and navigating it. Moreover, the applied Moodle tools turned out to be quite effective in delivering the course content for students with a low language level. Thus, the Page tool in LMS is considered to be an appropriate way of providing new learning material for students [15]. The Assignment tool is convenient for our course because the task or exercise comes in a form of a file that is uploaded to the system and course participants need to download it. Assessment procedure is performed via multiple-choice tests that considerably facilitate evaluation of students by automatic grading. The research has shown that such simple tools and clear course structure make it possible to deliver the learning material and improve language skills of students with a low level of language proficiency.

References

1. M. Simonson, S. Smaldino, S. Zvacek, *Teaching and learning at a distance* (IAP–Information Age Publishing, Inc, 2015)
2. N. Cavus, Procedia - Social and Behavioral Sciences **191**, 872-877 (2015) https://doi.org/10.1016/j.sbspro.2015.04.611

3. T. Markova, I. Glazkova, E. Zaborova, Procedia - Social and Behavioral Sciences **237**, 685-691 (2017) https://doi.org/10.1016/j.sbspro.2017.02.043
4. M. Dougiamas, A journey into constructivism, (1999), URL: https://www.researchgate.net/publication/20002240 4 A journey into constructivism (date of access: 10.12.2020)
5. A. Chavan, S. Pavri, Open-source learning management with moodle. Linux Journal **12**, 66-7 (2004)
6. R. D. Costa, G. F. Souza, R. Valentim, B. Castro, Cognitive Systems Research **64**, 134-145 (2020) https://doi.org/10.1016/j.cogsys.2020.08.00
7. B. Holmberg, M. Shelley, C. White, *Distance Education and Languages: Evolution and Change* (Multilingual Matters Ltd, Bristol, 2005)
8. I. Rymanova, N. Baryshnikov, A. Grishaeva, Procedia - Social and Behavioral Sciences **206**, 236-240 (2015) https://doi.org/10.1016/j.sbspro.2015.10.01
9. S. Tsvetkova, I. Malinina, (2019). Engineering education, 25, URL: http://www.ac-raee.ru/files/io/m25/art_8.pdf (date of access: 10.12.2020)
10. D. Keržič, A. Aleksander, N. Tomaževič, L. Umek, International Journal of Innovation and Learning **21 (2)**, 165-177 (2017) doi: 10.1504/IJIL.2017.10002132
11. N. V. Khudoley, RUDN Journal of Informatization in Education **15 (4)**, 410-32 (2018)
12. A.N. Bekhterev, A.V. Loginova, Open Education **4 (99)**, 91-97 (2013) https://doi.org/10.21686/1818-4243-2013-4(99)-91-97
13. C. Costa, H. Alvelos, L. Teixeira, Procedia Technology **5**, 334-343, (2012) https://doi.org/10.1016/j.protcy.2012.09.03
14. K. Lee, H. Choi, Y. H. Choy, The Internet and Higher Education **41**, 25-33 (2019) https://doi.org/10.1016/j.iheduc.2018.12.001
15. D. Yawson, F. Yamoah, Heliyon **6 (11)**, 1-9 (2020) https://doi.org/10.1016/j.heliyon.2020.e05519

Prospects of legal regulation and use of the digital rouble

Ekaterina Kostikova [1,*], *Diana* Alekseeva [2], *Svetlana* Rybakova[3], and *Anna* Savina[4]

[1] Russian State University of Justice, 69 Novocheremushinskaya St., Moscow, Russia, keg2003@bk.ru
[2] Financial University under the Government of the Russian Federation, 49 Leningradsky Prospekt, Moscow, Russia, Kutafin Moscow State Law University (MSAL), 9 Sadovaya-Kudrinskaya St., Moscow, Russia, alekseeva.dg@yandex.ru
[3] Saratov State Law Academy, 1 Volskaya street, Saratov, Russia, svrybakova@mail.ru
[4] Derzhavin Tambov State University, 33 Internatsionalnaya St., Tambov, Russia, anna.savina56@mail.ru

AbstractThis article considers the problems associated with the prospects of central banks issuing and circulating digital currency, in particular the digital rouble. The study found that at present there is virtually no need to issue such a currency as an alternative to cryptocurrency (virtual currency), because cryptocurrency has largely lost its properties as a means of payment with the acquisition of the qualities of an investment instrument. The advantages of a cryptocurrency over non-cash money, which is familiar and also has excellent qualities as a settlement tool, are also not obvious. The positioning of offline payments as a benefit of the digital rouble is questionable given the prospects of full Internet coverage in Russia and the significant technological and legal complexities involved in the organization and control of such payments. Building trust in the new form of the national currency, without which it will not be in demand, will require building a mechanism for legal protection of consumers' rights. There is also a need for a mechanism to address the risks associated with laundering proceeds of crime, financing terrorism and proliferation of weapons of mass destruction. It is difficult to formalise such a mechanism in the absence of international legal regulation

1 Introduction

Since 2018, several states have been developing and implementing projects to introduce central bank digital currency (CBDC) into money circulation.

In November 2018, the Riksbank of Sweden announced Europe's first trial of a sovereign digital currency, the e-Crown, which, according to Stefan Ingves, Governor of the Riksbank, is a pilot version of the Riksbank's new type of money [1]. In October 2020, China began testing China's national digital currency, DCEP (Digital Currency Electronic Payment). The project has been piloted in Shenzhen, Suzhou and Chengdu, as well as in the Xiongnan New Area. The digital yuan will be used to pay for transport fares and food. Testing of the DCEP is scheduled to continue at the 2022 Winter Olympics. The Bank of England also announced preparations for a digital currency in March 2020. It indicated that the digital currency, as an electronic form of money by the central bank, would be intended for payments by households and businesses and would not be a crypto-asset or cryptocurrency [2].

In October 2020, the Central Bank of the Russian Federation published a report on "The Digital Rouble" (report) (Central Bank of the Russian Federation [3]. The report invited the public to discuss the prospects of issuing the digital rouble as a new form of national Russian currency. The Russian central bank is positioning the digital rouble as a central bank digital currency (CBDC) that combines features of cash and non-cash forms of money: a unique digital code similar to the series and number on Bank of Russia banknotes and no physical carrier like non-cash money, allowing for remote payments. The digital rouble would also have other characteristics of a national currency: a monopoly state issuing centre and the status of legal tender. However, the digital rouble is not intended to replace cash and non-cash: it is seen as a third form of money.

2 Problem Statement

Initiatives to issue digital currency by national central banks emerged as a response to the creation and large-scale diffusion of a radical innovation generated by new Internet technologies [4] - cryptocurrency [5] (virtual currency). The first and most popular of such currencies is Bitcoin.

The monetary launch of central bank digital currencies is an offer to interested consumers from an official issuer of an analogue of "digital coins" released by decentralised issuance systems.

However, there are doubts about the feasibility of introducing a new form of money. This is evident from central bank projects to launch such a currency in Europe and China in test mode, as well as from a public

* Corresponding author: keg2003@bk.ru

discussion initiated in Russia. The Russian issuer has no understanding of the benefits of the digital rouble. There is no clear understanding of the legal issues associated with its issuance and circulation. It is impossible to shape the digital currency model of the Bank of Russia without it.

3 Research Questions

The raised issue requires research into the following questions:

1. The need to introduce central bank digital currency as a consumer-demanded alternative to cryptocurrency.
2. The advantages of the digital rouble compared to non-cash money
3. The legal problems related to the issuance and circulation of the digital rouble.

4 Purpose of the Study

Purposes of this study include:

- To determine the feasibility of introducing central bank digital currency, including the digital rouble, as a consumer-demanded alternative to cryptocurrency.
- To identify the advantages of the digital rouble compared to non-cash money.
- To identify the legal problems associated with the issuance and circulation of the digital rouble.

5 Research Methods

The authors used the following research methods: system analysis, formal-legal, and generalization.

6 Findings

6.1 Central bank digital currencies as an alternative to cryptocurrency

Cryptocurrency is a modern economic phenomenon. Researchers identified characteristics to consider cryptocurrency as a parallel to the national currency [6], an alternative form of liquidity to traditional monetary assets [7]. The European Central Bank in its February 2015 report "Virtual Currency Schemes - Further Analysis", in describing cryptocurrency, noted that in some circumstances it is feasible to use it as an alternative to money (European Central Bank [ECB], 2015) [8]. In October 2015, the Bank for International Settlements pointed out the monetary characteristics of cryptocurrencies in its Digital Currencies report [9]. At the same time, there was also a debate about the extent to which cryptocurrencies have inherent money functions [10]. Cryptocurrency is fundamentally different from national (fiat) money: it is not issued and guaranteed by any state [11], central bank, credit institution or electronic money institution [8]; its price process is not regulated or stabilised by third parties such as the central bank [12].

The cryptocurrency's meteoric success was due to its consumer properties as a means of payment. It is primarily about Bitcoin, whose market capitalisation of cryptocurrencies at the beginning of 2020 is over USD 194 billion, of which Bitcoin accounts for USD 124 billion [13]. Researchers noted the following advantages of the cryptocurrency compared to national (fiat) money: low inflation rate due to limited issuance (Bitcoin 21 million) [14]; transaction security ensured by special technologies [15]; accessibility of cross-border transactions [16] due to lack of connection to national borders [17]; anonymity [18].

However, the Bitcoin launch was based on idealistic ideas about the possibility of replacing existing monetary systems with an alternative system [18]. Today, cryptocurrency acquired, in addition to its properties as a means of payment (settlement), the properties of an investment instrument. The high volatility in the value of cryptocurrencies means that their inherent properties as a payment (settlement) instrument are replaced by those of an investment asset. This significantly reduces the competitiveness of cryptocurrencies regarding national currencies, whose payment properties are secured by currency regulation in the issuing countries.

Russian legislation reflected the dual nature of cryptocurrencies. The Federal Law of 31.07. 2020 No. 259-FZ "On Digital Financial Assets, Digital Currency and on Amendments to Certain Legislative Acts of the Russian Federation", calling cryptocurrency a digital currency, defined that it is a set of electronic data contained in an information system that is offered and may be accepted as (1) a means of payment that is not a monetary unit of the Russian Federation, a foreign state or an international monetary or settlement unit, and (2) as an investment; there is no obliged person to the holders of electronic data.

In addition, the payment properties of decentralized digital currencies are suppressed by prohibitions and restrictions on their use, including in payments.

In particular, in Russia, the above-mentioned federal law prohibits payments for goods, works and services in digital currency, as well as the dissemination of information on the possibility of such payments. This prohibition applies to persons who can be conditionally called Russian residents: legal entities, separate subdivisions of international and foreign organizations established in the territory of Russia, natural persons permanently residing in the Russian Federation (for at least 183 days during the next 12 consecutive months). This prohibition is reinforced by rules relating to the possession of the currency by Russian residents: judicial protection of the rights of such entities associated with the possession of the digital currency will only be possible if they inform the state of the facts of their possession, transactions and operations with the digital currency.

Thus, at the present stage, cryptocurrency is unable to compete with national currencies, including digital currencies, so the introduction of central bank digital currencies as an alternative to cryptocurrency is not appropriate. However, it is reasonable to give central bank digital currencies the properties of virtual money, which were incorporated when creating cryptocurrency and provided consumers with interest in it.

6.2 On the advantages of the digital rouble compared to non-cash money

Modern non-cash has excellent consumer properties (affordability of mobile payments, speed, low transaction costs). In this light, the desirability of central bank digital currencies should stem from their unique qualities.

The exceptional advantage of the digital rouble is the possibility to use it in offline mode, where consumers will be able to make payment transactions without access to the Internet. This new currency option is relevant for remote areas not covered by the Internet. However, the Bank of Russia expresses doubts about its future relevance, as over time the Internet will cover the entire country and users everywhere will be able to make the usual cashless transfers without hindrance. The Bank of Russia is planning to conduct a special study for forming an opinion on the feasibility of giving the digital rouble such a feature. A balanced approach is necessary because developing technology for offline payments (including storing digital currency on devices that support wireless services; synchronization with an online wallet) will incur significant costs. In addition, such settlements carry risks associated with the loss of the device on which they are stored (actually a wallet), violations of the requirements of Federal Law of 07.08.2001 No. 115-FZ "On Counteracting the Legalization (Laundering) of Proceeds of Crime and the Financing of Terrorism".

Another advantage of the digital rouble could be the availability of cross-border transfers, including reducing the transaction costs of such transactions. This quality is highly appreciated in cryptocurrency settlements; it was ensured by the absence of restrictions imposed by individual jurisdictions on such transfers. The accessibility of cross-border transfers in digital roubles can result not from overcoming jurisdictional restrictions, but from reducing the number of intermediaries in such transactions. This will require adopting and legislating a digital currency model with a minimum number of participants (intermediaries). The Bank of Russia will also need to ensure interaction with foreign payment systems, which is necessary for the conversion of the digital rouble.

A unique useful feature of the digital rouble compared to non-cash money is the ability to accurately identify participants in settlement transactions. However, transaction identification has risks associated with data protection issues, compliance with legislation on anti-money laundering, financing terrorism and proliferation of weapons of mass destruction.

Thus, at present, the advantages of the digital rouble over non-cash money are not obvious. And without such benefits, the Bank of Russia's digital currency may not be in demand, as the choice of money form for the consumer is voluntary (given the limitations of cash payments between economic entities) and cashless payments have become customary for the widest range of individuals. Therefore, the Bank of Russia needs to define more clearly the useful qualities of the new form of the national currency that should serve as a basis for the digital rouble model and the legal framework for its operation.

6.3 Separate legal issues related to the issuance and circulation of the digital rouble

Only high quality legal regulation serves as a criterion for public confidence in the financial and monetary system [19], including the digital rouble. Thus, it is important to identify the legal problems related to the issuance and circulation of the digital rouble.

Monetary circulation, including central banks' digital currencies, involves risks associated with laundering proceeds of crime, financing terrorism and proliferation of weapons of mass destruction.

The introduction of the digital rouble raises the question of establishing a mechanism to counter such threats, especially in terms of controlling monetary transactions.

In the current system, control functions are shared between the Bank of Russia and credit institutions, as well as other entities involved in cash transactions. The central bank believes that it would be efficient to retain the current functionality of these entities with regard to digital rouble transactions as well, upgrading it as necessary.

However, such an approach is fraught with legal complexities.

The Financial Action Task Force currently distinguishes between the virtual currency (cryptocurrency) and electronic money to assess and manage anti-money laundering, financing terrorism and proliferation of weapons of mass destruction [19]. There is no FATF-formulated regulation for central bank digital currencies, which will cause difficulties in reforming national legislation.

In particular, problems will arise in the development of the following documents:

- rules for opening and maintaining individual e-wallets for storing digital roubles in terms of identifying customers, customer representatives, beneficiaries and beneficial owners, storing and periodically updating information about these entities, and determining the grounds for refusing to open an e-wallet;
- internal control rules in terms of anti-money laundering, financing terrorism and proliferation of weapons of mass destruction related to digital currency transactions. In the current context of legal uncertainty regarding existing legal categories ("unusual", "suspicious", "questionable" and "transit" transactions), confusing grounds for mandatory and optional controls in this area, gaps and conflicts in relevant legislation, this task seems ineffective [20].

Another issue that is important to address is providing a consumer protection mechanism for the digital rouble. The financial literacy of the Russian population remains rather low, even with regard to financial products that have long been known and widely marketed. The emergence of digital rouble as a new financial product will inevitably entail the emergence of unfair practices (mis-selling) both in opening and maintaining individual e-wallets and in transactions and transactions with them. Therefore, it is necessary to develop and enshrine in law protective mechanisms. This is necessary to build public confidence in the digital rouble.

It is important to exclude the application of liability measures in transactions and operations involving the digital rouble to bona fide actors. This requires a clear formulation of the grounds for such measures, including clarification of the offences, since inaccurate and incorrect wording of the objective side of the offence often leads to unlawful prosecution even today [21]. The risks of criminal prosecution of innocent people will increase with the introduction of digital currency.

7 Conclusion

The authors conclude that doubts by the Bank of Russia about the feasibility of introducing the digital rouble into monetary circulation are reasonable: there is no need to offer consumers a governmental alternative to cryptocurrency; the advantages of central bank digital currency over non-cash money are not obvious; solving the legal problems associated with the issue and circulation of the digital rouble is difficult at the moment.

In such circumstances, the Bank of Russia needs to clearly define the public interests that will form the basis for the digital rouble model and the legal basis for the issuing and circulation of the national currency in the new form [23].

References

1. S. Ingves, The e-krona and the payments of the future (2018), URL: https://www.riksbank.se/globalassets/media/tal/engelska/ingves/2018/the-e-krona-and-the-payments-of-the-future.pdf (date of access: 16.03.20).
2. Bank of England (2020), URL: https://www.bankofengland.co.uk/-/media/boe/files/paper/2020/central-bank-digital-currency-opportunities-challenges-and-design.pdf?la=en&hash=DFAD18646A77C00772AF1C5B18E63E71F68E4593 (date of access: 16.03.20).
3. Bank of Russia (2020), URL: https://cbr.ru/analytics/d_ok/dig_ruble/ (date of access: 16.03.20).
4. E. Iidogan, Blockchain-based cryptocurrency equilibrium model (2018), URL: https://doi.org/10.2139/ssrn.3152803(date of access: 16.03.20).
5. S. Corbett & B. Lucy & A. Urquhart & L. Yarovaya, International Review of Finance **62,** 182 – 199 (2019).
6. L. Larue, Ecological Economics (2020), URL: https://doi.org/10.1016/j.ecolecon.2020.106823(date of access: 16.03.20).
7. R. Boehme, & N. Christine, & B. Edelman, & T. Moore Perspective **29 (2),** 213 – 238 (2015)
8. European Central Bank (2015). *Virtual currency schemes – a further analysis.* URL: https://www.ecb.europa.eu/pub/pdf/other/virtualcurrencyschemesen.pdf?d5b48352d3ce982d8c4507dcad038018 (date of access: 16.03.20).
9. Bank for International Settlements (2015). *Digital currencies.* URL: https://www.bis.org/cpmi/publ/d137.pdf (date of access: 16.03.20).
10. D. Ermak, Academic Press **2**, 31 – 43 (2015)
11. Bank of Russia (2017), URL: http://cbr.ru/content/document/file/36009/rev_ico.pdf (date of access: 16.03.20).
12. L. Schilling, & H. Uhlig, Journal of Monetary Economics, 16-26. (2019), URL: https://doi.org/10.1016/j.jmoneco.2019.07.002 (date of access: 16.03.20).
13. Y. Ma, F. Ahmad, & M. Liu, & Z. Wang, Technological Forecasting and Social Change. (2020), URL: https://doi.org/10.1016/j.techfore.2020.120265 (date of access: 16.03.20).
14. J. Papp, J. Pittsburgh, Technol. Legal Policy, 33-56 (2014)
15. S. Bayern, Wash. Lee Law Rev, 22–34 (2014)
16. V. Maese, The Banking Law Journal **133 (8)**, 468 (2016).
17. V.A. Mez, & A.V. Avery, & B.A. Naftalis, & S.P. Vink, & Yu.D. Valdez, The Banking Law Journal, **133 (8),** 468-471 (2016)
18. E. Sayedi, & A Brostrem, & F. Ruiz, Economy of small buses (2020), URL: https://doi.org/10.1007/s11187-019-00309-8 (date of access: 16.03.20).
19. M.K. Borodavina, & S.V. Rybakova, & A.V. Savina, & A.P. Khakhulina, Enforcement, 4, (1), 75-84 (2020)
20. Financial Action Task Force (2014). *Virtual Currencies - Key Definitions and Potential AML/CFT Risks.* URL: http://www.fedsfm.ru/content/files/documents/fatf/virtualnye_valyuty_fatf_2014.pdf(date of access: 16.03.20).
21. D.G. Alekseeva, Banking law **4**, 9-17 (2018)
22. D.G. Alekseeva, & I.E. Mikheeva, & T.P. Suspitsyna, All-Russian criminological journal **13 (6),** 962-970 (2019)
23. I.A. Tsindeliani, I.V. Bit-Shabo, A.D. Selyukov, M.M. Proshunin, E.G. Kostikova, S.S. Tropskaya, S.V. Rybakova, Journal of Legal, Ethical and Regulatory Issues **22 (5),** 1-12 (2019)

Human rights protection on the internet in Russia and abroad

Natalia Kovaleva*[1], *Svetlana* Kulikova[2], *Natalia* Zhirnova[3] and *Tatiana* Rudchenko[4]

[1] Saratov State Law Academy, Saratov, Russia, kovaleva.natalia@mail.ru
[2] Saratov Chernyshevsky National Research State University, Saratov, Russia, kulikovasveta@inbox.ru
[3] Saratov State Law Academy, Saratov, Russia, zhirnowanata@yandex.ru
[4] Saratov State Law Academy, Saratov, Russia, rudchenko27@yandex.ru

Abstract. The article focuses on the analysis of legal protection of personal, political, labor, educational and other rights, freedoms and legitimate interests of citizens on the Internet, as there are generally recognized problems related to their violation by individuals or various organizations.The authors analyze how citizens' rights, such as the right of access to the Internet (the right to the internet connection, the right to IT neutrality, the right to availability of internet service, the right to comfortable usage of internet services, the right not to be disconnected) the right to digital identity (the possibility to manage internet data of private persons), the right to anonymity (the possibility of carrying out certain actions under a false name without revealing personal data that would allow the person to be identified), the right to discard unwanted information (to protect oneself against spam), the right to be forgotten, right to safe internet use, both in the technological as well in the contextual meaming, are declared and implemented on the Internet.

1 Introduction

82 million people - 66 percent of Russia's population - use the Internet [1] An essential feature of modern social development is the general digitization of economic, political, communicative and other social relations which are carried out via the Internet. These relations of course require legal regulation that must take into account the specifics of the legal and technological nature of the electronic environment [2].

2 Problem Statement

A lot of personal, political, labor, educational and other rights and freedoms are implemented on the Internet which creates problems related to violation of citizens' rights and freedoms by individuals or different types of organizations that may pursue profit-seeking or destructive goals (e.g. spreading extremist ideas or recruiting individuals for terrorist groups) and be under the jurisdiction of various states [3].

The model law "On the Basics of Internet Regulation", which was adopted at the 36th plenary session of the Interparliamentary Assembly of the CIS member states on 16 May 2011 (revised on 25 November 2016) gives the following definition of the Internet: "Internet is a global information and telecommunications network that connects information systems and telecommunications networks from different countries via a global address system based on the use of IP complexes (Internet Protocol) and TCP complexes (Transmission Control Protocol) and enables the use of various forms of communication, including the ones that make information available to an unlimited number of people. "The rights and obligations of companies that are active on the Internet are defined in the Federal Act No. 149-Federal Law from 27 July 2006 "On Information, Information Technologies and Information Protection" (hereinafter: the Information Act). The aforementioned act defines the legal status of such companies as website owners, hosting providers, search engine operators, owners of news aggregators, owners of audiovisual services, etc. However, the status of a user on the Internet is not defined in the Information Act. The entirety of his/her rights and obligations is regulated by a large number of regulations, permissions and prohibitions, which are established to some extent in the legislation and govern relations in the field of information technologies [4].

3 Research Questions

The "Guidelines on Human Rights for Internet Users", adopted by the Committee of Ministers of the Council of Europe in 2014 are of certain interest. They attempt to create a set of rights for an ordinary person who uses various internet services and neither has a special legal status nor any legal, cybernetic or IT expertise.

The Guidelines name the following distinctive rights of internet users:

- the right to access the Internet,
- freedom of expression and information,

* Corresponding author: kovaleva.natalia@mail.ru

- the right of assembly, association and participation,
- the right to privacy and data protection,
- the right to education and literacy,
- the right to special protection of children and young people when using the Internet,
- the right to effective means of protecting their internet rights.

The analysis of modern Russian and foreign legislation, recommendations of international bodies for the protection of human rights, as well as the scientific positions of Russian and foreign scientists allow the conclusion that two groups of rights of internet users can be distinguished: first, new digital rights, the appearance of which is based on the legal nature the Internet and second, traditional rights, the implementation of which changes significantly with the transition to the electronic environment [5-6].

4 Purpose of the Study

The purpose of this study is to determine the informational and legal status of a person as a citizen on the Internet.

5 Research Methods

This study is an analysis of the Russian and international laws that regulate the contemporary legal issues arising from the information technologies playing an increasing role in today's world.

During this study the authors have analyzed over 20 sources about the realization of rights and legitimate interests of people on the Internet.

The authors have applied the following scientific methods:

- a formal-legal method for determining the features and analyzing the definition of a person's "right to access the Internet";
- a comparative method for comparing the laws on the realization of information rights of citizens in Russian Federation and other countries;
- a structural analysis method for analyzing the features of the term "right to internet access" as a structural element of a more general term "right to information";
- an informative approach - when analyzing any object, process or phenomenon in nature and human society the primary object is determining and analyzing its specific information aspects that strongly influence its current state and development.

6 Findings

The contemporary understanding of the right to internet access consists of:

- right to the internet connection: all citizens who wish to be connected to the Internet should have this opportunity. This right corresponds to the duty of the state to enable internet connection for all citizens regardless of their place or residence, so internet connection should be possible even in rural areas and remote regions of the country. According to Part 1 of Article 57 of the Federal Law "About Means of Communication" from 7 July 2003 providing access to the Internet is a universal connection service that must be provided by the state for all its citizens;
- right to IT neutrality or right to protection from technological discrimination: every citizen must have the opportunity to use the information technologies that correspond to his goals and interests and have guaranteed access to the information resources they need. Article 3 of the Russian Law about Information prohibits passing any legislation to grant privileges to certain information technologies compared to the others.

The issue of net neutrality has been actively discussed lately. Internet providers have the opportunity to enable faster connection for certain resources while slowing it down for others for different groups of consumers. Such actions are viewed by some experts as "discrimination of services with a goal to make them pay fees to gain priority connection" [7]. In the Netherlands the net neutrality is obligatory by law, in the USA the net neutrality is guaranteed in the Internet regulations issued by the Federal Communications Commission. In Russia this principle is stated in the recommendations of the responsible working group in the Federal Anti-Monopoly Agency (FAMA), in this document it is stated that the goal of setting this principle is to remove barriers to accessing the desired content for end users, to make discrimination in internet connection services impossible, to enable cooperation between different market players for improving the quality of services and to ensure the safety of citizens and the state [8].

- right to internet connection services also has an economical component. In the whole world connecting to the Internet and providing internet services are handled by private companies that provide these services in exchange for payment. The duty of the state here is to ensure that internet access is affordable at a reasonable price for a wide range of its citizens.
- the right to comfortable use of internet services implies that certain technological standards will be put in place to ensure the quality of these services. Such standards include the speed and stability of internet traffic, requirements for the architecture of and navigation on websites, automated information search, the possibility to determine the date and time of publishing the information, conditions for copying and storing information from the Internet by the user, availability of the English version of the website or tools for its automatic translation into the most common world languages, the possibility to contact the content the website creators online, a special option for using the internet service by people with disabilities [9]. Part 2 of Article 57 of the Federal Law "About Means of Communication" states that data should be transmitted to the user's equipment with a speed not below 10 (ten) megabit per second. In 2020 Chapter 7.1 was added to this law, this new chapter sets new requirements for

ensuring stable, safe and uniform functioning of the internet in the Russian Federation.

- the right not to be disconnected from the internet was first formulated in the verdict of the Constitutional Council of France on June 10, 2009. The Constitutional Council was reviewing the bill that proposed to punish the persons that on numerous occasions violated the copyright laws by distributing illegal copies of copyrighted materials on the Internet by disconnecting them from the Internet for a period from two months to one year without the right to get access from any other internet provider during this time [10]. The Constitutional Council declared that the bill was in conflict with the country's constitution since its creators hadn't considered the fundamental freedom of speech and communicating with others and the important role the telecommunication services play in democratic processes, so disconnecting a person from the Internet is a disproportionate punishment for such violations.

Later the right not to be disconnected from the Internet was enshrined in the above mentioned Human Rights Guide for internet Users (2014). It means that a person can't be disconnected from the Internet at will, such a decision can only be made by a court or by the internet provider in case of a contract obligations' violation. In this case providers are obligated to inform the users of the reasons and legal base of the disconnection, as well as procedures to appeal this decision and possibilities of applying for regaining full access to the Internet.

The right to internet access was further strengthened with the project "Accessible internet". On 15 January 2020 the Russian president Vladimir Putin in his annual address to the Federal Assembly made it a goal for the country's government to provide free internet access to socially relevant Russian internet services everywhere in the country. The four biggest telecommunication services providers in Russia supported the president's initiative. The Russian Ministry of Communications issued a decree with conditions for the experimental realization of this project and a list of 391 free internet resources [11]. The decree recommends the internet service providers "not to terminate internet services contracts even if the users haven't been paying for these services, but instead only give them access to the free socially relevant services". In other words, even if users stop paying for internet services they will still have limited access to these services in order to use the socially relevant internet resources.

The Russian laws provide liability for obstructing the work of internet websites or any actions aimed at a knowingly illegal restriction of access to websites (Article 13.18 of the Code of Administrative Offenses of the Russian Federation from 30 December 2001 no. 195-Federal Law).

So, even though the right to internet access is a new right, its realization is to a certain degree already regulated by Russian law and can be defended on the administrative level or in court.

Another right that is predicated on the emergence and development of the Internet is the right to digital identity. It is connected with the possibility to manage the internet data of individuals.

The personal data that users submit to different service providers on the Internet, as well as the so called digital traces – information about the user's actions on the Internet – forms a digital portrait of the user, a projection of the real person in the virtual environment. This phenomenon is called a digital identity. In Law this phenomenon corresponds to the following two features: 1) the originality, individuality of each person; 2) digital form of data presentation; 3) the ability to "recognize" the user in the virtual environment, without confusing him with others, to be sure that the virtual projection of this person belongs to "him/her" and not to someone else.

The right to digital identity includes three aspects: its forming, use and protection.

The forming of the digital identity presumes the possibility for the users to know who and how collects their data and what data have already been gathered about them. Contemporary studies show that the internet users data that is being collected can be divided into different layers, and the data that people consciously and of at their own will provides to the information mediators or publish on blogs, social networks, mobile apps are only the first layer. But the information technologies nowadays allow collecting a lot of other data about users' behavior. For example, about their current location, the content they view, which keys they press on their keyboard, about their values, political views and consumer preferences, even about their psychological and emotional state. According to A.M. Kondakov and A.A. Kostylyova use of different algorithms for analyzing the user's web searches allows to find out everything about that person that he/she wouldn't voluntarily reveal about themselves: "their weaknesses, their psychometric profile and IQ level, family situation, addictions, illnesses, small obsessions (i.e. games) and their serious commitments (i.e. business projects)" [12].

This data can later be used by commercial companies for marketing their goods and services to that specific user or for social control like with the Social Credit System in China which uses information technologies to automatically assess the citizens' behavior and give them points based on this assessment. Citizens with a low social rating face restrictions for purchasing tickets, the rating is taken into account when a person wants to enter a university or get a job.

Protection of digital identity is especially necessary because of the rising number of personal data theft cases. A report of the International Electronic Communications Union with the telling name "Digital Life" contains results of a study analyzing the consequences of violating the right to digital identity. The victims of personal data theft (they are called "those who were stolen from themselves") lost their jobs, couldn't get student loans, mortgages or car loans and in some cases were even arrested for crimes they didn't commit. According to the data cited in the report the victims on average had to devote about 600 hours of their time to recover their digital identity, reputation and credit history. The emotional impact of this on the victims is comparable with that of violent crimes [13].

The General Data Protection Regulation (GDPR) that was put in place by the Resolution no.2016/679 of the European Parliament and the Councils of the European Union on 27 April 2016 and came into force in 2018 contains fundamentally new provisions for the protection of personal user data.

For example, it differentiates between the data that users submit by themselves (Article 13 of the Regulation) and the data about the user that are gained not from him/her (Article 14 of the Regulation). And the key is enabling the user to manage that latter set of data. In particular, the Regulation enshrines the user's right to inquire about what data, from which sources and for which purposes were gathered about him/her, about the legal base for using his/her data and about whether or not the gathered data will be used for automated decision making as well as the user's right to recall his/her consent to use of their data and the user's right to file a complaint with the overseeing authority.

In Russian Federation some aspects of the right to digital identity are regulated by Law on Information and the Federal Law "On Personal Data" from 27 July 2006 No. 152-Federal Law (hereinafter referred to as the Law on Personal Data), which establishes the requirements for receiving and use of personal data, as it is stated in the Law, "among other things in information and communication networks" (Art. 1). The Russian legislation establishes the requirement for the mandatory consent of the individual to the processing of his/her data.

The right to anonymity is one of the most controversial rights exercised on the Internet. The right to anonymity (from lat. "without a name") on the Internet means the ability to perform certain actions under a false name (nickname) without disclosing personal data that would allow the individual to be identified. Such actions usually include anonymous communication on social networks, commenting on articles, blogs, posts, sending private messages via instant messaging services, making online payments, distributing creative products (produced by themselves or others).

To date there is no binding international legislation securing the right to anonymity on the internet. Although the Report of the Special Expert on Ensuring and Protecting the Right to Freedom of Opinion and Expression of the UN Human Rights Council, David Kaye, dated 22 May 2015, that he presented at the 29th session of the UN Human Rights Council, advocates the idea that anonymity is necessary for exercising freedom of opinion in the digital environment and therefore requires international protection, but this report is only advisory in nature [14].

Russian legislation establishes a number of rules that prevent users from maintaining anonymity. Article 10 of the Law on Information establishes that information disseminated without using mass media must include reliable information about its owner or another person disseminating this information, in such form and amount that is sufficient to identify this person. In 2016, the Russian parliament adopted the so-called Yarovaya's legislative package aimed at de-anonymizing users.

According to the new edition of Parts 3 and 3.1 of Article 10.1 of the Law on Information, the organizer of the information dissemination on the Internet is obliged to store on the territory of the Russian Federation information on the facts of reception, transmission, delivery and (or) processing of voice recordings, written text, images, sounds, video or other electronic messages of internet users and information about these users for one year from the end of the implementation of such actions, and the messages themselves - for six months. The organizer of the information dissemination on the Internet is obliged to disclose this information to the authorized state bodies when it is needed for an ongoing investigation or ensuring the security of the Russian Federation.

Another quite controversial issue are anonymous comments. If an Internet resource on which a comment containing signs of abuse of freedom of the media is posted is registered as a mass media, then Roskomnadzor (Federal Service for Supervision of Communications, Information Technology, and Mass Media) which monitors legislative compliance in this area will contact that media's editorial office and demand to delete the comment or edit it. Within one day, the editorial office must perform the required actions. Otherwise, Roskomnadzor has the right to issue a warning to that media. Two warnings within one year may result in the media being closed.

If an Internet resource is not registered as a mass media, but has a different legal status, the responsibility for anonymous messages on its website that violate the rights of third parties may be assigned to the site owner, since he/she is the domain administrator and therefore the person who has provided the necessary conditions and technical capabilities to visitors of his/her internet resource.

Next we consider the right to decline unsolicited information.

With the development of digital technologies, information messages can be repeatedly transmitted to a large number of recipients. This raises the question of protecting the user from information that he has not requested and does not want to receive. The right to reject unwanted information could be defined as the right to be protected from spam.

Electronic submissions, that were not approved by the user, are prohibited in accordance with Article 18 Part 1 of the Federal Act "On Advertising" of March 13, 2006 No. 38-Federal Law, which provides that the distribution of advertising via telecommunications networks, including the use of mobile phone, fax and cellular telephone communications, are only permitted to receive advertising with the prior consent of the subscriber or recipient.

In this case, the advertisement will be recognized as distributed without the prior consent of the subscriber or recipient, unless the merchant can prove that such consent has been obtained.

The advertising distributor is obliged to immediately stop distributing advertising to the address of the person who made such a request.

In 2014, a provision was introduced in the Information Act obligating the organizer of the information distribution to ensure the technical ability of users of the instant messaging service to refuse to receive electronic messages from other users.

Currently the register of organizers of information distribution, which is maintained by Roskomnadzor includes the social networks Vkontakte, Moy Mir, Odnoklassniki, the Yandex service complex, the Mail.ru mail service, Telegram and Sberbank messengers, cloud storages, online publications, electronic magazines, news agencies, forums, blogs that provide services for the exchange of text messages, audio, photo and video information in real time between clients.

Therefore, a binding regulation applies to all mailings, regardless of their content and the electronic platforms used, in order to guarantee the possibility of rejecting the information sent.

The introduction of the right to oblivion can be traced back to the development of digital technologies for finding and storing information on the Internet. Back in the day, before the Internet, if information about events, phenomena or actions of an individual was published, for example, in the media, after a while it lost all its relevance, copies of newspapers or television programs were transferred to the archives, the event itself was forgotten, where in the conditions in which information is displayed in Internet resources, it is always available (i.e., its distribution is no longer one-time event) and can be easily found using search engines. In academic literature, the right to be forgotten is linked to the constitutional principle of humanism, the ability of a person to receive forgiveness, and the right to further personal development [6].

The origins of the right to be forgotten are associated with the decision of the European Court of Justice (ECJ) of 13 May 2014 in the case of Gonales v. Google. In 2010, Spanish citizen Costeja Gonales filed a complaint against Google Span with the National Data Protection Authority which stated that when entering his name into the Google Group search engine, he received links to the 1998 newspaper page, which published the advertisement about the sale of his house at auction in connection with the collection of debts for non-payment of contributions to the Social Security Fund. The applicant requested that the ad be removed and that this information no longer be included in search results. The agency denied him a request to remove the ad in reference to the legal validity of such a post, but supported the request to Google to remove the corresponding links. The company tried to challenge the Agency's decision before the Supreme Court of Spain, which, when examining the case, referred the matter to the EU Court of Justice. In its decision, the Court of Justice of the European Union recognized the right of a natural person to demand to stop the publication of links containing harmful information about them [15].

Subsequently, the right to erasure was enshrined in Article 17 of the aforementioned General Data Protection Regulation (GDPR), where it is referred to as "the right to be forgotten". Part 1 of this GDPR article enshrines the subject's right to demand the erasure of his own data in cases where: the purposes for which they were collected and processed have been exhausted; the consent to data processing has been revoked; the data has been processed unlawfully, etc.

A.A. Antopolsky points out that "the right to be forgotten is protected as part of the right to respect the private life, provided for in Article 8 of the Treaty.

At the same time, it frequently conflicts with the freedom of expression, but the first steps have already been taken to achieve a "fair balance" between the protection of each of these rights " [16].

In Russia, the right to be forgotten was enshrined in law earlier than in many other European countries.

In 2015, Article 10.3 was added to the Information Act requiring the operator of the search engine, at the request of the citizen, to stop providing information on a website index on the Internet, allowing access to information about the applicant.

Links must be disabled when the information:

- is distributed in violation of Russian legislation;
- is unreliable;
- is outdated, has lost its meaning for the applicant due to subsequent events or actions of the applicant.

The Internet is not a law-free space; on the contrary, today it is already a regulated and controlled communication space. However, most of the Internet has market-based Amor-regulation, its "code is law" tool. New Network architecture, new search algorithms, and new types of information selection and connection are emerging faster than ever. Therefore, it is important to guarantee an effective and open to innovation communication-free legal framework, which is the task of a democratic legislator to create [17].

7 Conclusion

Hence, the new digital rights have already been largely consolidated both in international legal acts and in Russian legislation. The mechanisms for their implementation and protection have been developed by national legislators to varying degrees and need to be further improved

References

1. Y. Sinelschikova, 10 facts about the Russian Internet (2016), URL://https://www.rbth.com/science_and_tech/2016/01/11/10-facts-about-the-russian-internet_558165 (date of access: 16.03.21)
2. A.T. Karasev, O.A. Kozhevnikov, & V.A. Meshcheryagina, Antinomies **19 (3)**, 99-119 (2019)
3. Shah, The Internet and Human Rights. Global Issues. Web. 02 Nov. 2020. URL:https://www.globalissues.org/article/151/the-internet-and-human-rights(date of access: 16.03.21)

4. M.S. Salikov, & S.E. Nesmeyanova, Russian law, education, practice, science **1**, 5-13 (2019)
5. E.V. Talapina, Law of the Russian Academy of Sciences **14(3)**, 122-146 (2019)
6. E.A. Voinikanis, Pravovedenie 3, 70-89 (2016)
7. Institute for Internet Research. Network Neutrality or Long Distance Internet (2020) URL: // https://internetinstitute.ru/ (date of access: 16.03.21)
8. Federal Antimonopoly Service of the Russian Federation. Basic Document on Net Neutrality (2020) URL: // http://fas.gov.ru/upload/other(date of access: 16.03.21)
9. V.N. Sereda, & M.Yu Sereda, *Protection of human and civil rights and freedoms on the Internet* (Voronezh. Russian Federation: Publishing and Printing Center "Scientific Book", 2013).
10. L.B. Sitdikova, A.L. Shilovskaya, M.A. Volkova, R.R. Lenkovskaya, & N.A. Stepanova, Mediterranean Journal of Social Sciences **6(6),** 163-169 (2015).
11. Ministry of Digital Development, Communications and Mass Media of the Russian Federation On conducting an experiment on the provision of communication services to citizens on a pro bono basis for the transfer of data and for providing access to the information and telecommunication network "Internet" on the territory of the Russian Federation for the use of socially significant information resources in the information and communication network "Internet": order of the Ministry of Digital Development, communications and mass communications of the Russian Federation from 31.03.2020, No. 148 (2020)
12. A.M. Kondakov, & A.A. Kostyleva, Vestnik RUDN. Series: Informatization of education. 16 (3), 210 (2019)
13. International Telecommunication Union, Digital Life (2006) URL: https://www.itu.int/osg/spu/publications/digitalife/docs/digital-life-web.pdf (date of access: 16.03.21)
14. David Kaye, Report of the Special Rapporteur on the promotion and protection of the right to freedom of opinion and expression (2016) URL: // https://ru.euronews.com/2016/07/05/un-denounces-disruption-of-internet-access-as-human-rights-violation (date of access: 15.04.21)
15. A.A. Chebotareva, Legal support of personal information security in the global information society (Dissertation for the degree of Doctor of Law), (Moscow, Russian Federation, 2017)
16. A.A. AntopolskyHuman Proceedings of the Institute of State and Law of the Russian Academy of Sciences **14(2),** 159-185, (2019)
17. M. Kall, Impulspapiere. Münster **2**, 1-86 (2011)

The influence of digital platforms and algorithms on legal regulation of competition

Marina Kozlova[1]*, *Dmitriy* Kozhemyakin[2], *Olga* Sergacheva[3], and *Alexandr* Bortenev[4]

[1] Volgograd Institute of Management, The Russian Presidential Academy of National Economy and State Administration, ul. Gagarina,8, Volgograd, Russia, ORCID 0000-0002-6651-4392, kozlova-my@ranepa.ru
[2] Tomsk State University, 36, Lenin Avenue, Tomsk, Russia, ORCID 0000-0002-9722-4136, dv-k@outlook.com
[3] Volgograd Institute of Management, The Russian Presidential Academy of National Economy and State Administration, ul. Gagarina,8, Volgograd, Russia, ORCID: 0000-0002-7419-3277 sergacheva-oa@ranepa.ru
[4] Volgograd Institute of Management, The Russian Presidential Academy of National Economy and State Administration, ul. Gagarina,8, Volgograd, Russia, ORCID: 0000-0002-0239-554X bortenev-ai@ranepa.ru

Abstract. The antitrust regulation faces challenges in the context of digitalization and algorithmization; several of them are analyzed in the article. The authors explore the influence of digital platforms and pricing algorithms on competitive environment, the practice of their application in view of monopolizing the market and the possibilities to resist such practice from the point of view of the antitrust legislation. They examine the approaches to regulating digital instruments of pricing in order to ensure fair competition. One of the approaches is based on the fact that there is no need to develop specific regulatory instruments due to the lack of information on the consequences of algorithmic pricing applied to competition. Another one implies checking every new pricing algorithm in order to establish its influence on competitive environment and allowing or prohibiting it based on that. The third approach suggests further control using new legal standards that differ from the current ones. Banning algorithmic pricing is viewed as unreasonable as the pricing itself becomes more economically effective. The question of responsibility is raised for the cases when cartels are formed as a result of the activity of computer software. It is concluded that the owners of the software must be held accountable for any actions of the algorithm, as no program can act of its free will, they are controlled by the creator or the right holder.

1 Introduction

The changes in the structure of markets, the emergence of new business models under the influence of the digitalization and algorithmization processes lead to the necessity to ensure and protect competition on digital markets that have several specific features compared to the traditional ones.

For instance, the defining factor in this sphere is building a market around the competition of digital platforms that form the infrastructure for providing services and selling goods. The peculiarity of the way digital platforms function is the application of algorithms in order to process big volumes of data accumulated there in order to coordinate the activities of the platform's users, regulate their behavior, form a pricing policy.

Currently application of the business model based on digital platforms goes beyond providing services and selling goods on the Internet, as it recently was. Traditional offline services such as passenger and freight transport can also be managed based on platform decisions and algorithm application. In this case the Internet does not act as the environment for providing services like it does in case of digital content but rather as a necessary technological element of the platform's functioning.

Thus, the platform decisions and the technologies of algorithmization applied by them are spreading outside digital markets influencing the competition on the traditional ones. The application of business project algorithmization creates new possibilities for coordination of the activity of entities, and also new ways for monopolization. Organization for Economic Co-operation and Development [1] notes that one of the key challenges for competition policy connected to spreading digital data is using digital algorithms (formalized computer procedures) and establishing cartels using them.

2 Problem Statement

The development of economic relations, their digitalization and algorithmization lead to the emergence of new ways of doing business and to the development of smart technologies and gadgets. Using new technologies and business models (e.g. services for sharing resources such as houses, cars), creates previously unknown possibilities of collecting and

* Corresponding author: kozlova-my@ranepa.ru

processing data about consumers, algoritmization (pricing algorithms, risk evaluation and prediction algorithms) is beneficial for the market participants; however, it is a challenge for the legislator. Law is unable to regulate these relations with traditional methods, they are intended for a different economical reality.

The questions related to digitalization and algorithmization as challenges to antitrust regulation are being discussed in the legal and economic literature. The issue of antitrust regulation in the digital epoch is undergoing integrated monographic research [2]; specific aspects of digitalization and its influence on competition legislation in particular countries are being studied as well [3].

The possibilities of using big data are analyzed; in this context probable competition issues are detected, as well as their potential influence on the performance of instruments of competition assistance and on the main spheres of work of the competition authorities [4].

The questions of the influence of computer technologies on competition environment are studied [5], as well as the possible application of the EU law on competition on digital markets [6].

The issues of influence of algorithmic pricing on the antitrust policy [7], of expansion of the term "collusion" considering the possibilities of computer technologies [8] are indicated.

However, there is no common understanding of how the competition authorities should respond to the challenges of the modern digital economy, whether they should use traditional instruments or develop new specific methods.

3 Research Questions

The research analyzes the practice of applying digital platforms and algorithms for monopolizing the market and to form approaches of the legislation to resist negative influence of such practice.

4 Purpose of the Study

The research is aimed at identifying the optimal approach for the competition authorities to react to the challenges of the digital age, such as using digital platforms and algorithms.

5 Research Methods

The methodology is based on general cognition methods: dialectical and systemic. Not only the general scientific methods of analysis, synthesis, induction and deduction were applied in the research, but also specific legal ones, the comparative legal method in particular.

6 Findings

The issue of digital platforms in competition practice is widely known and well-studied. It is based on the effect of the right holder of the platform controlling the goods and services market that appears due to the fact that the digital platform is the entrance point to the markets for both consumers and suppliers. Since the platform itself is a communicative service that lets a consumer find a supplier and vice versa, the platform's right holder controls not only the market of their own services but also all other markets where communication between contractors is conducted through using the platform [9].

Undoubtedly, control of this kind only becomes attainable in absence of sufficient amount of competing platforms and when the vast majority of consumers and suppliers concentrate around the monopolist platform. However, this exact situation can be seen at most digital markets.

Two factors lead to the emergence of monopolistic platforms. The first one is connected to the fact that developing, introducing and promoting digital platforms requires a lot of economical and organizational resources that new technological companies generally do not have, while technological giants do. Thus, the emergence of a lot of competing digital platforms is impeded by the high price of entering the market of the services provided. The situation is exacerbated by the fact that a lot of services provided by the platform right-holder are free, which excludes the possibility of rapid payback of investments.

The second factor impeding the emergence of a big amount of platforms are so-called network effects. Being a communicative service providing the connection of contractors at the market, digital platforms acquire value and significance only in case they have a lot of users. However, the more users the platform has, the more valuable it becomes for the more new users it attracts. It is understandable that the participants of the market that have a loyal audience that uses their other products are more likely to enjoy the advantages of the network than new market participants.

The network effect also intensifies due to circumstances preventing the clients from leaving the platform. The obstacles are connected to the fact that the clients get used to certain interfaces and possibilities and, consequently, they do not want to study new instruments provided by rival platforms.

Thus, the emergence and dominance of a limited amount of digital platform s can be considered a natural consequence of the process of their development and functioning.

However, the naturalness of the limited number of platforms does not mean that the competition authorities are to stop controlling the competition of such platforms with each other and their influence on related markets. For instance, the dominance of a digital platform can lead to the introduction of unreasonably high fees for using the platform for suppliers of services and goods. High fees on digital platforms along with the platform's price regulation can lead to decline in quality or safety of services, deterioration of working conditions in certain sectors of economy as it did in the taxi services market.

The market influence of digital platforms can also lead to takeovers and displacement of independent suppliers and to establishing discriminating conditions for using the platform. Providing suppliers with the instrument of selling and providing services the platform owners can start competing with suppliers using the platform directly, driving them out of the market after gathering enough data.

In August 2020 the Federal Antimonopoly servic e of Russia has examined the case upon the complaint by "Kaspersky Lab" on Apple Inc. connected to abusing the dominant position on the market of mobile devices functioning on iOS. Apple Inc. has forced Kaspersky to significantly reduce the functionality of their parent control application in order to promote its own similar (competing) application "Screen time". The Federal Antimonopoly Service committee has admitted it to be abuse of dominant position in sense of part 1 of article 10 of the Federal law "On protection of competition" the actions of Apple Inc. on setting conditions for the developers of B2C applications to access App Store; they let the right holder decline or not admit any app of any developer for any reason. The actions of Apple of normative and behavioral character were also considered abuse of dominating position, and they have led to a considerable decline in functionality of parental control applications [10].

One of the effective instruments of maintenance of competition environment in digital markets is preventing the digital platforms from merging; it is performed via preliminary checks of merges takeovers [11].

One of the examples of successful application of this instrument is the refusal of the Federal Antimonopoly service to approve the agreement of joint action between groups of companies owning "Yandex. Taxi" and "Vezyot" taxi aggregators [12]. As it was concluded after the analysis conducted by the Antimonopoly service , in case the deal is settled, the market share will increase up to 70% within the borders of the Russian Federation, more than 80% within 19 regions of the Russian Federation and more than 50% within 32 regions of the Russian Federation.

However, the examples listed above of a successful response of the antimonopoly service and of prevention of monopolization of the market using digital platforms are more of an exception than a rule.

For instance, Tim Wu states that the antitrust legislation per se is not suitable for regulating the information industry, as its application is limited by cases of price manipulation and other specifically listed kinds of violations that do not necessarily characterize the activity of companies owning digital platforms [13].

A. Dotsenko and A. Ivanov note that economic concentration and the market power strengthening of digital platforms is mainly connected to collecting and concentration of a lot of data about the users. However, traditional antitrust regulation approaches do not regard such information as an asset and mainly focus on prices, which becomes pointless considering the fact that a lot of services are free on the platforms [9].

Application of algorithms for collecting and analyzing data can be used not only for the creation of digital platforms and monopolization of digital markets through them, but also for the coordination of activity of entities both in traditional and digital markets. For instance, algorithmization can be used for the elucidation of the price level and detection of demand. This process facilitates collusions and forming of cartels. There is problem of detection and of establishing responsibility for using algorithms for competition authorities.

An "algorithm" can be defined as a way (a program) of solving calculation and other tasks, exactly prescribing how to and in what sequence achieve a result clearly defined by the source data [14]. This term is used in different sciences, and has received a special meaning for business processes and relationships on competition protection. Price algorithm is a software product used to determine prices on goods using which estimation and (or) determination and (or) monitoring of prices are executed based on the parameters set by the algorithm's user.

Recent researches focus on the necessity of full analysis of pricing mechanisms in the digital age, as the sphere is of pronounced interdisciplinary character (economics, law, informatics, deontological ethics, etc.)

In a brief description of the pricing phenomenon and the behavior of entities in goods markets caused by it generally three types of pricing are distinguished in the literature [15]:

1. Dynamic pricing that happens in real time, a reaction on the situation in the market with undefined demand;
2. Personalized pricing that is also called price discrimination of the first degree, the price varies depending on the consumer's personality;
3. Algorithmic pricing is a result of rapid digitalization that has the features of both dynamic and personalized pricing; it's main characteristic is the fact that it is based on algorithms as key components of digitalization.

The consequences of application of mechanisms of algorithmic pricing by the entities present the most interest in this research.

A lot of companies use the potential of algorithmization in their businesses based on the analysis of a lot of data. Pricing algorithms dominate online selling of goods for optimization of the price based on the supply available and demand; they are also used for booking hotels and travels, in commerce, sports and entertainment [5].

A business based on algorithms includes predictions of demand, price changes, behavior and other preferences of clients, risk assessment and assumptions on upheavals that can influence the market environment. What is more, prediction can be used to optimize the decision taking process, development of new services.

Moreover, algorithms can be used to optimize business processes through the reduction of expenses, segmentation of consumers, setting optimal prices [16].

Using algorithms gives an undoubted competitive advantage in business; however, it can limit competition.

As it is noted in the literature, decisions about the prices are mostly handed to algorithms. Scientists and competition authorities express concerns about the fact that these algorithms can more effectively lead to collusions than people making decisions [7].

One of the first cases where a price-fixing cartel using algorithms was discovered was the one investigated by the United States Department of Justice about a collusion on prices on posters sold via the Amazon Marketplace. Participants of the cartel have agreed upon prices on particular posters; specific software was used to set prices according to the agreement [17]. However, there are examples of using algorithms for collusion in the "pre-digital" age; for example, the case of leading airlines of the USA agreeing on ticket tariffs using the database of the Airline Tariff Publishing Company. The case was not tried in court, the Department of Justice and the airlines have entered a settlement agreement [18]. It is emphasized that the influence of pricing algorithms is estimated differently in the literature. There are two groups of views: the most positive one states that the influence of pricing algorithms is exaggerated and the human factor influences most cartels more. The others say that the anti-competition potential of pricing algorithms is not fully realized yet; it will be realized as the algorithms are still at an early stage of development and the antimonopoly authorities are not yet capable to counteract misuse of digital instruments of this kind since it does not possess the necessary technical equipment [7].

American researchers Ezrachi and Stuke distinguish four main categories of cartels that are possible with the help of the algorithms — Messenger, Hub and Spoke, Predictable Agent and Digital Eye [5].

The first category of cartel agreements is based on the application of messengers. A cartel agreement is achieved and the data analysis carried out by software is used for creating, monitoring and controlling the cartel. A traditional collusion model is applicable to such agreements.

Competitors may agree on information exchange on prices and on application of a specific behavior model; for instance, setting a specific price, applying discounts etc. An algorithm can control the enforcement of the agreement. Limiting competition happens when the participants of the agreement cannot set their own prices.

The practice of the Federal Antimonopoly Service of the Russian Federation can serve as an example. For instance, a company that was the only one importing and selling LG smartphones on the territory of the Russian Federation, set the recommended retail prices on smartphones and then controlled if the retailers followed the guidelines. A special price algorithm was applied. When a deviation from the recommended prices was detected, the company put pressure on resellers applying sanctions (shipment termination) [19]. The company was found guilty in concluding anti-competition agreement, administrative measures were taken.

The second category of illegal agreements is based on using one algorithm for determining the market price by several subjects. One vertical agreement does not necessarily have an anti-competition effect; however, several vertical agreements with several subjects can lead to the creation of a classic branched cartel that leads to an increase in prices.

The third category (predictable agent) implies that on a complex market a situation can occur when different subjects use their own algorithms programming them in a certain way to react to changes in market situation. In this case the introduction of similar algorithms into the whole sphere can lead to an anti-competition effect due to the creation of interdependent actions. In order to facilitate the usage of price mechanisms the firms increase transparency which makes the market more sensitive to collusions due to which prices start growing. Importantly, the price increase is not a result of an explicit collusion, it is rather a natural consequence of a tacit collusion.

Conscious parallelism occurs when an algorithm of one entity interacts with other algorithms; for instance, under certain conditions competition ceases.

Using an algorithm one can also watch the prices of other entities not only for objective analysis but for coordination as well.

Using auction bots in e-tendering can lead to the establishment of such cartels.

The fourth category called Digital Eye implies that the competitors unilaterally create and use computer algorithms for achieving the set goal; for example, profit maximization.

The fact of using price algorithms in activities of entities is not viewed as violation of competition law, since it is actually a trade instrument. However, illegal coordination can be performed using the algorithms, or they can lead to cartel agreements.

There are other repercussions for the development of competition applying mechanisms of algorithmic pricing.

Calvano, Calzolari, Denicolo and Pastorello emphasize that using algorithmic pricing significantly expands the possibilities for price discrimination. Even though direct influence of price discrimination is generally not apparent, the prices set under the influence of pricing algorithms can depend not only from relatively "innocent" information such as the time of purchase or the remaining capacity of the firm, but also the complete purchase history of the buyer. Such pricing "under condition" can lead to so-called "poaching" of consumers or to using exclusivity or discount in order to get a share of the market, both of these consequences can have an anti-competition effect [7].

Currently the spread and development of digital platforms and the application of technologies of data processing outside exclusively digital markets compel to review traditional approaches to regulating competition.

For instance, the usual markers of market power abuse such as setting monopoly high or low prices can be ineffective in regard to the assessment of digital platforms, the expansion of market power of which is connected to the concentration of a lot of user data and the network effect connected to it.

Speaking of the phenomenon of algorithmic pricing and its influence on the state of competition, three main approaches are to be noted; they concern how the

https://doi.org/10.1051/shsconf/202110901020

application of digital pricing instruments should be regulated in order to provide fair competition.

The first approach suggests not to start developing additional normative instruments and observe, since the exact influence of algorithmic pricing on the state of competition is not fully studied yet.

The second approach implies the regulation of introducing the pricing algorithms ex ante just like commercialization of new medicine: any new pricing algorithm should be tested by a regulating authority in order to determine whether it can potentially provoke creation of a cartel (in case it is to be banned by the regulator) or cannot (in this case it is sanctioned).

The introduction of such preliminary regulation is supported in research of Ezrachi and Stucke who note the possibility of creating a "digital sandbox" where pricing algorithms could be modeled and studied, creating their regulation as a result of such empirical studies [5].

Finally, the third approach lies in ex post regulation, as it usually happens in competition law; however, it implies using new legal standards. In particular this approach requires reassessment of the disputable question of the tacit collusion ban.

A fourth approach can be found in literature, it implies a total ban of algorithmic pricing; however, in our view, such measures are unreasonable. Evidently, using digital instruments in competition law benefits the markets in pricing since the pricing becomes more economically effective.

We agree with Calvano and Calzolari who note that legal regulation of applying the instruments of algorithmic pricing confronts insufficient elaboration of their essence and peculiarities of their application. In particular, features of collusion of pricing algorithms can depend on what other algorithms they interact with. Let us assume that an algorithm A was approved based on the proof that it does not tend to collude with the existing algorithms B, C and D. Suppose that a new algorithm E tends to collude with A but not with B, C and D. Which algorithm should be banned? E is better than A, so on grounds of effectiveness E must be approved and A must be banned. However, A was approved from the beginning, so banning it at a later stage may turn out to be quite costly [7].

Nevertheless, extremes in setting regulating measures are evidently ineffective as well. Practicing specialists must review proactive approaches to providing consumers and authorities with possibilities to check algorithms and using their personal information on the spot in order to enable the assessment based on influence [15]. Such algorithmic audit can be accompanied by a system of algorithmic certification, as well as with certification of developers based on ethics codes [20].

Describing what approaches to regulating the application of pricing algorithms are to be put into practice, M. Gal gives the following interesting example: "A commissioner of the Federal State Commission Maureen Ohlhausen has suggested a simple test that records a lot of these simple cases: if the word "algorithm" could be replaced with a phrase "a guy called Bob" then one can work with algorithms just like with traditional agreements". In this case it comes to a situation when there is a cartel collusion between entities and algorithms just serve as instruments for realization of such agreement. However, another option is possible, when algorithms serve as self-learning programs that under certain conditions regulate the prices by themselves based on each other's figures, interacting in order to adjust the price. It is noted that right holders using algorithms in such situations can be viewed as a cartel, except for the cases when such use causes "significant compensatory profits" for consumers [21].

However, there is still no definitive answer to the question of how to qualify coordination of prices when there is no agreement on such coordination. In other words, when several entities independently develop pricing algorithms that are based on the principles of autonomous functioning; the price increases as a result of their application.

The mere use of algorithms does not prevent the application of legal liability to their developers or users. As a competition European Commissioner has stated, "entities should be accountable for the consequences of algorithms they had chosen to use". In order for the legal liability to appear, the developer or the user should know about the price effects being created. This could be illustrated with an example of European Eturas [22]; 30 Lithuanian travel agencies have participated using the same online booking system. The system operator has programmed the algorithm is such a way so that the agencies would not be able to offer more than 3% discounts and has notified the agencies about this restriction through the inner messaging system. The agencies have used the algorithm. The question was whether these events imply that there was an agreement between the agencies on changing the algorithm and reducing competition. The European court has admitted the awareness about the changes in the algorithm to be a necessary condition for identifying a cartel. Underestimating the possible effects of algorithms under certain circumstances can be sufficient for proving awareness. What kind of awareness is required in cases when the algorithm designed for autonomous identification of decision taking parameters contributes to collusion [21]. In literature this phenomenon is called "cartel of the machines" [8].

7 Conclusion

In our opinion, the following approach is logical; since any software even while performing defined, already set functions, does not act upon its will and is controlled by its creator and/or right holder, the actions of "cartels of the machines" concerning pricing violating competition are the direct consequence of behavior of such creators (right holders). As a result, it is logical that for any actions of the machines, even relatively autonomous and based on autonomous calculations of the algorithms, the program owners should be held accountable (since even inaction (intended non-interference to the activity of the program) can serve as indirect evidence of coordinated anti-competition practice).

References

1. K. Doi, OECD Discussion on Digital Diffusion Problems, 11 (2018), URL: https://fas.gov.ru/pages/international-partnership/oesr/oesr-gvx-rczk/informaczionnyij-byulleten-oesr-gvx-rczk/ (date of access: 01.12.2020)
2. A. Tsarikovsky, E. Voynikanis, A. Ivanov, *Antitrust in the Digital Age: How to Protect Competition in the Context of Globalization and the Fourth Industrial Revolution* (National Research University Higher School of Economics, 2019) https://doi.org/10.17323/978-5-7598-1984-4
3. P. Honoré, R. Verzeni, Italian Antitrust Review, **4 (2)** (2018) https://doi.org/10.12870/iar-12853
4. S. Hutchinson, Law. Journal of the Higher School of Economics **1**, 216–245 (2017) https://doi.org/10.17323/2072-8166.2017.1.216.245
5. A. Ezrachi, M.E. Stucke, SSRN Electronic Journal (2015) https://doi.org/10.2139/ssrn.2591874
6. A. Ezrachi, SSRN Electronic Journal (2018) https://doi.org/10.2139/ssrn.3191766
7. E. Calvano, G. Calzolari, V. Denicolò, S. Pastorello, Review of Industrial Organization **55 (1)**, 155–171 (2019) https://doi.org/10.1007/s11151-019-09689-3
8. A. Heinemann, A. Gebicka, Journal of European Competition Law & Practice **7 (7)**, 431–441 (2016) https://doi.org/10.1093/jeclap/lpw036
9. A.V. Dotsenko, A.Y. Ivanov, Law **3**, 31–45 (2016)
10. AD/74808-DSP/20 (2020), URL: https://br.fas.gov.ru/ca/upravlenie-regulirovaniya-svyazi-i-informatsionnyh-tehnologiy/b70e9d96-8d39-46f5-9d7d-342da95b354b/?query=App%20Store (date of access: 01.12.2020)
11. Competition Issues in the Digital Economy (TD/B/C.I/CLP/54) (2019), URL: https://undocs.org/ru/TD/B/C.I/CLP/54 (date of access: 01.12.2020)
12. IA/49910/20, URL: https://br.fas.gov.ru/ca/upravlenie-regulirovaniya-svyazi-i-informatsionnyh-tehnologiy/2b16e00b-8fd3-4df7-a551-a46d814248ee/ (date of access: 01.12.2020)
13. T. Wu, The Master Switch: The Rise and Fall of Information Empires (Vintage Books, 2011)
14. Great Encyclopedic Dictionary **1** Soviet encyclopedia (Algorithm, 1979)
15. P. Seele, C. Dierksmeier, R. Hofstetter, M.D. Schultz, Journal of Business Ethics (2019) https://doi.org/10.1007/s10551-019-04371-w
16. Algorithms and Collusion: Competition Policy in the Digital Age, URL: www.oecd.org/competition/algorithms-collusion-competition-policy-in-the-digital-age.htm (date of access: 01.12.2020)
17. Former E-Commerce Executive Charged with Price Fixing in the Antitrust Division's First Online Marketplace (2015), URL: https://www.justice.gov/opa/pr/former-e-commerce-executive-charged-price-fixing-antitrust-divisions-first-online-marketplace (date of access: 01.12.2020)
18. S. Borenstein, Rapid Price Communication and Coordination: The Airline Tariff Publishing Case 19 (1994)
19. 4-14.32-340/00-22-18 (2018), URL: https://fas.gov.ru/documents/ats-46961-18 (date of access: 01.12.2020)
20. J. Beranger, The Algorithmic Code of Ethics: Ethics at the Bedside of the Digital Revolution. Wiley-ISTE (2018)
21. M.S. Gal, Illegal pricing algorithms. Communications of the ACM **62 (1)**, 18–20 (2018) https://doi.org/10.1145/3292515
22. C-74/14 (2016), URL: http://curia.europa.eu/juris/document/document.jsf?text=&docid=173680&pageIndex=0&doclang=EN&mode=lst&dir=&occ=first&part=1&cid=523288 (date of access: 01.12.2020)

Selected aspects of legal regulation of e-justice in Russia

Natalya Kvitsiniya[1], *Elvira* Osadchenko[2*], and *Olesya* Belova[3]

[1] Volgograd State University, 100 Prospect Universitetsky, Volgograd, Russia e-mail: gimchp@volsu.ru
[2] Volgograd State University, 100 Prospect Universitetsky, Volgograd, Russia e-mail: elviraosa@rambler.ru
[3] Volgograd Institute of Management, branch of the Russian Presidential Academy of National Economy and Public Administration, 16 ul. Gagarina, Volgograd, Russia e-mail: belova.oa@list.ru

Abstract. The article is dedicated to the research of how information and telecommunication technologies are used in the legal procedure and of problems of development of e-justice in Russia. What is more, the topic of additional opportunities for the trial parties in civil law cases is covered as well. The authors have identified several problems connected to the development of e-justice and have suggested solutions, using video conferences in court proceedings. Recommendations are given in the article, as well as possible solutions to problems emerging in the process of development of e-justice in Russia.

1 Introduction

Among the aims of creating the electronic justice system are the improvement of its availability and the transparency of administration as well. Availability of justice is generally understood as the feasibility of receiving legal protection, the effectiveness of justice mechanisms at all stages of its realization, the pretense of actual guarantees of getting a fair and effective ruling upon recourse to the court.

The European community recognizes the need of introducing electronic elements to the functioning of public authorities, including the judiciary. Electronic institutes should naturally complement the traditional institutes of democracy by ensuring the fullest realization of fundamental human rights and freedoms [1]. While uncovering the problem of legal regulation of e-justice in Russia, the authors have sought to analyze the current state of electronic digital law enforcement.

2 Problem Statement

It must be noted that one can easily forget about the risks innovations cause while being immersed in them. For instance, the electronic workflow includes more safety risks than the habitual paper one. The possibility to amend electronic documents calls into question their credibility as evidence when assessed by the court and creates additional requirements for identification of persons submitting such documents [2].

3 Research Questions

The term “e-justice” has been understood ambiguously in scientific literature for a long time. In the absence of a normative definition different scientists have implied a wide range of phenomena starting from the activity of judiciary ending with the order of submitting electronic documents to court [3].

The Concept of development of court computerization until 2020 contains the first legal definition, according to which e-justice is a method and a form of actions established by courts based on applying the information technologies system to their work, including the electronic interaction between courts, citizens and organizations [4].

Despite the significance of e-justice and the rise of scientific activity on the problem in recent years, the autonomous nature of the research object is currently still in question. Despite the particularity of such a form of realization of the right to judicial protection, using it changes neither aims nor the essence of justice, as it is a law enforcement activity that has some distinctive features of technical and procedural character.

The application of information technologies is aimed at increasing the transparency and availability of justice, the realization of rights to receiving reliable and actual information. All these goals do not stand out of the general conjuncture of justice administration, so, in our opinion, the interest in researching e-justice and its technical and procedural peculiarities is caused by its relative novelty. Once e-justice becomes an established institute in civil law and other spheres of procedural law, it will stop being a special research subject.

However, considering the tendencies of e-justice formation in Russia, one can assert with certainty that this process is far from over.

The active stage of e-justice formation in Russia started in 2000 after the All-Russian Congress of Judges V. At the same time, the first program of information management of courts of law was created. The next milestone of the process was the adoption and functioning during 2002-2006 of the federal program

* Corresponding author: elviraosa@rambler.ru

dedicated to the development of the judicial system in Russia.

The first practically important achievement in the e-justice sphere was the creation of "Justice", an electronic court filing system which has formed a united information space of courts of law and has provided not only the electronic support of the procedure but also the information about it to the citizens and organizations on the Internet.

It was clear that despite the progress made, the system was far from perfect, qualitative changes were needed, as well as a broader distribution. In this regard the next program of the Russian judicial system development for the period of 2007–2011 was adopted, during which the "The register of judgements of State commercial courts" information system was created in 2008.

The changes introduced to the Code of Arbitral Procedure of the Russian Federation in 2010 were a breakthrough. Due to them, the sides have been provided with the opportunity to submit procedural documents via the Internet, it became possible for the sides to participate in judicial proceedings not in person but using the means of teleconferencing applications.

The next Federal target program was the program of development of the judicial system during 2013-2020, according to which improvement of the e-justice system was one of the key tasks, as it made it possible to ensure the availability of justice, the trust of the citizens and the protection of personal data of the participants of the process.

It must be noted that the legal positions of the Supreme court of the Russian Federation have become more profound and better developed over the years. The ruling of a plenary session of the Supreme court of the Russian Federation of 09.02.2012 №3 "On altering some orders of plenary sessions of the Supreme court of the Russian Federation", the ruling of a plenary session of the Supreme court of the Russian Federation of 29.09.2015 №43 "On some issues connected to applying the rules on the statute of limitations of the Civil code of the Russian Federation" have regulated only some aspects of e-justice; for instance, informing the parties of judicial proceedings via SMS and the moment of submitting the statement of claim in electronic form [5], [6]. In 2016 the Department of Justice of the Supreme court of the Russian Federation has established the order in which the electronic documents should be submitted to courts of law [7]. In 2017 the plenary session of the Supreme court of the Russian Federation published a fundamental rule №57, regulating the aspects of e-justice not covered by the law [8].

In conclusion, during the years of its existence e-justice in Russia has undergone long and complicated development; however, the process is not over yet.

4 Purpose of the Study

Let us address some practical problems that arise in the field studied due to the faults of technical equipment and procedural norms, namely the problem of using teleconferencing systems in arbitral courts and courts of law to provide participation of the sides in judicial proceedings in the process of justice administration.

Digital technologies are developing by the requests of society. A video conference is a technology letting subjects who are physically distant to interact and exchange information in real time.

The emergence of this system has significantly reduced the money and time spent on proceedings.

As S. V. Vasilkova notes, e-justice aims at improving the quality of judicial paperwork services and the quality of workflow in general by using information and communication technologies [9].

The introduction of the rule on judicial proceeding using videoconferencing systems to the legislation is caused by several factors. Firstly, this way the time spent on procedural actions, such as questioning the witnesses, interviews, expert consultation, etc., is reduced. Secondly, the availability of justice increases. Thirdly, the necessity to use video calls can be caused by such situations when the parties are located in different regions for objective reasons. There are also cases when it is difficult to present certain evidence in court.

According to part 1 of article 155.1 of the Civil Procedural Code of the Russian Federation, the provision of a video conference to the parties, their representatives, experts and other specialists is only possible upon submission of a motion [10].

As demonstrated by judicial practice, such a motion must contain good reasons for the possibility to have a video conference. For instance, one ruling of cassation contains an indication to the fact that the possibility to participate in judicial proceedings by using the means of video conferencing is a right and not an obligation of the court. The motion submitted was not granted, as the court had not found reasons for it, namely, there was no reason to grant the motion that could have influenced the decision. Thus, the appeal was investigated without the representative taking part [11].

As stated in point 8.1.5 of the order of the plenary session of the Supreme Arbitration Court of the Russian Federation of 2013, before appointing a judicial proceeding using the video conferencing systems in courts of law, one has to consider the following circumstances: timezone of the region, "video conference plan" created with the help of the United corporate portal of arbitral courts created for convenient planning of session time [12]. In this regard, conducting a court session via a videoconference is possible under several conditions, namely the same timezone and the technical possibility to do so in courts.

According to article 153.1 of the Code of Arbitration Procedure of the Russian Federation, there are two grounds for refusal: the necessity to conduct a closed session and no technical possibility to provide a video conference [13].

According to part 5 of article 159 of the Code of Arbitration Procedure of the Russian Federation, not granting a motion on a video conference can be based on the fact that the motion had not been submitted in time and when the submission of such notion delays the process and creates obstacles for proceedings [10].

It is worth paying attention to the fact that organizing a video conference shortly before a session does not seem to be possible in case the motion had been submitted on short notice.

5 Research Methods

The main method used was dialectical; it has made it possible to identify the purpose of the study. To solve the problems of this research the authors have used various specific cognition methods.

6 Findings

Experience has shown that in civil proceedings a motion for a side to take part in a session via a video conference can be denied for various reasons that depend upon a judge. We suppose that this problem requires a solution on the legislative level by amending the Civil Procedural Code of the Russian Federation.

Clear grounds in which the court must refuse granting the motion on making it possible to take part in the court session via video conference and cases in which it cannot be denied must be introduced to the Russian Procedural codes (article 153.1 of the Arbitration Procedure Code of the Russian Federation, article 155.1 of the Civil Procedure Court of the Russian Federation, article 142 of the Code of Administrative Procedure) [13], [10], [14].

The researchers Bochkarev Y. A. and Shapovalov S. I. note that the question of the absence of the possibility to appeal against the decision of the court on refusing the usage of a videoconferencing system should be taken into consideration as well [15].

Thus, addressing the gaps will establish justice with no violations from the court. It must be noted that using a videoconferencing system in court during proceedings has its advantages: it saves time, decreases the duration of proceedings and expenses necessary to commute to the court from where the person is.

An issue of organizational character is the fact that in courts there are no hardware specialists that could ensure continuous work of video conferences. Such a person must be responsible for the maintenance of video calls. The introduction of such specialists can improve the efficiency of using modern digital technologies in judicial proceedings.

7 Conclusion

Thus, using video conferences in judicial proceedings is a mechanism as important and necessary in today's circumstances as the possibilities to initiate proceedings online, submit documents in electronic form, record videos of court sessions and keep digital case files.

References

1. J.A. Gavrilova, N.V. Kvitsinia, N.A. Kalashnikova, *Competitive Russia: Foresight Model of Economic and Legal Development in the Digital Age. CRFMELD 2019. Lecture Notes in Networks and Systems* **110** (Springer, Cham, 2020) https://doi.org/10.1007/978-3-030-45913-0_44
2. V.I. Tsyganov, A.A. Demin, E.O. Osadchenko, *Competitive Russia: Foresight Model of Economic and Legal Development in the Digital Age. CRFMELD 2019. Lecture Notes in Networks and Systems* **110** (Springer, Cham, 2020) https://doi.org/10.1007/978-3-030-45913-0_69
3. R.V. Kostenko, Y.V. Komissarova, V.N. Perekrestov, *Competitive Russia: Foresight Model of Economic and Legal Development in the Digital Age. CRFMELD 2019. Lecture Notes in Networks and Systems* **110** (Springer, Cham, 2020) https://doi.org/10.1007/978-3-030-45913-0_48
4. The council of Judges, *The concept of information policy of the legal system in 2020-2030* (Moscow, 2019)
5. Plenum of the Supreme court of the Russian Federation, *Ruling (April, 2012, N 4)* (Moscow, 2012)
6. Plenum of the Supreme court of the Russian Federation, *Ruling (September 29, 2015, N 43)* (Moscow, 2015)
7. The Supreme Court of the Russian Federation, *Ruling (December 26, 2016, N 251)* (Moscow, 2016).
8. The Supreme Court of the Russian Federation, *Ruling (December 26, 2017, N 57)* (Moscow, 2017)
9. S. V. Vasilkova, *E-justice in civil process. (Dissertation of candidate of law)* (Saratov, 2018)
10. Russian State Duma, *Civil Procedural Code of the Russian Federation (November 14, 2002, N 138-FZ)* (Moscow, 2002)
11. Kemerovo regional court, *4G-72/2015* (2015), URL: http://oblsud.kmr.sudrf.ru/modules.php?name=sud_delo&srv_num=1&name_op=doc&number=907510&delo_id=2800001&new=2800001&text_nu (date of access 01.12.2020)
12. Plenum of the Supreme Arbitration Court of the Russian Federation, *Ruling (December 25, 2013, N 100)* (Moscow, 2013)
13. Russian State Duma, *Code of Arbitration Procedure of the Russian Federation (July 24, 2002, N 95-FZ)* (Moscow, 2002)
14. Russian State Duma, *Code of Administrative Procedure (March 8, 2015, N 21-FZ)* (Moscow, 2015)
15. Y. A. Bochkaryov, S. I. Shapovalov, Bulletin of International Law Institute **3 (32)**, 21-25 (2017)

https://doi.org/10.1051/shsconf/202110901022

Legal regulation of investment activity in the republic of Kalmykia in the context of the digital economy development: the civil aspect

Ochir Minkeev [1,*], and *Sergey* Sharonov[2]

[1] Government Office of the Republic of Kalmykia, Elista, Republic of Kalmykia, Russia, ochir91@mail.ru
[2] Moscow University of the Ministry of Internal Affairs of Russia named after V.Ya. Kikot, Moscow, Russia, sharonov345@mail.ru

Abstract. The article uses the means of civil science to identify the problems of applying the normative legal acts of the Republic of Kalmykia regulating the relations of investment activity in the context of the digital economy development. Within the framework of this study, the author systematizes three groups of regulatory legal acts dealing with the relations of investment activity in the Republic of Kalmykia, each containing certain problems. The key problem of the first group, consisting of civil legislation and other acts containing civil law norms, is the lack of a unified approach to regulating investment activities. The second group, which is a set of separate Laws and other regulatory legal acts of the Republic of Kalmykia, dealing with the issues under the joint jurisdiction of the Russian Federation and the Republic of Kalmykia, contains a problem expressed in the lack of proper legal regulation of investment activity in the republic. The problem of the third group, consisting of municipal acts providing legal regulation of relations arising from the management and disposal of municipal property objects that can act as objects of investment activity, is the lack of a legislative basis for the implementation of the public function.

1 Introduction

The relevance of the topic is due to the rapid development of the digital economy and its penetration into investment activities. Currently, the priority state task is the process of "launching a new investment cycle", which allows significant investment increasing in various segments of "industry, agriculture, and services" [1]. Attention should be paid to the fact that investing in mentioned segments is most widespread in the Republic of Kalmykia territory (hereinafter – RK).

However, as the practice in several countries shows, the future of the Russian economy is not possible without the digitalization of investment activity and its legal regulation, which currently lags behind the needs of civil turnover.

Currently, the digital transformation of the national economy is carried out within the framework of the national project "National Program "Digital Economy of the Russian Federation", aimed at, among other objectives, increasing domestic costs for the development of the digital economy from all sources of funds [2].

To date, the results of these policy documents implementation are the introduction of amendments to civil legislation in terms of adding digital rights to the list of civil rights objects [3], the adoption of a Federal Law regulating the process of investing and attracting investment using investment platforms [4]. Additionally, these acts do not actually detail the legal regulation specifics of investment activities in the RK, which, definitely, reduces its effectiveness for the civil turnover participants.

2 Problem Statement

In the RK the implementation of this national project is carried out through the execution of several regional programs that are correlated with federal projects: digital technologies, digital public administration, personnel for the digital economy, information infrastructure, information security [5].

In this regard, it should be noted that measures aimed at regulating the digital environment are carried out exclusively at the federal level, which, without taking into account the specifics of investment activity in the RK, is not always effective.

This circumstance is explained, among other things, by the fact that, according to paragraph "o" of Article 71 of the Constitution of the Russian Federation, civil legislation is under the jurisdiction of the Russian Federation.

* Corresponding author: ochir91@mail.ru

However, the current state of investment activity in the Republic of Kazakhstan requires a comprehensive approach to its legal regulation. In practice, this means the application of the norms not only of civil legislation but also of other branches of law, which norms directly affect the legal relations developing in certain areas of investment activity carried out in the RK. Also, it should be noted that this regulation has problems that do not allow achieving the goals of the national project.

3 Research Questions

In the question of studying the problems of legal regulation of investment relations that are developing in the RK, in the context of economic processes digitalization, it is necessary to take into account its complex nature, paying attention to the possibility of its regulation by the norms of both private and public law.

4 Purpose of the Study

Thus, the research goal of the publication is to identify the problems of complex legal regulation of investment activities in the RK in the context of economy digitalization. To meet this goal, it is necessary to solve two tasks: first, to argue that the complex nature of the regulation under study actually means the application of the norms of other branches of law, which objects are objects of investment activity (land, housing, etc.), and the relations developing in this case, along with civil legislation, can be regulated by acts of other sectorial affiliation, the right to adopt these acts is granted to the relevant bodies of the RK; second, based on the analysis of regulatory legal acts, to identify and resolve problems arising in the process of their application in the RK.

Let us sequentially consider the solution of the tasks set, paying attention to the features of the methodology designed by the authors: the identification of problems in the studied area (the second task) will be carried out sequentially in the context of solving the first task of the publication.

In the context of solving the first task, it should be noted that according to the investment strategy of the RK, investment activities in the RK are carried out in several economic sectors of the RK, which study on investment attractiveness requires the application of cluster analysis [6]. As a result of cluster modeling, the first cluster includes two sectors – "Agriculture" and "Education"; the second cluster – "Industry"; the third cluster – "Construction", "Trade", "Transport and Communications"; the fourth cluster – "Hotels and restaurants".

It is possible to achieve the objectives of the RK investment strategy, in particular through improving the legislation under the joint jurisdiction of the Russian Federation and the RK, including budget legislation, legislation of the RK on taxes and fees, land legislation, etc.

We will evaluate mentioned types of legislation while focusing on the classification of the relevant regulatory legal acts and the degree of their impact on the essence of the activity under study.

According to the results of the analysis, the regulatory legal acts dealing with the investment activity relations in the RK can be systematized into three groups. Thus, the first group comprises civil legislation, which includes codified acts, laws of the Russian Federation and federal laws, and other acts containing the provisions of civil law, including Decrees of the President of the Russian Federation, RF Government Regulations and regulatory legal acts of federal executive bodies. The second group comprises the legislation of the RK, containing the laws of the RK, decrees of the Head of the Republic of Kalmykia, RK Government Regulations and regulatory legal acts of the RK executive authorities. The third group comprises the decision by the representative local bodies and municipal legal acts of the RK municipalities.

The first group of classification should include the Civil Code of the Russian Federation, which structure largely coincides with the content of investment activities in the Russian Federation as a whole and the Republic of Kalmykia in particular. This leads to the conclusion that the Civil Code of the Russian Federation is the fundamental act regulating relations in the studied area. For example, the legal capacity of citizens and legal entities that can act as both investors and customers is determined by subsection 2 "Persons". The norms of Chapter 9 "Transactions", as well as the norms of Section III "The general part of the law of obligations" can be fully used to regulate the relations under study, since the basis for the emergence of investment activity and the means of its legal regulation can be a contract, which in its legal essence is a type of transaction and obligation.

The first group also includes several federal laws, dealing with the relations developing within the process of capital investments, concession agreements, agreements on public-private, municipal-private partnership, in attracting investment using digital platforms, in the issuance, accounting and circulation of digital financial assets, which enter into force on January 10, 2021.

The analysis of the above-mentioned federal laws discloses the key problem of the first group, which is expressed in the application of regulatory legal acts only separately to each type of investment activity, which indicates the lack of a unified approach to regulating investment activity. The need to supplement this regulation with such acts, which could be adopted by the authorized bodies of the RK, is therefore acute. In addition, it is necessary to take into account the current trends of legal regulation, in particular, the penetration of digitalization elements into this process [7].

The content of the second group is represented by the Laws of the Republic of Kalmykia "On the tax on the property of organizations", "On the investment fund of the Republic of Kalmykia", "On state support and protection of the rights of investment entities in the Republic of Kalmykia", "On the establishment of a reduced rate of corporate income tax credited to the budget of the Republic of Kalmykia for certain categories of taxpayers", "On the application of the investment tax deduction for corporate income tax in the territory of the Republic of Kalmykia", Decrees of the Head of the Republic of Kalmykia "On approval of the investment strategy of the Republic of Kalmykia until 2025", "On approval of the investment declaration of the Republic of Kalmykia", Government Regulations of the Republic of Kalmykia "On approval of the procedure for support of investment projects implemented and/or planned for implementation on the principle of "one window" on the territory of the Republic of Kalmykia", "On approval of the rules for capital investments in state-owned objects of the Republic of Kalmykia at the expense of the republican budget", "On approval of the procedure for making decisions on the compliance of large-scale Investment projects with the criteria established by Article 4 of the Law of the Republic of Kalmykia "On the establishment of criteria that must meet the objects of socio-cultural and communal purposes, large-scale investment projects, for the placement (implementation) of which it is allowed to provide a land plot that is in state or municipal ownership for rent without bidding", etc.

The formation of acts of the second group is dictated by the fact that in accordance with Article 72 of the Constitution of the Russian Federation, the joint jurisdiction of the Russian Federation and the Republic of Kalmykia includes the administrative, administrative procedure, labor, family, housing, land, water, and forest legislation; legislation on subsoil and environmental protection. For the legal regulation of investment activity in the RK, this has the following meaning: 1) the RK has the right to carry out state regulation of investment activities carried out in the form of capital investments by approving inter-municipal investment projects and investment projects for state-owned objects of the RK, establishing the procedure for conducting expert reviews of investment projects, approving the procedure for involving temporarily suspended and abandoned construction projects and objects owned by the RK in the investment process; 2) the possibility of legal regulation that determines the legal status of individual entities and the legal regime of investment objects. Thus, the subjects of investment activity in the RK can be, including the public legal entity – the Republic of Kalmykia represented by the state authorities of the RK [8], and in turn, the objects of investment activity in the RK - objects of housing, land, water and other relations.

The analysis of the acts of the second group shows that some of them are not coordinated with each other. This circumstance reveals the main problem of the group under consideration, which is expressed in the lack of proper legal regulation of investment relations in the RK. In this regard, it is proposed to adopt the Law of the Republic of Kalmykia "On Investment Activity in the Republic of Kalmykia", establishing the powers of the RK state authorities in regulating the relations of investment activity in the RK, defining the list of subjects and objects of investment activity in the RK, the procedure for concluding investment agreements to which the public legal entity – the Republic of Kalmykia represented by the RK state authorities is a party, as well as aimed at solving other problems in the field under study. Also, attention should be paid to the fact that "digitalization of the regional economy...should become the main tool..." for its effective development [9].

The third group is due to the amendments to the Constitution of the Russian Federation, according to which local government bodies shall be included in a single system of public authority of the Russian Federation in order to maximise their efficiency in fulfilling tasks in the interests of the people living in the corresponding territory [10].

Thus, local government bodies, within the framework of constitutional provisions implementation in the field of investment activity, have the right to act as its subjects and, through the management of municipal property, have the right to carry out legal regulation of relations concerning the objects (buildings, constructions, structures) of culture and sports, educational, industrial, social infrastructure, etc., which have the potential to act as objects of investment activity.

As rightly pointed out by V.N. Pleskach, the development and concretization of constitutional provisions in legislation is of paramount importance for the implementation of the Constitution of the Russian Federation [11].

In addition, the analysis of the third group leads to the conclusion that the legal regulation of investment activities in the RK by local government bodies is limited to a certain extent due to the lack of a legislative basis for the implementation of public functions in terms of regulating investment activities carried out in the form of capital investments.

5 Research Methods

Dialectical, logical, historical, systematic and structural, and other methods of scientific cognition were used in the study.

6 Findings

Summarizing up the results of this study, we can draw the following conclusions.

First, the legal regulation of investment activity in the RK is of complex nature, which is manifested in its impact on the norms of various branches of law, while combining private law and public law relations for the implementation of investment activities.

Secondly, the core issue of legal regulation dealing with investment activity in the RK identified in this publication is the lack of conceptual legal regulation both at the level of civil legislation and at the level of the RK legislation.

Thirdly, at present, proper legal regulation of investment activities in the RK is becoming impossible without the digitalization process, its legal registration should be reflected in the proposed bill.

7 Conclusion

Thus, based on the analysis of public and private legislative means of regulating investment relations, the classification of problems allows us to identify and systematize issues to be addressed in the area under study.

A number of problems concerning legal regulation of investment activity in the RK can be solved by improving the RK legislation through the development and adoption of the Law of the RK "On Investment Activity in the Republic of Kalmykia", making up the legal regulation system of the studied activity and providing provisions on its digitalization.

References

1. V.V. Putin, Rossiyskaya Gazeta **7** (2020)
2. L.S. Gorskina, & O.V. Propp Russian Journal of Innovation Economics **9(2),** 275-286 (2019)
3. I.M. Konobeevskaya, Izvestiya of Saratov University. New Series. Series Economy. Management. Law **19 (3),** 330-334 (2019)
4. Yu.V. Pavlova, Bulletin of the Russian University of Cooperation, **3 (37),** 78-84 (2019)
5. T.N. Mikheeva, Courier of Kutafin Moscow State Law University **9 (61),** 114-122 (2019)
6. A.A. Mantsaeva, & T.G. Delikova, Bulletin of the Institute for Integrated Studies of Arid Territories **1(32),** 6-19 (2016)
7. V.P. Kamyshansky, Vlast Zakona **1**, 14-18 (2019)
8. O.D. Minkeev, Bulletin of the Volga Region Institute of Administration **3 (54),** 127-131 (2016)
9. D.E. Bekbergeneva, *Collection of scientific articles on the results of the fourth round table with international participation* **2**, 27-28 (2020)
10. M.M Mukhlynina, Gosudarstvennaya sluzhba i kadry **2,** 30-33 (2020)
11. V.N. Pleskach, Carskosel'skie chtenija, **III (XX),** 89-99 (2016)

 https://doi.org/10.1051/shsconf/202110901023

The application of digital technologies in the conditional agreements

Irina Mitrofanova [1*]

[1] Volgograd institute of Management of the Russian Presidential Academy of National Economy and Public Administration, Civil law Chair, Volgograd, Russia, advance100@inbox.ru

Abstract. This article is devoted to the study of some theoretical and practical problems of concluding a conditional agreement in the form of a smart contract. The article defines and characterizes a conditional agreement and describes its differences from ordinary transactions. The author analyzes the possibility of expressing conditional agreements in the form of a smart contract, taking into account the features of setting program code, as well as features of the agreement (a condition that the special nature of the rights and obligations of the parties activity). As a result of the research, it was concluded that the conditional transaction structure is ideally suited to the elementary program code for creating a smart contract. However, the condition contained in the transaction must be strictly formalized and expressed in clear values, such as numbers. The main problems that arise when creating a smart contract are also identified: the complexity of getting data from the real world by the system, weak state control, and possible data leakage from the system. Thus, the use of digital technologies is a prospective direction for the development of conditional agreements, but this requires solving both legal and some technical problems.

1 Introduction

The whole world is currently in a state of tremendous changing because of transition to the informational society. Access to the global information space, the increasing informatization of the society and the main role of information technology changes not only economic, but also social reality. Digital technologies are increasingly used in the regulation of legal relations. The legal system must respond to changes in a timely and comprehensive manner. But the opportunities offered by modern technologies do not always succumb themselves to adequate legal regulation. This is a completely new area of legal control, which, so far, is quite difficult to implement.

2 Problem Statement

This article is devoted to challenges faced by contract law. Now more and more attention of researchers is attracted by the so-called smart contracts. Such contracts exist in a digital reality, so there are many advantages of their using, but many problems too.

The author's research interests include conditional agreement. For this reason transition process of conditional agreements into the digital environment and the potential for their existence in digital form are of considerable interest.

3 Research Questions

3.1 What is a smart contract?

In the last few years concepts such as "bitcoin", "blockchain", "cryptocurrency", etc. is firmly established in everyday life. The Blockchain technology made it possible to create not only digital "currency", but also smart contracts.

A smart contract is a self-executing contract written in software code based on Blockchain technology [1]. This is a contract that is written in a programming language (artificial language) and sealed with an electronic digital signature of each of the parties (or certified with a special key). Such contracts are concluded in a decentralized environment with using a specific protocol (for example, the Ethereum project), which allows you to exchange assets without intermediaries.

Many big companies already sign contracts in this format. For example, Alfa-Bank and S7 have already realized a financial transaction using blockchain in Russia [2]. They used two smart contracts at once – for opening and executing (closing) a letter of credit. It can be argued that there is a need to legally regulate this type of legal relationship, because the number of such transactions will increase.

* Corresponding author: valentinatsybaneva@yandex.ru

3.2 Features of conditional agreement

According to the article 153 of the Civil code of the Russian Federation, transactions are actions of natural persons and juridical persons aimed at establishing, changing or terminating civil rights and obligations. Conditional agreements are interesting because they have quite specific effects. The essence of a conditional agreement determines by the circumstance on the occurrence (non-occurrence) of which the occurrence (termination) of civil rights and obligations depends.

For example, an agreement made under a suspensive condition exists in some hypothetical form before the condition happens. The moment of making a deal has happened, but the consequences will only come when condition will occurs. We can't accurately determine the duration of the legal relationship if agreement includes a subsequent condition. The legal relationship may terminate at any time with the occurrence of a condition. Thus, the conditional agreement has a number of additional properties. The parties can foresee negative consequences for themselves and neutralize them. At the same time, new risks appear, which can be expressed both in the occurrence and non-occurrence of a condition, because nobody knows if condition occurs [3].

G.F. Shershenevich wrote that a condition is the occurrence or non-occurrence of a known circumstance in the future [4]. N.L. Duvernua argued that the use and meaning of the condition in the civil relationship based in its future character [5]. Any circumstance can be as a condition, for example, reaching a certain currency exchange rate, obtaining a license, and so on. The condition may absolutely not depend on the will of the parties to the contract, or it may depend to some extent (for example, to get a loan, a person must do certain actions).

V.S. Em the author believes that conditional agreement is a transaction where the parties make the occurrence or termination of rights and obligations dependent on some circumstance that may or may not occur in the future [6].

Some authors, such as Ivanova E.V. [7], Zenin I.A. [8], give the concept of conditional agreement in accordance with the position set out in the Civil code of the Russian Federation. In other words, a transaction is considered through specific concepts – a transaction made under a suspensive and subsequent condition.

The legal purpose of transactions made under the condition is to obtain the economic and legal result that is characteristic of transactions of this type, as well as the creation of conditional rights and obligations, creating a state of uncertainty that is good for parties of the contract.

As a result of a conditional transaction, there are two types of consequences. First, such transactions serve as the basis for the emergence of rights and obligations not in their classic version, but conditional rights and obligations. Secondly, from the moment of the transaction and until the condition occurs, the parties must maintain the possibility of consequences that will lead to the dynamics of legal relations: do not perform actions that make it impossible for legal consequences to occur, take measures to ensure the safety of the property and its quality, inform the counterparty of significant changes in the condition of the property, etc.

The features of conditional rights and obligations which arising from suspensive conditional agreement are: 1) such rights and obligations should be sufficiently defined both by subject and by content at the time of conclusion of such an agreement; 2) they do not imply actual performance, but are a unique model of the rights and obligations that will arise with the occurrence (non-occurrence) of a suspensive condition [9]. The legal relationship that has arisen on the basis of such a transaction will not yet be either obligation, real, or corporate, but a special relative legal relationship.

The feature of conditional rights and obligations which arising from subsequent conditional agreement is that the parties of the agreement start their execution from the moment when the transaction is considered concluded. The legal relationship will not be considered conditional, but the continuation of its existence.

The occurrence of a condition is necessary for the dynamics of the legal relationship. This is a situation in which a circumstance (condition) occurs, or disappears (it is not carried out as it was provided for in the transaction, or it becomes obvious to everyone that such a circumstance can't occur ever. If the condition is suspensive, then its occurrence leads to the beginning of the performance of rights and obligations characteristic of this type of transaction. If the condition is subsequent, then its occurrence leads to the termination of rights and obligations and the transaction itself as a whole. The occurrence of a subsequent condition terminates the transaction for the future. Previous relationships as they were formed before the onset of the conditions are not changed. The onset of the condition has no retroactive effect.

4 Purpose of the Study

The purpose of the study is to research the possibility of using digital technologies in the conclusion of conditional agreements, as well as to identify problems that may arise in this case.

5 Research Methods

The methodological basis of research is the dialectical method of cognition, which presupposes the all-roundness, objectivity and interconnectedness of the phenomena under study; general scientific methods of cognition (analysis, synthesis, hypothesis, analogy, etc.); comparative-legal, functional methods.

6 Findings

6.1 Technology for including a condition in a smart contract

In order to find out whether a conditional agreement can be expressed in a smart contract, we need to understand the following. Conditional agreement — this is a legal structure, a complex legal composition consisting of two legal facts: a transaction that entails to rights and obligations, and the fact of occurrence or non-occurrence of a condition, which is expressed in the form of a legal action or legal event.

The structure by which the program code is set (when a smart contract is forming) must include certain reference points, after which events develop in one way or another. As a rule, the program is built on basic conditions, initially set parameters. After that, the algorithm of actions is set, and these actions are not linear, the options are calculated (yes/no, 1/0). That is, when a certain event occurs, the program performs one order of actions. When another event occurs, the order of actions may be completely different. When the third event occurs, the program code ends, and then nothing will happen.

The occurrence of consequences for a conditional agreement is caused the occurrence of certain circumstances. This is embedded in the source code of such a contract. The formula can be briefly described as: if…then... otherwise. Any modern schoolchild understands how to set program code using operations «If... then...» [10]. From the point of view of Russian civil law, the formula may well describe a conditional transaction, or a contract in which the performance of one party is conditional on the performance of the other's duties[1].

It should be noted that a smart contract has a number of distinctive features that will produce a positive effect in conditional agreements. A smart contract is a sequence of commands – this is the program code that is embedded in the Blockchain [11]. Therefore, the independence and self-enforceability of the terms of such a contract is ensured in the event of circumstances that are determined in advance.

Both parties of contract are in an equal position relative to each other in terms of available information because all the terms of the contract are formalized in advance and nothing "outside" can change them [12].

This is especially true for conditions that absolutely do not depend on the will of the parties. It is an event on the stock exchange, for example. The system can track such facts by itself [13]. In real life, one of the parties that does not benefit from the occurrence of the condition may not notify the other party about the occurrence of the condition. Therefore, the other side should always monitor the situation so as not to miss this moment. The system has a very important advantage. It is objectivity. In any case, the system will detect the occurrence of the condition and start acting on a specific way. None of the parties will be able to influence this process.

It can be concluded that conditional agreements in which the condition is objective (that is, absolutely independent of the actions of the parties) can exist in the form of smart contracts. There will be the positive aspects: the absence of intermediaries, automatic execution (which does not depend on the will of the counterparty, who may, for example, delay the payment or not make it at all), strict regulation and interpretation of the specified conditions, the inability to interfere with the specified code from the outside, etc. Smart contracts allow to transfer everything that is of any value, whether it is money, shares or data, without any intermediaries [14]. If we consider the" ideal" concept of a smart contract, its using can significantly reduce not only the organization's costs for concluding a contract, conducting transactions, etc., but also the number of legal disputes.

However, not everything is so simple and there are many problems for the legislator and those who would like to use this type of contract.

6.2 Problems

The weak point of a smart contract is its connection to the real world [15]. The system must receive data from the outside about the occurrence of certain conditions. So-called Oracle programs are designed to perform this function. However, experts note the lack of development of such programs. Smart contracts exist in their own environment – the environment of the blockchain platform. Accordingly, they always have access to data that also exists within the platform. For example, the fact of making a cryptocurrency transaction. But getting data from the outside is still not so easy. Moreover, this data must not only be received quickly in the system, but also be reliable. Who should verify the accuracy of this data and how?

In the context of conditional agreements, the method of obtaining information can be both a strong and a weak side of the contract. The system can objectively evaluate information that it can get from outside. On the other hand, if this information is not strictly formalized – for example, it is not expressed in specific and accessible figures, but represents a specific action of the party to the agreement, the system will not record the occurrence of the condition. Also, the system cannot assess the unfair behavior of a party that prevents or facilitates the occurrence of a condition.

There is also a problem of state control. The blockchain platform guarantees the anonymity of participants, as well as absolute independence. However, this indicates a lack of state control. Although the entire legal system, the existence of norms of human behavior is an expression of the power of the state. If the blockchain gets out of the control of the state, then there can be no legal regulation.

There is more than one platform in the world where you can make such a deal. Due to their decentralization, there is no "superuser" who would provide the state with information about all contracts concluded. In this case, there is a risk that the parties may have to prove in court the fact of concluding the contract (if it is not duplicated on paper).

Blockchain platforms are decentralized and, for the most part, open. Accordingly, there may be a risk of data leakage (including, given the increased attacks on blockchain platforms). For many large companies, data

https://doi.org/10.1051/shsconf/202110901023

privacy may simply not allow them to use smart contracts and "open" their data to the system. Everyone can track the chain of transactions made, how much cryptocurrency was transferred, etc.

Therefore, despite the fact that digital technologies are already used for making transactions, there are still many legal and technical problems.

If there are such problems, a conditional transaction can be expressed in a smart contract only if the condition is an objective event that the system can get information about. If the condition contains a subjective component, then human participation in the process is necessary, that is, it is impossible to achieve full automation of such transactions today.

7 Conclusion

Technology does not stand still and the legal reality must change at the same pace. So far, one thing can be stated – smart contracts can be widely used as transactions with simple, measurable conditions. At the same time, transaction costs will disappear, and the behavior of the parties will not affect the execution of the conditional agreement. On the other hand, the legislator cannot yet interfere in the sphere of digital contracts without restricting the freedom of this activity. But freedom is the highest value of blockchain technology. Control should be associated with the recognition and legalization of a new reality, since it is already obvious that the digital economy is real. This is a completely new challenge to both the state and the legal system. It is quite possible that in the foreseeable future, lawyers will have to get a basic programmer's education in addition to knowledge of the law in order to fit into the new legal reality.

Over time, conditional agreement can become the very legal structures that can be implemented through smart contracts in everyday reality. The potential of conditional agreements can be revealed through a smart contract, as well as their development and application in the future.

Today, in our opinion, smart contracts can be an organic addition to traditional contracts, including conditional agreements, and simplify turnover by automating certain processes.

References

1. A.I. Savelev, Bulletin of civil law, **3**, *32-60* (2016)
2. O. Ovechkin, Edition Rubase, (2016). URL: https://rb.ru/news/sseven-blockchain (date of access:15.02.2021).
3. A.G. Karapetov, Bulletin of The Supreme Arbitration Court of The Russian Federation, **7**, 28-93 (2009)
4. G.F. Shershenevich, *Textbook of Russian civil law* (Moscow, Russia: Statut, 2005)
5. N.L. Duvernua, *Reading about the civil law. Vol. 2: The doctrine of things. The doctrine of a legal transactio.* (Moscow, Russia: Zertsalo, 2004).
6. E.A Sukhanov, *Civil law: textbook. Vol.1.* (Ed Moscow, Russia: VoltersKluver, 2008).
7. E.V. Ivanova, *Civil law. General part: textbook and workshop for applied bachelor course* (Moscow, Russia: Urait, 2018).
8. I.A. Zenin, *Civil law* (Moscow, Russia: Urait, 2018)
9. V.V. Vasnev, *The nature of the conditional obligation* (Dissertation of Candidate of Law) (Saint-Petersburg, Russia: St Petersburg University, 2013).
10. N. Szabo, A Formal Language for Analyzing Contracts (2002), URL: http://szabo.best.vwh.net/contractlanguage.html(date of access:15.02.2021).
11. A.I. Savelev, Bulletin of civil law, **3**, 32-60 (2016)
12. A. Vashkevich, *Smart contracts: what, why, and how* (Moscow, Russia: Simployer, 2018)
13. Vashkevich, A. *Law.ru - the first social network for lawyers* (2017), URL: https://zakon.ru/blog/2017/12/27/pyat_vyvodov_o_smart-kontraktah (date of access:15.02.2021)
14. Journal Forklog (2017), URL: https://forklog.com/chto-takoe-smart-kontrakt (date of access:15.02.2021).
15. V. Elkina, Edition Rubase (2017), URL: https://rb.ru/story/ethereum-in-business (date of access:15.02.2021).
16. E. Krauzova, Forbes (2016) URL: https://www.forbes.ru/tekhnologii/idei-i-izobreteniya/320325-slozhno-svyazat-kod-na-blokcheine-s-realnym-mirom-intervyu-s (date of access:15.02.2021).

Digital rights under the civil law of the Russian Federation

Armine Mograbyan[1*]

[1] Volgograd Institute of Management - a branch of the Russian Academy of National Economy and Public Administration under the President of the Russian Federation, 400066, st. Gagarin, 8, Volgograd, Russia, mograbyan-as@ranepa.ru

Abstract. The article discusses a new object of civil rights, which appeared as a result of the addition of the Civil Code of the Russian Federation with article 141.1 "Digital rights". The features of the definition of digital rights in Russian civil law are revealed. Particular attention is paid to the formulation of the concept of digital rights contained in the Civil Code of the Russian Federation. The author draws attention to the fact that according to Russian civil law, digital rights include only those that are directly named as such in the law, which indicates a legislative limitation of their turnover. In addition, the adopted laws on crowdfunding and digital financial assets were reviewed, as well as utilitarian digital rights and digital rights that relate to digital financial assets were analyzed. Attention is also drawn to other problems raised in the science of civil law regarding the modernization of Russian civil legislation in the field of digital rights. The author emphasizes the positive nature of the appearance in the Russian law of norms on digital rights, as an indicator of a legislative response to the digitalization of the economy and law, and a necessary prerequisite for further regulation of civil law relations changing under its influence.

1 Introduction

In the era of comprehensive digital transformation and the penetration of digital online platforms into all sectors of society [1], changes have affected not only the economy, but also various sectors of law, including Russian private law. This is evidenced by the latest changes in the civil legislation of the Russian Federation. So, from October 1, 2019, Art. 141.1 of the Civil Code of the Russian Federation (State Duma of the Federal Assembly of the Russian Federation [2], 1994), which is called "Digital Rights". This article was introduced by federal law No. 34-FZ of March 18, 2019 [3] in order to adapt the current civil legislation of Russia to the changing public relations for digitalization, in particular, for the adoption of digital financial assets [4] and on attracting investments through an electronic platform [5]. Supplementing the Civil Code of the Russian Federation with the provision on digital rights of witnesses "on the legislative recognition of digital rights as a kind of civil rights from the point of view of civil legislation" [6]. Corresponding changes were made to Article 128 of the Civil Code of the Russian Federation "Objects of Civil Rights" in accordance with various digital legal means, including non-cash funds and uncertified securities. At the same time, it should be noted that the chosen legislative formulation of the concept of digital rights has caused an emergence of numerous discussions in the science of civil law.

2 Problem Statement

Most researchers rightly point out that the introduction of digital rights into the Civil Code of the Russian Federation as one of the objects of civil rights raises many questions and problems, some of which were noted by researchers at the stage of discussing draft laws [7-9]. In particular, L. Novoselova rightly criticizes the term "digital law" used in the Civil Code of the Russian Federation, since it is used in the meaning of the technical term "token to which information about property rights is tied"[10] without explaining the content of the law itself. An interesting position is S. Sarbash, who, noting the controversial nature of the concept of digital rights in civil science, believes that it is not new for civil law science. In his opinion, "Digital law, in fact, is only a form, that is, the registration of any known law: proprietary, obligatory, corporate, exclusive, personal. Digitalization as such and digitalization of the economy are only a method of registration, but not the content of property relations" [10]. Perhaps it is difficult to disagree with this. M.A. Rozhkova is also in solidarity with the opinion that the concept of digital rights introduced into the Civil Code of the Russian Federation in the existing edition is insolvent: "Digital rights" have obviously become superfluous" [11].

It should also be noted that the concept of "digital rights" used in the Civil Code of the Russian Federation is rather narrow, and, judging by the legal definition, it

* Corresponding author: mograbyan-as@ranepa.ru

 https://doi.org/10.1051/shsconf/202110901024

includes only "obligations and other rights named in such a capacity in the law, the content and conditions of implementation of which are determined in accordance with the rules of an information system that meets the characteristics established by law." [2] In this definition, a number of restrictions on the classification of rights as digital can be distinguished. First, they must be named as such in the statute. Secondly, their content and conditions for implementation must be determined in accordance with the rules of the information system. Thirdly, the named information system must meet the criteria established by law. In addition, the law establishes that the implementation and disposal of digital rights, as well as the limitation of the order, are possible only in the information system, while contacting a third party is not required for this [2]. As a general rule, the holder of a digital right is a person who can dispose of it in accordance with the rules of the information system. The exceptions are cases provided by law and certain grounds when another person is recognized as the owner of a digital right [2]. At the same time, in the case when the basis for the transfer of digital rights is a transaction, then the consent of the person obliged under such a right to transfer is not required [2]. Thus, an analysis of Article 141.1 of the Civil Code of the Russian Federation allows us to speak of a legislative limitation in the turnover capacity [12] of digital rights as objects of civil law.

To date, only one legislative act has come into force, which mentions digital rights - Federal Law of August 2, 2019 No. 259-FZ "On attracting investments using investment platforms and on amending certain legislative acts of the Russian Federation" (hereinafter - Crowdfunding Law). In particular, the digital rights provided for by the law include the "utilitarian digital rights" [5] named in Article 8 of the Crowdfunding Law: demand the transfer of things or exclusive rights to use them, as well as demand the performance of work and (or) the provision of services. At the same time, these rights are recognized as digital only if they originally arose as such on the basis of an agreement for the acquisition of utilitarian digital rights, which was concluded using an investment platform, in accordance with the rules established by law. At the same time, the right to claim property, the rights to which or transactions with which are subject to state registration or notarization, cannot be utilitarian digital rights [5]. Analysis of Article 8 of the Law on Crowdfunding suggests that along with state registration and notarization, another form of certification of property rights has appeared in the law - digital. At the same time, this concept is not consistent with the design of Article 128 of the Civil Code of the Russian Federation, in which digital rights are named as one of the varieties of property rights, along with non-cash funds and uncertified securities. In this regard, the opinion of A. Asoskov is correct, who believes that the legislator has formulated a new, along with uncertified securities, method of certifying property rights [10].

In addition to the Law on Crowdfunding, as part of the implementation of Article 141.1 of the Civil Code of the Russian Federation, another legislative act was adopted - the Law on Digital Financial Assets, which comes into force on January 1, 2020, with the exception of certain provisions for which the law provides for a different period. The said Law provides for one more, in addition to utilitarian, type of digital rights - digital financial assets, which are understood as digital rights, including monetary claims, the possibility of exercising rights under equity securities, the right to participate in the capital of a non-public joint stock company, the right to demand the transfer of equity securities, which are provided for by the decision on the release of digital financial assets in the manner prescribed by law, the release, accounting and circulation of which is possible only by making (changing) entries in the information system based on the distributed register, as well as in other information systems (State Duma, 2020). As you can see, the legislator has distinguished between utilitarian digital rights, which cannot be for monetary claims, from digital financial assets, which cannot be for things, works, services and results of intellectual activity. At the same time, in accordance with the new edition of Article 8 of the Law on Crowdfunding, from January 1, 2021, it will be provided as an opportunity to acquire and dispose of digital financial assets, as well as digital rights, including both utilitarian digital rights and digital financial assets, in the investment platform; and the possibility of executing transactions with utilitarian digital rights through the operator of the exchange of digital financial assets. But the Law on Digital Financial Assets contains restrictions for credit institutions, in particular, the latter cannot perform the functions of operators of investment platforms. Thus, "the legislation of the Russian Federation allows, based on the functioning of utilitarian digital rights and digital financial assets in one complex information system, to organize the circulation of a wide variety of digital assets, including those tied to monetary claims and physical assets" [13].

Another problem in the field of civil legal regulation of digital rights, which has been repeatedly raised by civil lawyers, is seen from the analysis of Article 141.1 of the Civil Code of the Russian Federation and Article 8 of the Law on Crowdfunding and is associated with the restriction in terms of contacting a third party. In particular, we are talking about the possibility of exercising, ordering and restricting the disposal of digital rights only in the information system without contacting a third party. In this regard, absolutely fair questions are posed by S. Sarbash: "If some entity in circulation is ready to provide this or that benefit both by referring to the information system, and without referring to it, then the right, formalized in digital, becomes something else? I have already cited an example with a bank account agreement, where a person has an alternative: either interact with the bank through appropriate programs, or go to a bank branch. Do my rights in the second case stop being digitalized? After all, in the bank, the owner will not require anything except a passport, but the rights, as they were in digital form, remain. Why does the legislator limit our opportunities? Or is that not what you mean?" [10].

Civilists also identify a number of other problematic issues in the area under study. So, for example, an important remark regarding the legal definition of digital rights was made by A. Semenov, who draws attention to the separation of the authority to order from the exercise of digital law, "which suggests a certain doctrinal turn of codifiers to the relationship of these concepts, since earlier the exercise of the right included number and disposal of them." [10] As you can see, the addition of civil legislation with a rule on digital rights has led to the emergence of numerous discussions and questions to be answered by the legislator, including after a collision with the practice of resolving disputes arising in the new digital reality.

3 Research Questions

The article is aimed at solving the following research questions:

- What is the specificity of the definition of digital rights in Russian civil law?
- Has the introduction of norms on digital rights into the Civil Code of the Russian Federation solved the challenges of reality in the field of regulation of private-law relations changing in the context of the digitalization of the economy and society?

4 Purpose of the Study

On the basis of a comprehensive analysis of theoretical provisions and legal regulation, identify the specific features of the legal definition of digital rights in the civil law of Russia, and assess the emergence in the legislation of the new concept of "digital rights" for Russian civil law.

5 Research Methods

Both general scientific and sectoral research methods were used as the main ones, in particular, analysis, synthesis, comparative legal, formal legal, logical, systemic and others.

6 Findings

Given the speed of change in public relations, it is difficult for a legislator to keep up with it, and such a restriction will generate many problems associated with the unsettledness of changing civil law relations. A specific feature of digital rights in Russian civil law is the fact that the law significantly restricts the range of rights that can be classified as digital, in comparison with foreign legal systems [14,15], which, inevitably, gives rise to legally unregulated public relations in the private law sphere, which, in turn, gives rise to corresponding problems in practice. On the other hand, the emergence of the very concept of "digital rights" in the civil legislation of Russia is a positive factor on the way to the settlement of private law relations arising in the new reality.

7 Conclusion

Thus, despite the well-founded criticism of the legal definition of digital rights and its content, the very fact of the appearance in the Civil Code of the Russian Federation of the concept of "digital rights", which means the private law regulation of this concept, deserves positive attention, since it indicates the desire of the legislator, albeit not always ahead of reality and timely, to promptly respond to objective changes in law and economy in the era of global digitalization.

References

1. J. Van Dijck Computer Law & Security Review **36,** 105377 (2020)
2. State Duma of the Federal Assembly of the Russian Federation *Civil Code of the Russian Federation. Part one* (November 30, 1994 No 51-FZ) (as revised July 31, 2020). (Moscow, Russia, 1994)
3. State Duma of the Federal Assembly of the Russian Federation *Federal Law "On Amendments to Parts One, Two and Article 1124 of Part Three of the Civil Code of the Russian Federation"* (March 18, 2019 No. 34-FZ) (Moscow, Russia, 2019a)
4. State Duma of the Federal Assembly of the Russian Federation *Federal Law "On digital financial assets, digital currency and on amendments to certain legislative acts of the Russian Federation"* (July 31, 2020 No. 259-FZ) (Moscow, Russia, 2020)
5. State Duma of the Federal Assembly of the Russian Federation *Federal Law "On attracting investments using investment platforms and on amending certain legislative acts of the Russian Federation"* (August 2, 2019 No. 259-FZ) (as amended on July 20, 2020). Moscow, Russia, 2019b)
6. D.A Petrov, V.F. Popondopulo, E.V. Silina, *Digital law* (Moscow, Prospect, 2020)
7. A. Savelyev, Computer Law & Security Review **34(4),** 863-869 (2018)
8. A.M. Erdelevsky, About digital rights (2019), **URL:**http://www.consultant.ru/cons/cgi/online.cgi?req=doc&base=CJI&n=122169#07570494312440597(date of access: 16.03.21).
9. R.B. Golovkin, O.S. Amosova, Bulletin of the Vladimir Law Institute **2(51),** 163-166 (2019)
10. L. Novoselova, A. Gabov, A. Savelyev, A. Genkin, S. Sarbash, A. Asoskov, A. Semenov, R. Yankovsky, A. Zhuravlev, A. Tolkachev, A. Kamelkova, M. Uspensky, R. Krupenin, V. Kislyi, M. Zhuzhzhalov, V. Popov M. Agranovskaya, Law. **5**, 31-54 (2019)
11. M.A. Rozhkova, Economy and law **10,** 3-12 (2020).
12. L.V. Borisova, Actual Problems of Russian Law **1(74),** 126-131 (2017)
13. A. Lisitsyn, Settlements and operational work in a commercial bank **4**, 4-17 (2020)
14. E.M. Weitzenboeck Computer Law & Security Review **29(3)**, 293-295 (2013, June).

15. M.A. Rozhkova Digital rights - what are they and are they needed in the Civil Code? [Blog post] (2020) URL: https://zakon.ru/blog/2020/08/17/cifrovye_prava_digital_rights__chto_eto_takoe_i_nuzhny_li_oni_v_grazhdanskom_kodekse (date of access: 16.03.21)

The economic approach to designing anticorruption institutions

Viktor Molokanov[1*], *Elena* Malysheva[2], *Ekaterina* Chumakova[3], *Vadim* Yakovenko[4] and *Andrey* Zelensky[5]

[1] Volgograd Institute of management – branch of The Russian Presidential Academy of National Economy and Public Administration, 8 Gagarin street, Volgograd, Russia, vikmol@inbox.ru
[2] Volgograd Institute of management – branch of The Russian Presidential Academy of National Economy and Public Administration, 8 Gagarin street, Volgograd, Russia, decanatec@vlgr.ranepa.ru
[3] Volgograd Institute of management – branch of The Russian Presidential Academy of National Economy and Public Administration, 8 Gagarin street, Volgograd, Russia, chumakova-ea@vlgr.ranepa.ru
[4] Volgograd Institute of management – branch of The Russian Presidential Academy of National Economy and Public Administration, 8 Gagarin street, Volgograd, Russia, yakovenko-vv@ vlgr.ranepa.ru
[5] Volgograd Institute of management – branch of The Russian Presidential Academy of National Economy and Public Administration, 8 Gagarin street, Volgograd, Russia, zelensky-ag@vlgr.ranepa.ru

Abstract. Contemporary economic analysis of corrupt behaviour requires practical consideration of the issues concerning projecting of social anti-corruption institutions. The economic science has accumulated sufficient practical experience of institutional construction in society, for instance approaches provided by K. Sunstein's and E. Ostrom. The requirement of reasonability in managing economic interactions makes the society refuse from the dichotomy of market (anarchical, spontaneous) and non-market (hierarchical) institutions. The same tendency can be observed in arranging anti-corruption struggle in state management. Both approaches - aimed at creating fixed vertical hierarchy as well as at creating competition at the lowest bureaucratic level - are inappropriate. That is why anticorruption institutions arranging should be based on economic models of rational criminal and citizens involvement in law enforcement activity. These models enable to harmonize state dirigisme related to the state management ex ante with basic provision of laissez-faire doctrine – personal motives of people's behaviour. Standard economic supply and demand model has shown inelasticity of corruption crimes supply in Russia. Non-price determinants (cultural norms, "tabu") seem to be more significant in motivating corruption behaviour than price determinants (seriousness and probability of punishment). Economic characteristics of elasticity and inelasticity of supply and demand in the model of involvement into law enforcement activity enable to create strategy of developing social anti-corruption institutions.

1 Introduction

In mainstream economic theory there are two approaches explaining allocation of economic goods. Goods in the market of production factors can be allocated either by means of the market itself or by the state. The economic science offers theories explaining social choice of allocation of world rare goods.

However, this institutional polarization (either market or the state) typical for mainstream science is subject to more and more critics by scientific society. Ideologically approved constructions under pressure of experience accumulated by humanity have been replaced with opportunist approach. If the truth can exist only as "useful delusion" rather than reflection of "objective reality" the person's intellectual activity must be directed at transformation of experience itself (reality) rather than systematic perfection of it in all spheres of human life [1].

The most severe criticism against polarization of market and non-market institutions is mostly directed at blurring of line between anarchic and hierarchical orders.

One of the most significant works in this respect is the book "Economic Institutes of Capitalism" by O Williamson [2], in which the author discloses a convincing theory of effective opportunity management in market economy. The idea that making a deal requires appropriate structure of management enabled to prepare the most significant construction of market economy of anarchic order. The spontaneous order of the market covers only part of all possible deals in the economy. The other part but not the least requires creating long-term sustained structures of managing the deals with established hierarchical relations where the state can take its place as well. Uncertainty of the future makes

* Corresponding author: vikmol@inbox.ru

entrepreneurial activity highly-risky if it is connected with rarely taken opportunities related to specific assets. The practice of overcoming this uncertainty demonstrates high relevance of law as the most significant state institute.

Another significant work postulating opportunism approach to managing economy is the work by E. Ostrom «Governing the Commons. The Evolution of Institutions for Collective Action" [3]. In this paper the author reinterprets one more dogma of mainstream economic theory, so called tragedy of commons. The tragedy of commons reveals significance of property rights in fulfilling the key goal of economic theory - effective usage of rare resources. Any resource without property title is subject to non-productive usage. In case of tragedy of commons the standard recommendation is privatization or nationalization of the resource. On the basis of multiple examples E. Ostrom demonstrated that the system of common property can be effective. Irrigation systems of Spain and California, cattle grazing and timber cutting in the mountains of Switzerland and Japan, fishing in Turkish Alania are the examples of successful cooperation of close communities having common economic activity, possessing rare, scarce resources without private or state property. The book gives description of the rules of constructing institutions enabling effective allocation of scarce resources in conditions of communal, common property, shows the consequences of managing common resources without following agreed rules. Wide implementing of the mechanisms of private and public partnership into Russian practice of municipal management is the reflection of ideas described by E. Ostrom [4]. The problems and perspectives of such reforming of municipal management are touched upon in many articles devoted to this subject [5].

Strong arguments against ideologization of economic science can also be found in R. Thaler and K. Sunstein's works [6]. R. Thaler, a Nobel laureate on Economics in 2017, suggested a curious construction - "libertarian paternalism" [7]. This oxymoron discloses Thaler's opposing the ideologeme that the public paternalism is always an evil. As R. Thaler and K. Sunstein write the liberal economy advocates' prejudices base on "false assumption and two misconceptions" [8]. The false assumption implies that proponents of free market economy argue that every person knows better than others what is the most preferable for him. It is obvious that the right choice is possible only in case of having full information, possessing appropriate competences and having enough experience. The belief that every person knows everything in every sphere of human existence characterizes quite naïve researcher's position.

The first misconception implies that there can be unconditioned choice. The theoretics of behavioural economics call conditioned choice the framing effect. And if it is impossible to isolate human choice from the context why not use conscious constructing of social institutes to make the human choice more conditioned, based on knowledge and experience.

The second misconception implies assurance in forced character of any paternalism. It's the reason why Thaler's paternalism is called libertarian as it doesn't imply any bans or enforcement. The state can and must construct institutes orienting the human to make "the right choice" although he can make "the wrong choice" as well.

The above-represented review of worldviews against extra ideologization of approaches to managing economy doesn't support both extreme statist as well as anti-statist views. Opportunism as the principle of achieving effectiveness in organization of social economy enables to release from ideology the process of developing every day instruments for reforming modern society.

As it was already highlighted there is not only one way of allocating resources in the society. Resources can be allocated by means of the market system or political system. Moreover, allocation mechanism of the political system is not subordinate or optional in relation to the market allocation. Over half of the national income practically in all developed countries including the USA is allocated through the public structures. That is why the effectiveness of the political market is in focus of researchers-economists. One of the main problems determining efficiency of the state impact on the economy is corruption.

2 Problem Statement

The economist's standard approach to the problems of corruption is the analysis of the phenomenon according to the algorithm described in agency theories. The logics of this algorithm is simple. The Consumer (the Principal) concede some of its functions to the Agent. The Agent is understood as rational subject isolated from society and making decisions if it is necessary to leap to opportunistic behaviour or not, basing only on balance between benefits and losses. The Agent leaps to opportunistic behaviour only in case when his benefit from it more significant than punishment inncured [9]. Legislative power is the Agent of voters. The Supreme executive power (the Government) is the Agent of a legislative body. Ordinary officials can be agents of the ministers and heads of governmental agencies. Corrupt actions of public authorities (such as Parliament and Government) are naturally limited by periodicity of voting cycle. It enables to limit time and size of corruption. The institute of political reputation and derived from it requirement of publicity of the authority effectively prevent corruption at top authority levels. Lower officials' corruption requires another approach to analysis.

If scientists-criminologists try to explain criminality by means of unique psychological and physiological traits of criminals economists consider these properties as given and they believe that criminals as any other economic agents react to incentives and make rational choice of the best available alternative [10]. Thus, economists-theorists researching criminality by means of economic approach base their research on "the common assumption of individual rational behaviour" according to which the criminal action is committed only when the general benefit from it taking into consideration sanctions and other costs higher than from legal alternatives. Criminal's behaviour in economic theory is mostly considered as ordinary human behaviour rather than deviant behaviour with uncommon motivation while criminality is considered as a common trade similar to doctor's or teacher's profession [11].

3 Research Questions

During the research it was supposed to get answers to the following questions:

1. Is it possible to combine public enforcement and motivation of citizens' behaviour while working out anti-corruption policy?
2. What economic models can be used in analysis of corruption behaviour?
3. How does the standard model of demand and supply disclose peculiarities of corruption services market?

4 Purpose of the Study

The lack of attention on the part of Russian legal science to the opportunities of economic theory determined the goal of the research which implies development of theoretic approach enabling to combine public enforcement institutes with personal motivation of citizens to analysis of corruption behaviour. In order to reach this goal it necessary to fulfil the following objectives:

1. to reveal inefficiency of taking only administrative anti-corruption measures;
2. to present economic models enabling to consider both citizens' inner motivation as well as administrative factor;
3. to demonstrate how demand and supply model discloses the character of corrupt services market.

5 Research Methods

Within the present research classic general scientific methods of research and processing information were used such as: comparison and analogy, analysis and synthesis, induction and deduction, geographic method, comparative statics.

6 Findings

The analysis of low level corruption requires formalized model of criminal behaviour. Institutional economic theory has got the model of rational criminal [12]. Let's suppose that x is the number of definite crimes committed, i/e/ the number of times when the criminal takes possession of a definite sum of money . Then y is the criminal's compensation for committed crimes. The function of criminals compensation will increase with increasing x, i.e. with increasing number of crimes $y=y(x)$. Let's introduce the function of criminal's punishment f. The function of punishment will increase with increasing number of committed crimes as well $f=f(x)$. Punishment for crimes is of probabilistic nature, that is why we will introduce p-parameter in the model as probability of punishment for committed crimes. The function of punishment probability also increases in case of increasing x ($p=p(x)$). The rational criminal choses the optimal number of definite crimes being conducted by the classic principle of maximizing net advantage $\partial(y(x)-p(x)f(x))/\partial x=0$.

If x is the number of definite crimes committed by one individual then $X=\sum x$ is the total number of crimes committed in a society. According to the model of rational criminal the number of crimes in the society is inversely related with increasing probability and severity of punishment. Now we can draw the supply curve of definite crimes in a definite society (see figure 1).

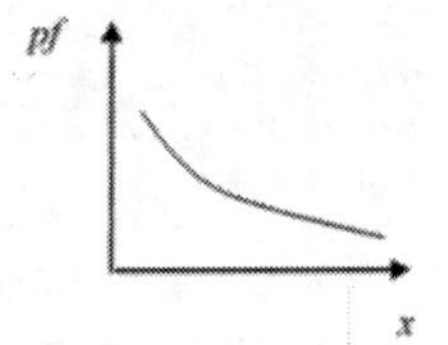

Fig. 1. Crime supply.

This curve enables to formulate the first law of sustaining: the more the probability to be caught or the severer the punishment the fewer the number of committed crimes. The slope of crime supply curve enables to come down to the quality conclusion about the character of criminality including corrupt activity. The more elastic is the crime supply curve the stronger its connection with the price of these crimes the brighter is the dependance described in the first law of sustaining. And vice versa, the less elastic is the crime supply the weaker manifestation has the above-mentioned law (see figure 2).

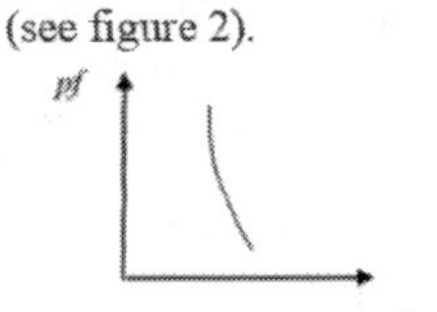

Fig. 2. Non-elastic crime supply.

Anti-corruption fighting in public management in RF has been significantly strengthened for the last years. There are institutes that were created to prevent negative tendences of corruption character that certainly led to the increasing probability of being caught by law enforcement agencies, increasing penalties and sentences for corruption crimes. However, there cannot be observed qualitative shift in decreasing the number of committed corruption crimes. One of the possible explanations of this situation can be non-elasticity of corruption crimes demand in Russia. And if it is so then the factors described in the model of rational criminal (probability and severity of punishment) don't play significant part in motivating corrupt officials. There occur the phenomenon that economist call non-price factors of supply and in this case public moral principles, cultural norms, "tabu", the quality of education are the factors determining the character of criminal activity of

Russian corrupt officials. The embeddedness and wide spread of corruption make us to assume that they cannot develop without creating definite institutional environment, asserts famous scientist and popular former official of the federal level G.A. Satarov [13].

There can be distinguished three models of making decisions by bureaucrats: fragmentarily sequential hierarchical and disorganized [14]. In fragmentarily sequential model decisions are made at several autonomous stages. Applicants request is divided into parts and every part undergoes its special procedure of confirmation. In hierarchical model (traditional bureaucracy) decisions made at the lower level can be reconsidered at any moment at the higher level of management. These models differ by incentives of corrupt behaviour of lower level official. Honest official at the higher level of the bureaucratic ladder will stimulate corruption at lower levels in fragmentarily sequential model and decrease corruption in hierarchical model. And vice versa dishonest official at the higher level of the bureaucratic ladder stimulates corruption in hierarchical models and discourages corruption at lower levels of in fragmentarily sequential model. Disorganized model is typical for bureaucratic structures, where criteria of decision making are spontaneous and the structure of management is subject to constant changes.

In fighting against corruption it is commonly recommended to change the structure of decision making in order to decrease authority of the level involved in corruption activity. And thus as routine corruption is mass at the lowest level than the most populist decision will be shifting authority to the officials of the higher level of the hierarchy. The lowest level officials just accept the requests of the applicants while the final decisions are made by the superior. However, centralization doesn't lead to decreasing of corruption. The corruption simply shifts from lower to upper levels. If the lower level official does not make decision and transfers requests to the upper level without processing the volume of superior's work can significantly increase. It will influence the quality and first of all the quantity of decisions made and in this case the bribe becomes the institute that makes the superior's work easier.

Rejecting of centralization of management stimulates corruption at lower levels. The standard reaction of the economist fighting against routine corruption is creating competition among officials. However, advantages from bureaucratic competition often do not compensate advantages from specialization and monopolization of managerial functions. Consequently, reforms aimed at shifting officials' authority to different levels of management appear to be counterproductive. Approaches related to centralization and decentralization of management appeared to be senseless in the sphere of anti-corruption fighting[15].

If corruption has been deeply established in Russian culture then anti-corruption fighting should be concentrated on breaking norms approving corruption activity. This can be realized only by means of wide involvement of the population of the country in anti-corruption activity. The economic science can provide its theory of population involvement to fight against criminal activity.

Undoubtedly for an ordinary person any help rendered to law enforcement agencies causes costs: spending private time, taking efforts, waiving personal comfort and safety. Let's consider the model with the vertical axe OY showing costs related to providing help to law enforcement agencies and horizontal axe OX showing the share of population ready to incur such costs. There is no need to prove inverse relation between these parameters. (see Figure 3)

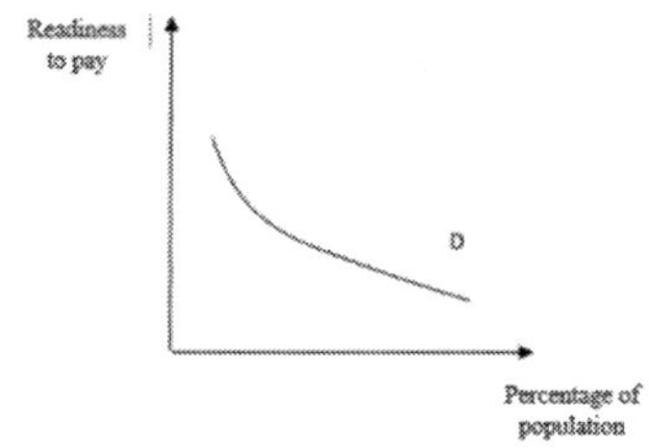

Fig. 3. Readiness to pay for participation in law enforcement.

The cost value of every person's involvement in fighting against criminality decreases due to psychological effect of "comfort behaviour" with growth of the number of people providing help to law enforcement agencies. (see Figure 4). The growth of popularity of a definite action among citizens decreases involvement costs.

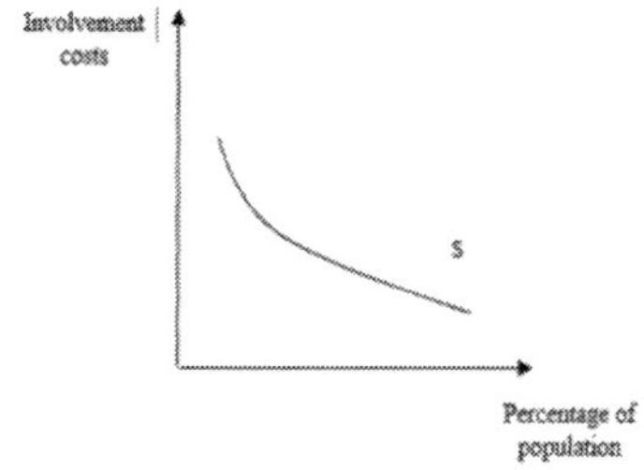

Fig. 4. Costs of public involvement in law enforcement activity.

The share of citizens that will provide help to law enforcement agencies in anti-corruption fighting in this model is determined by intersecting of the demand curve that implies «readiness to pay» (Figure 3) and supply curve that implies «readiness to incur costs» (Figure4). As the result we get the standard set of instruments used by economists - the demand and supply model in the market of citizens involvement in law enforcement activity (see Figure 5)

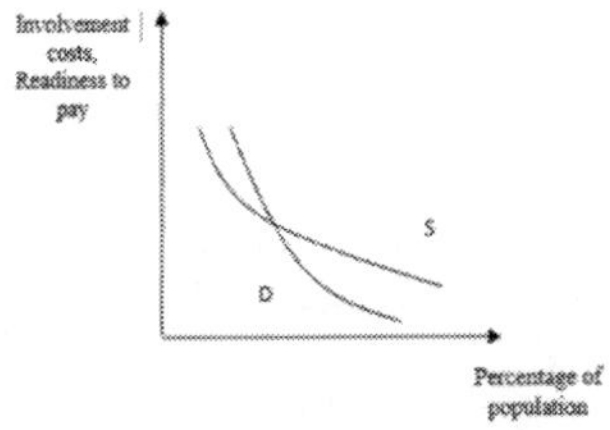

Fig. 5. Equilibrium in the market of involvement in law enforcement activity.

At that there can be two states of the market : a) supply curve S is more elastic than demand curve D (see Figure 5); b) demand curve D is more elastic than supply curve S (see Figure 6).

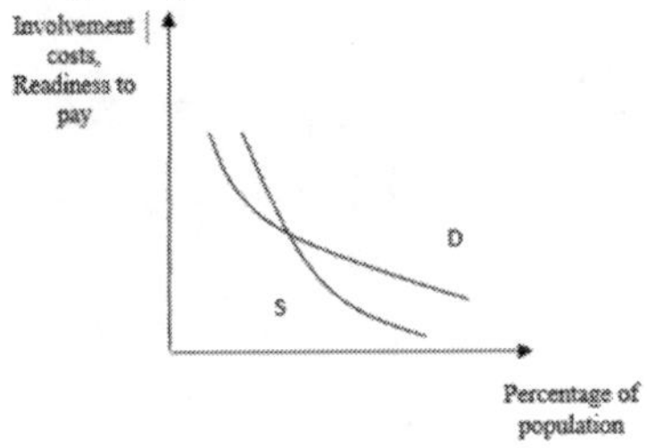

Fig. 6. Equilibrium in economy of scale.

Figure 5 illustrates the opportunity of the only equilibrium in full conformity with the theory of demand and supply. If in the market demand is higher than supply it leads to deficiency People are ready to pay for participation in law enforcement activity more than it is required by social costs of involvement , that leads to the growth of the number of participants in this market by the level of the intersection point. If in the market the supply exceeds demand then there is a surplus of involvement in law enforcement activity and the share of population involved in this market will decrease by the level of intersection point because social costs of involvement in law enforcement activity are higher than the human desire to incur such costs.

From the perspective of developing legal norms the situation described in Figure 6 seems to be more interesting. Here the surplus is created to the left from the point of intersection and the deficiency is formed in the right part. The surplus stimulates decreasing participation in law enforcement activity and consequently the equilibrium will tend to reach zero share of involvement while the deficiency stimulates the growth of participation in law enforcement activity and consequently the equilibrium will tend to reach 100 % of involvement. In Figure 6 there are two equilibriums and both of them appear at the intersection of demand and supply curves. Such situation occurs when social approvement of participation in law enforcement activity sharply decreases involvement costs. This dynamics of costs is interpreted by the economist as the economy of scale.

7 Conclusion

The analysis of the market of public involvement in law enforcement activity by means of supply and demand curves determines the algorithm of fighting against corruption directed at breaking cultural norms approving corrupt behaviour. The strategy aimed at increasing the share of the population participating in law enforcement requires creating deficiency states both to the left and to the right of the intersection of demand and supply curves. It means that at first it is necessary to provide conditions described in the left part of the figure 5 and then reflecting the situation in the right part of the figure 6.

Realizing managerial decisions in modern conditions it is necessary to avoid polarization of approaches typical of theoretical opposition of market and public institutes. Reaching economic effectiveness including one in public reforming requires using more opportunistic methods combining both hierarchical as well as anarchic structures of management. Fighting against corruption by means of shift of power at different levels of managerial hierarchy is non-productive in theory as well as in practice. It is impossible to create competition in the market of bureaucratic services due to significant benefits from specialization and monopolization of managerial functions. Character of Russian corruption enables to come down to the conclusion about its non-price nature. In case the probability and seriousness of the punishment for corruption crimes are not determining it is necessary to focus on reviewing accepted in society cultural norms approving corruption behaviour. It requires in its turn wide public involvement in law enforcement activity. Supply and demand model in the market of public involvement in law enforcement activity enables to develop growth strategy of the population involved in fighting against corruption

References

1 D.V. Semenov, Contemporary Scientist **4,** 234-245 (2017)
2 O. Williamson, *The Economic institutions of capitalism. Firms, Markets, Relational Contracting.* (Saint Petersburg, Russia: Lenizdat; CEV Press, 1996)
3 E. Ostrom, *Governing the commons: the evolution of institutions for collective action* (Moscow, Russia: IRISEN, Thought, 2010)
4 S. I., Shishkin, & M. A. Lavrik, Siberian Legal Bulletin **3**, 30-36 (2018)
5 U.A. Aparina, & D.A. Obozny, Transport Industry in Russia **2**, 82-87 (2019)
6 R. Thaler, & C. Sunstein, *Nudge: Improving Decisions about Health, Wealth, and Happiness.* (Moscow, Russia: Mann & Ferber, 2017)
7 R.I. Kapelyushnikov, *Aroung Behavioural Economics: a Few Comments about Rationality and Irrationality* (Moscow, Russia: Publishing House of Higher School of Economics, 2018).
8 R.I. Kapelyushnikov, *Behavioural Economics and New Patternalism.* (Moscow, Russia: Publishing House of Higher School of Economics, 2013)

9 Yu. I. Budovec, & K.N. Lebedev, *Basics of the Theory of Agents* (Krasnodar, Russia: Publishing House Scientific and Research Institute of the Economy of the South Federal Region, 2019)
10 M.I. Levin, & E.V. Pokatovich, Economics and Mathematical Methods **48(1),** 44-55 (2012)
11 A. Bystryakov, & N. Epifanova, State Service **6,** 62-65 (2016)
12 P. Cooter, & T. Ulen, *Law & Economics* (Moscow, Russia: Publishing House «Business» RANEPA, 2018)
13 G.A. Satarov, *Russian Corruption: Level, Structure, Dynamics. Tests of Sociological Analysis.* (Moscow, Russia: Foundation «Liberal Mission», 2013)
14 M.I. Levin, E.A. Levina, & E.V. Pokatovich, *Lections on Corruption Economics* (Moscow, Russia: Publishing House of Higher School of Economics, 2011)
15 M.I. Levin, & G.A. Satarov, Issues of Economics, **10**, 4-29 (2012)

Digitalization of the decision-making process in criminal proceedings

Nadezhda Muratova[1*], *Natalya* Solovyeva[2], *Vladimir* Shinkaruk[3], *Victor* Rudkovskiy[4], and *Pavel* Fantrov[5]

[1] Kremlevskaya Street, 18, Kazan, Russia, muratowan@mail.ru
[2] Universitskiy Prospekt, 100, Volgograd, Russia, solovieva_na@volsu.ru
[3] Universitskiy Prospekt, 100, Volgograd, Russia, shinkaruk@volsu.ru
[4] Universitskiy Prospekt, 100, Volgograd, Russia, rudkovskiy@volsu.ru
[5] Universitskiy Prospekt, 100, Volgograd, Russia, pavelfantrov@volsu.ru

Abstract. The law enforcement behavior of the subject of decision-making in criminal proceedings has developed a certain algorithm in the system of criminal procedure law. Many probabilistic bases existing in the Russian legislation for making procedural decisions, rating expectations of the variable outcome of a procedural decision in a criminal case - all this provides for the need to introduce mathematical methods for researching the actual circumstances of a criminal case, their classification structure and logical justification. The purpose of the research is to identify the features of digitalization of the decision-making process in criminal proceedings. The research used a system of scientific methods, which is based on the principle of consistency, an institutional approach and sociological methods, which made it possible to evaluate the institutional basis for the introduction of mathematical and information and communication methods in the decision-making process in criminal proceedings. In the context of this problem, the authors pay special attention to the blockchain technology, this is a continuous sequential chain of blocks built according to certain rules, containing information, access to which excludes data theft, fraud, violation of property rights, and invasion of the system of making criminal procedural decisions. In the course of the research, they show the main institutional directions in the process of algorithmization of criminal procedure decisions. The article concludes that by applying a universal mathematical language in describing legal phenomena, quantitative values and statistical indicators, the subject of law is able to calculate the optimal correct decision in specific conditions in a criminal case.

1 Introduction

The modern system of criminal procedure acts is based on the conceptual foundations of the domestic adversarial criminal procedure law. The study of the system of law is multidimensional in its essence and content. The genesis of the idea of an optimal, fair and correct decision in the conditions of possible coercion dates back to ancient times. Novgorod birch bark manuscript No. 531 (XI c.) describes the trial, and the manuscript becomes the first official decision in the judicial process of Ancient Russia. Models of procedural decisions were built logically in accordance with the socio-political events in Russia and the stages of development of domestic legislation. First of all, the legal field was filled with ancient judicial decisions ("ordalia", "general search", "written certificates", "princely letters").

The differentiation of decisions in criminal proceedings under adversarial conditions acquired symbolic significance in accordance with the Statute of Criminal Proceedings of 1864, according to which there were proceedings in cases of crimes and misdemeanors: private complaints, decisions of the judicial investigator, the prosecutor's indictment, the sentences of magistrates.

The theoretical and legal basis for the application of mathematical and information and communication methods in the research of the system of criminal procedure decisions is based on the general scientific term "system". The explanatory Dictionary of the Russian language by V. I. Dal (2002) does not contain the concept of this term. The encyclopedic dictionary "World History of State and Law", edited by A.V. Krutskikh (2001), also does not contain this term. In the encyclopedic dictionary "Political Science" under the general editorship of Yu. I. Averyanov (1993), there is no commentary on this term. In the encyclopedic dictionary of Brockhaus and Efron, "system" (philosophical) is the combination of homogeneous knowledge into one whole, with a common idea and with the purpose of cognizing phenomena or the entire universe.

Sociology offers two concepts of "social system" and "sociotechnical system": the first term is a certain holistic education, the second term is a system education with the inclusion of a technical and technological

* Corresponding author: muratowan@mail.ru

subsystem and a system of roles and functions of service and management personnel [1]. The Legal Encyclopedia provides the concepts of "system of legislation", "system of law", "certification system". It is noted that, first of all, it is a set of phenomena, but, for example, the system of legislation, it is not just a set, but also a differentiated system [2]. The definition of the system of law proposed by R. L. Khachaturov is interesting, as objectively determined by the internal structure of law, expressed in the objectivity and arrangement of normative material in a certain sequence [3].

The final face of any "system" is entrusted to the philosophical and cognitive meaning of this term, which will make it possible to actualize the problem of forming the Concept of using mathematical methods and algorithms in the process of making criminal procedural decisions. The philosophy proceeds that the "system" (greek. Systema is made up of parts, connected - it is a set of interrelated elements that form a whole, unity. The semantic field of the "system" is also indicated: the terms "connection", "element", "whole", "unity", "structure".

The Code of Criminal Procedure of the Russian Federation in Article 5 defined a number of concepts – "procedural decision" (paragraph 55 of Article 5), "court decision" (paragraph 53.1 of Article 5),"final court decision" (paragraph 53.2 of Article 5),"interim court decision" (paragraph 53.3 of Article 5). The last three concepts are included in the Code of Criminal Procedure of the Russian Federation Federal Law No. 443 of 29.12.2010.

The quantitative indicators of the subjects of law authorized to make these decisions are obvious: there are 34000 judges in Russia, 22000 investigators of the Investigative Committee, and 65000 lawyers [4]. The law enforcement behavior of the subject of decision-making in criminal proceedings has developed a certain algorithm in the system of criminal procedure law. The court, as the third power in the state, takes verdicts in the conditions of an adversarial process, "cutting the knot of the right dispute" and, making optimal and correct decisions, which are announced to the public by posting on the information and telecommunications network of the Internet, claim to rating expectations of the variable result of the procedural decision in a criminal case. Here we can recall the traditional judicial statistics – "stability of sentences (court decisions)". Only digital indicators reveal a qualitative indicator of the work of a judge: if a lot of his decisions are canceled and changed, then it is bad, if they are not canceled, then it is good.

If mathematical science turns to abstract categories, it is difficult not to feel a certain temptation to perceive the actual circumstances of the case as a certain "image", "model", "truth". However, the Code of Criminal Procedure of the Russian Federation, when formulating the basis for making variable decisions, defines a certain limit – a concept – a pre-assumed and definite meaning in making a decision. Thus, a number of presumptive bases for making a decision strengthen the arguments of opponents that mathematical abstraction, which is the result of mental activity, is not applicable when applying decisions based on presumptive data: "data giving bases to suspect" (Part 2 of Article 91 of the Code of Criminal Procedure of the Russian Federation), "the presence of sufficient data to believe" (Part 1 of Article 182 of the Code of Criminal Procedure of the Russian Federation), "if there is sufficient evidence giving base for an accusation" (Part 1 of Article 171 of the Code of Criminal Procedure of the Russian Federation), "sufficient evidence giving basis for an accusation" (Part 1 of Article 223.1 of the Code of Criminal Procedure of the Russian Federation), "arise a doubt" (paragraph 3, Part 1, Article 196 of the Criminal Procedure Code of the Russian Federation).

D. N. Gorshunov believes that, unlike mathematical abstraction, legal abstraction "does not occur through the sequential decision of the problem, the derivation of the formula, but through the detection of signs and the degree of sufficiency" [5].

Thus, the formation of the Concept of applying mathematical and information and communication methods and algorithms in the process of making criminal procedural decisions involves: firstly, the formulation of the concept of a procedural decision; secondly, the definition of a system of probabilistic grounds for making procedural decisions; thirdly, the establishment of rating criteria for the optimal procedural decision; fourthly, the definition of an algorithm for the variable result of a procedural decision in a criminal case; fifthly, the study of the monitoring effect of judicial errors.

2 Problem Statement

The problems of the research are consonant with a number of new terms in law: mathematical methods, artificial intelligence, electronic evidence, cloud storage of information, computer technologies. The array of scientific research extends to many scientific branches: information technologies, information security [5], sociology, legal theories, criminal law, criminal procedure, criminalistics.

Taking into account the purposes of the stated topic, we focus on mathematical methods, electronic and information technologies for making criminal procedural decisions, as well as on the prospect of creating artificial intelligence (robot judge, robot lawyer, robot secretary, robot assistant).

3 Research Questions

1. To characterize the problems of applying mathematical methods in law.
2. To identify gaps in Russian legislation and identify ways to optimize the institutional principles of procedural decision-making and the use of mathematical methods and information technologies in criminal procedure.
3. To determine the main institutional directions in the process of algorithmization of criminal procedure decisions.

4 Purpose of the Study

 https://doi.org/10.1051/shsconf/202110901026

Identification of the features of the introduction of mathematical methods in the decision-making process in criminal proceedings.

5 Research Methods

The methodological basis of the research is a system of principles and methods of scientific cognition, which were used to evaluate the institutional basis for the introduction of mathematical methods in the decision-making process in criminal proceedings. The following scientific methods were used in the research process:

- general scientific methods (rise from the abstract to the concrete, analysis, synthesis, induction, deduction, modeling, comparison, etc.);
- system method – in a comprehensive research of the process of algorithmization of criminal procedure decision-making.
- institutional method, necessary to describe the key processes of digital transformation (strategic data management) in criminal proceedings.
- sociological methods that allow evaluating the effectiveness of the implementation of mathematical methods in the decision-making process in criminal proceedings on the basis of quantitative and qualitative data.

6 Findings

Currently, the institutional bases for the introduction of mathematical methods in criminal proceedings is being investigated in the following areas [7]. Firstly, it is necessary to use mathematical (formulas, tables, numerical sequences, geometric images), simulation and heuristic models in the investigation of corruption-related official crimes. Secondly, as a general education discipline, mathematics makes it possible to process digital data, use mathematical logic, combinatorics, mathematical statistics and methods of probability theory.

In the study of social phenomena and processes in the legal sciences, it should be noted the effective use of: probability theory, mathematical statistics, mathematical logic, information theory, operations research, and others [8]. In relation to the research topic, we can, for example, talk about the application of probability theory in rating expectations of the variability of the outcome of a court decision. Thus, in Federal Law No. 400-FZ of 02.12.2019 "On Amendments to the Federal Law "On Legal Practice and the Bar in the Russian Federation", paragraph 8 regulates the rule according to which the agreement on the provision of legal assistance may include a condition on the amount of remuneration paid by the trustee, depending on the result of the provision of legal assistance by the lawyer (with the exception of legal assistance in a criminal case and an administrative offense).

When applying the institute of rehabilitation, when claiming compensation for property damage (Part 2 of Article 135 of the Code of Criminal Procedure of the Russian Federation), mathematical calculations are required for wages, pensions, lost profits and other funds (Paragraph 1 of Part 1 of Article 135 of the Code of Criminal Procedure of the Russian Federation); property confiscated or converted to state income by a court decision (paragraph 2 of Part 1 of Article 135 of the Code of Criminal Procedure of the Russian Federation); fines and procedural costs (Paragraph 3 of Part 1 of Article 135 of the Code of Criminal Procedure of the Russian Federation); amounts paid to rehabilitated persons for providing legal assistance (Part 4 of Article 135 of the Code of Criminal Procedure of the Russian Federation). And if there is a right to partial rehabilitation (termination of criminal prosecution in part or acquittal under a number of articles of the charge), as stated by the legal position of the Constitutional Court of the Russian Federation (ruling of 18.07.2006 No. 279-O) and the Supreme Court of the Russian Federation (Resolution of the Plenum of 29.11.2011 No. 17), then it is necessary to submit to the court an opinion on the amounts calculated by the rehabilitated, taking into account the principles of reasonableness, justice and humanity. Thus, in the literature, in particular, A. M. Erdelevsky offers a table and formulas for calculating the amounts of compensation for moral damage: the death of the victim is 216-576 minimum wage or 720 minimum wage*0.3-0.8; bringing an innocent person to criminal responsibility is 360-720 minimum wage or 720 minimum wage*0.5-1, etc. The authors also propose an optimal formula that can be used to calculate the amount of compensation for damage to the victim. For example, in the dissertation research of R. G. Khasanshina, the application of civil law norms in proving the amounts of compensation to the victim when making procedural decisions is justified. Therefore, she developed a "formula for calculating the amount of lost earnings in compensation for damage to life and health and a formula for determining the amount of compensation for moral damage to the victim".

It should be agreed that when studying mathematical methods in legal practice, statistics, information processing, and the ability to identify a reliable forecast based on available statistical material plays an important role [9]. Criminological researchs also do not deny the use of mathematics in law, where the process of formalization of knowledge is possible [10]. You should pay attention to the judicial statistics. Thus, analyzing these data, it should be noted that the total amount of damage from crimes under judicial acts issued in 2014 is 29 billion rubles and this amount is growing exponentially – there are 43 billion rubles only for the first half of 2015. The amount of the compensated damage is 9.4 billion rubles (2014), against 4 billion rubles (2015). The greatest damage from crimes was caused to the personal property of citizens – there are 10.9 billion rubles or 37.6% in the structure of direct damage from crimes determined by judicial acts; state property is 9.3 billion rubles or 32.1 %; private property of legal entities is more than 5 billion rubles or more than 17.3 % of the total amount of damage. The statistics do not take into account suspended and unsolved criminal cases. Thus, we can conclude that there are low rates of compensation for damage to victims [11].

Thus, the bibliographic data on the research problem indicate a high interest of scientists in the application of mathematical methods in jurisprudence. Lawyers in the context of optimizing the system of procedural decisions in criminal proceedings will inevitably accompany the process of making criminal procedural decisions with innovative technologies, electronic resources, mathematical classification basis and programming.

First of all, it is necessary to find out the gaps in the legislation and identify ways to optimize the institutional principles of procedural decision-making and the use of mathematical methods and information technologies in this process.

After the researcher's affirmative answer about the use of mathematical methods in making procedural decisions in criminal cases, the problem of creating an algorithm for making them both individually and as a whole system arises [12].

The formation of the Concept of applying mathematical methods and algorithms in the process of making criminal procedural decisions involves: firstly, the formulation of the concept of a procedural decision; secondly, the definition of a system of probabilistic bases for making procedural decisions; thirdly, the establishment of rating criteria for the optimal procedural decision; fourthly, the definition of an algorithm for the variable result of a procedural decision in a criminal case; fifthly, the study of the monitoring effect of judicial errors; sixthly, anticipating mistakes in making procedural decisions.

In the process of implementing modern criminal procedure norms, law enforcement officers enter into extensive legal relations [14]. The criminal procedure is permeated with a system of procedural decisions - this is the relationship of the participants in the criminal procedure: investigator-prosecutor, investigator-lawyer, prosecutor-lawyer, investigator-victim, court-parties. Most criminal procedural relations have traditionally been typified and regulated for years by procedural procedures and a system of procedural decisions – stages are a vivid example of this.

Here we can formulate a rhetorical question: "Is such cumbersome, multivariate, dispositive principles of criminal justice justified in modern, intersectoral jurisprudence?". What is the difference between the sheets of paper formed into volumes in a criminal case in the XXI century and the same sheets of paper in a criminal case in the XIX-XX centuries? The answer is clear - there is practically nothing. And it's sad. Yes, there are computers, printers, video-film-photo shooting, video-conferencing, that is, modern technologies for creating the text of a procedural document - the result of criminal procedural activity. In this connection, there are prerequisites for the formulation of a legislative initiative on the use of modern digital technologies in making criminal procedural decisions.

P.A. Lupinskaya seems to have laid the institutional bases for the introduction of mathematical methods and algorithmization in the decision-making process, defining, firstly, the stages and phases of law enforcement decision-making, secondly, the legal models of decision-making, and thirdly, the criteria for choosing a decision and the form of the decision [14]. In another research, P. A. In another research, P. A. Lupinskaya set her task to "consideration the general model of decision-making in criminal proceedings", based on the fact that "the expected result is achieved not only by making the right decision, but also by its appropriate and timely execution" [15]. This, in our opinion, is the purest mathematical formula - to set a problem – to solve a problem – to get a result (answer).

Thus, it is possible to study the decision-making process "from the inside", without researching the procedural procedures according to this formula, only with a certain degree of probability, abstraction or simplification.

The influence of the general theoretical institutional foundations of law, certainly, should oblige the researcher to take individual institutions of criminal procedure and show its system in numerous procedural decisions [16]. The pace in the research is set by the principles of the criminal procedure. Thus, Part 4 of Article 7 of the Criminal Procedure Code of the Russian Federation states that court rulings and decisions of competent officials must be lawful, justified and motivated. None of the lawyers doubt this formula. Yes, this is an algorithm for the internal content of the decision. But any mathematician, programmer, specialist in the field of IT-technologies will ask – and what is it? where are the criteria? - discretion, dispositivity, the subject of law with his life experience. And is it possible for the "artificial intelligence" to ask a number of questions to the subject of the application of the law himself, so that he answers? And at the end of the technology, he would get an answer-clear and specific-about making the only correct decision from multi-variant situations combined, for example, by the stage. And then-the law enforcement officer, according to their knowledge of the law, will determine the possibility of making this decision. Our heuristic model can be implemented only in the conditions of joint efforts of many professions – "heuristic justice".

For example, the model of the validity of decisions in the criminal procedure of Ukraine is given by V. S. Zelenetsky and N. V. Glinskaya. They believe that the means used to confirm the adequacy of the validity of the decision are divided into four groups – factual, normative, logical and linguistic [17]. Indeed, all the elements as a means of ensuring validity are in a complex interaction and are combined into a single whole. Thus, the institution of criminal proceedings, which has now expanded its term to 30 days, is a traditional, historically established stage of the domestic criminal procedure. It would seem that there are only three decisions-in accordance with the list of decisions taken based on the results of consideration of a report on a crime (Article 145 of the Criminal Procedure Code of the Russian Federation).

Nevertheless, at this stage, numerous procedural decisions are made by the body of inquiry, the investigator, the head of the investigative body, the interrogator and the investigator, the prosecutor, the court, and other participants in the process when appealing against the actions of the competent authorities of the state. The algorithm for making procedural decisions applies: firstly, to the entire system of decisions of this classification group; secondly, it provides a decision of the procedural problem under the conditions of precisely specified procedures by law; thirdly, it provides a sequence of actions that are aimed at achieving the final (optimal) decision in effective ways. For example, digital data when making a criminal procedure decision (terms, dates, quantity, options) have legal significance and ensure the legitimate interests of participants in the criminal procedure. Thus, Federal Law No. 411-FZ of 12.11.2018 made significant changes to the institution of termination of a criminal case. Cancellation of a decision on the termination of a criminal case or criminal prosecution after one year from the date of its issuance is allowed on the basis of a court decision (Part 1.1. Article 214 of the Code of Criminal Procedure of the Russian Federation). The procedural algorithm of this rule is established by a new article - 214.1 of the Code of Criminal Procedure of the Russian Federation. "Judicial procedure for obtaining permission to cancel a decision on the termination of a criminal case or prosecution".

In December 2018, the Council of Europe's European Commission on the Effectiveness of Justice (CEPEJ) adopted the GEPEL European Ethical Charter on the Use of Artificial Intelligence (AI) in Judicial Systems and their Environment, which sets out five basic principles that apply to the processing of court decisions and technical data based on artificial intelligence methods.

The European practice of using artificial intelligence has shown that everything is not so simple and accessible. So, the Court of the Netherlands limited the Government in the use of artificial intelligence. This applies to the use of an automated social security fraud system when a version of the risk modeling algorithm violates the right to privacy. The court ordered to stop using the system-SyRI, developed in 2014, the essence of the technology of which is to model the probability of fraud on the part of persons claiming or receiving benefits. Christian van Veen, who is the director of "The Digital Welfare State and Human Rights" project at New York University School of Law, explained that "SyRI is not a unique system: many other states are experimenting with automated decision-making systems".

In November 2017, the Government's Project Activities Department addressed D. A. In November 2017, the Government's Project Activities Department addressed D. A. Medvedev with proposals to digitalize rule-making and law enforcement: to create "electronic codes", to generate standard court decisions using artificial intelligence, and to create an automated system for monitoring judicial practice. Documents will become easier to write, they will become predictable, and the overall legal technique will improve. An automated legal decisions support system will be able to check decisions for errors and corruption [18].

The Supreme Court of the Russian Federation at the forum "Improving the effectiveness of judicial protection and the legitimate interests of business" noted the demand for the introduction of artificial intelligence in the judicial system in the context of digitalization, which will ensure the uniformity of judicial practice, professional development and quality of justice, analysis of corruption risks.

It is interesting to think that artificial intelligence can be used as a mediator of the judiciary and increase the level of efficiency and fairness of court decisions and even predict them. The Supreme Court of the Russian Federation has formulated a number of proposals for the modernization of judicial proceedings. Thus, Chairman of the Council of Judges V. A. Momotov on April 4, 2019, spoke at the plenary session of the U1 Moscow Legal Forum "The Russian legal System in the context of the Fourth Industrial Revolution" and expressed the opinion that modern technologies will give "a new impetus to such basic principles of judicial proceedings as the independence of judges, legality, accessibility of justice, consideration of cases within a reasonable time, competitiveness, openness and transparency".

The report highlighted the technologies of automated case distribution, video conferencing, electronic document management, electronic form of document exchange between courts, as well as the possibility to apply blockchain technologies [19]. But it is "at least premature" to introduce a neural network instead of a judge, i.e. to automate the entire process of justice and replace the judge with a computer program (neural network) [20].

The essence of digitalization or the key process of digital transformation is the strategic management of data or virtual or real sectors of companies [21]. The virtual sector includes telecom operators, mass media, online stores, and aggregators. The main purpose is to increase the efficiency of using customer data, knowledge about customers. In the real sector, the purpose is to create digital doppelgangers of physical world assets, the technology of which will allow you to use the results of analysis and checks to improve productivity, optimize and implement predictive analytics.

There is a problem of discussing predictive judicial analytics (whether it is possible to know the judge's decisions before they are made) [22]. Rating expectations of the variable outcome of a procedural decision in a criminal case have long been known to practitioners, since the legislative options for the decision-making grounds are different. A comparative legal analysis of the criminal procedure legislation of Russia and foreign countries on the issues of making procedural decisions and the use of modern technologies in their adoption showed the following. For example, weight is a property of evidence when it is evaluated in the evidentiary law of the states of the Anglo-Saxon legal system [23]. The weight of evidence in the United States has relationships with the admissibility and relevance of evidence [24]. Although in the Code of Criminal Procedure of the Russian Federation, the legislator does not express an attitude to either weight or strength, one can only say about the evidentiary value of information on electronic media when they are withdrawn and copied, and here, perhaps, the legislator meant direct evidence, according to Article 164.1 of the Code of Criminal Procedure of the Russian Federation.

M. S. Sergeyev in his dissertation work studied the array of foreign legislation and found out that if the US legislation can be obtained secretly (also under the Federal Law of 12.08.1995 No. 144-FZ), then it is logical to talk about the following implemented programs: "CO-Traveler", "BoundlessInformant", "Dropmire", "X-Keyscore", "Tempora". In the context of this problem, attention should be paid to the mechanism of the blockchain – built according to certain rules, a continuous sequential chain of blocks containing information, access to which excludes data theft, fraud, violation of property rights, intrusion into the decision system. The user can only supplement the information, and only the person who owns the key and for whom the information is intended can take it back.

7 Conclusion

In the course of the research, the following results were obtained:

1. Classification bases of variable legislative types of decisions involve the identification of historical traditions in the algorithmization of the formation of procedural decisions.

2. The formation of the Concept of applying mathematical and information and communication methods and algorithms in the process of making criminal procedural decisions involves: firstly, the formulation of the concept of a procedural decision; secondly, the definition of a system of probabilistic bases for making procedural decisions; thirdly, the establishment of rating criteria for the optimal procedural decision; fourthly, the definition of an algorithm for the variable result of a procedural decision in a criminal case; fifthly, the study of the monitoring effect of judicial errors.

3. Bibliographic data on the research problem indicate a high interest of scientists in the application of mathematical methods in jurisprudence [25]. Lawyers in the context of optimizing the system of procedural decisions in criminal proceedings will inevitably accompany the process of making criminal procedural decisions with innovative technologies, electronic resources, mathematical classification basis and programming.

4. It is possible to study the decision-making process "from the inside", without researching the procedural procedures according to this formula, only with a certain degree of probability, abstraction or simplification. Our heuristic model can be implemented only in the conditions of joint efforts of many professions – "heuristic justice".

5. The algorithm for making procedural decisions applies: firstly, to the entire system of decisions of this classification group; secondly, it provides a decision of the procedural problem under the conditions of precisely specified procedures by law; thirdly, it provides a sequence of actions that are aimed at achieving the final (optimal) decision in effective ways.

6. The use of key digital transformation processes (strategic data management) in criminal proceedings will ensure the effective use of information and digital data about the system of procedural decisions and about each individual, using blockchain technology.

7. Algorithmization in the process of making criminal procedural decisions means: firstly, creating a consistent continuous chain of files from the actual circumstances of the case; secondly, creating conditions for the confidentiality of the user of the entire system of this chain of procedural decisions; thirdly, creating a ban for any user to intrude in order to change both an individual decision and the entire system; fourthly, the procedural transparency of making a procedural decision can be provided with the possibility of changing it only to the subject of managing the entire chain.

References

1. D.M. Gvishiani, N.I. Lapin, *A Brief Dictionary of Sociology* (Politizdat, Moscow, 1988)
2. M.Yu. Tikhomirov, L.V. Tikhomirova, *Legal encyclopedia* (Yurinformtsentr, Moscow, 2001)
3. R.L. Khachaturov, *Legal encyclopedia* (VUiTpublishinghouse, Togliatti, 2005)
4. N.S. Manova, V.M. Shinkaruk, P.V. Solovyeva, Studies in Computational Intelligence **826**, 77-83 (2019) doi: 10.1007/978-3-030-13397-9_9
5. D.N. Gorshunov, *Uchenyezapiski of Kazan State University* **5**, 27-34 (2008)
6. A.K. Bekishev, A.G. Kan, A.S. Ibrayeva, S.S. Dyussebaliyeva, N.S. Ibrayev, Journal of Advanced Research in Law and Economics **3**, 711-716 (2019) doi: 10.14505/jarle.v10.3(41).03
7. N.V. Olinder, V.V. Romanova, I.Z. Tovysheva, S.V. Yunoshev, E.A. Gambarova, Amazonia Investiga **19**, 491-499 (2019)
8. M.P. Basilio, V. Pereira, G. Brum, Data Technologies and Applications **3**, 333-372 (2019) doi: 10.1108/DTA-12-2018-0109
9. I. Ilyin, Young scientist **17-1**, 89-91 (2016)
10. S.V. Maksimov, Y.G. Vasin, K.A. Utarov, Russian journal of criminology **12 (4)**, 476-484 (2018) doi: 10.17150/2500-4255.2018.12(4).476-484
11. N.G. Muratova, R.G. Khasanshina, *The essence and significance of compensation for harm to the victim when making procedural decisions in criminal cases* (Yurlitinform, Moscow, 2016)
12. Y.V. Frantsiforov, N.A. Solovyeva, Studies in Computational Intelligence **826**, 67-76 (2019) doi: 10.1007/978-3-030-13397-9_8
13. K. Almazkyzy, Y.N. Esteusizov, Journal of Advanced Research in Law and Economics **3**, 834-841 (2018) doi: 10.14505/jarle.v9.3(33).05
14. P.A. Lupinskaya, *Decisions in criminal proceedings: theory, legislation and practice* (Jurist, Moscow, 2006)

15. P.A. Lupinskaya, *Decisions in criminal proceedings: theory, legislation, practice* (Norma: INFRA-M, Moscow, 2010)
16. R. Romashov, E. Bryleva, Studies in Computational Intelligence **826**, 411-417 (2019) 10.1007/978-3-030-13397-9_48
17. V.S. Zelenetskiy, N.V. Glinskaya, *Theory and practice of substantiating decisions in the criminal process of Ukraine* (Stride, Kharkov, 2006)
18. A.V. Groshev, A.G. Saprunov, A.V. Shulga, R.R. Galiakbarov, A.A. Tushev, International Journal of Engineering and Technology **3.15 S15**, 240-248 (2018)
19. J.A. Gavrilova, O.Y. Rybakov, E.A. Antonyan, N.V. Sharueva, Lecture Notes in Networks and Systems **129**, 1136-1142 (2020) doi: 10.1007/978-3-030-47945-9_123
20. A.V. Shulga, V.A. Yakushin, Lecture Notes in Networks and Systems, 61-67 (2020) doi: 10.1007/978-3-030-32015-7_8
21. S.V. Zuev, Studies in Computational Intelligence **826**, 419-425 (2019) doi: 10.1007/978-3-030-13397-9_49
22. N. HuyBinh, International Journal of Advanced Trends in Computer Science and Engineering **2**, 1909-1912 (2020) doi: 10.30534/ijatcse/2020/154922020
23. K.F. Gutsenko, L.V. Golovko, B.A. Filimonov, *Criminal process of Western states* (Zertsalo-M, Moscow, 2001)
24. S.V. Shcherbakov, Russian Judge **4**, 27-30 (2010)
25. G. Cereda, R.D. Gill, Entropy **4**, 439 (2020) doi: 10.3390/E22040439

"Elsevier inc. v. sci-hub": some aspects of copyright infringement in digital space

Mikhail Naumov*[1]

[1] Volgograd Institute of Management, branch of the Russian Presidential Academy of National Economy and Public Administration, 8, Gagarina Street, Volgograd, Russia, michael_naumov@mail.ru

Abstract. Many scientists are faced with the problem of finding the necessary scientific papers to use their work. In this regard, most of them resort to the opportunities provided by «scientific pirates». In connection with the digitalization of science, many scientists have the opportunity to download paid articles on specialized pirate sites. Such piracy activity is illegal and violates copyright. At the same time, neither legislation nor specific court orders can stop the illegal activities of such sites. In this article, using the example of one of the most famous copyright infringement case "Elsevier Inc. v. Sci-hub" conducted a comparative legal analysis of the copyright laws of the United States (a developed country) and Russia (a developing country), revealed weaknesses in the laws of both countries and concluded that the current state of copyright does not keep pace with the digitalization of science, and such inflexibility adversely affects not only the prestige of copyright itself, but also the research and investment climate in the state.

1 Introduction

Civil Code of the Russian Federation in Chapter 70 identifies three main objects protected by Russian copyright law, namely: works of science, literature, and art. Russian legislator widely uses this formulation to show the homogeneity of social relations formed around works of science, literature and art, as well as the absence of the need to differentiate these objects.

At the same time, it should be noted that in recent years, among all violations of copyright and related rights, the greatest public response was caused by a number of cases of copyright infringement of scientific works.

This situation is associated, first of all, with the fact that, despite the homogeneity of legal regulations and the similarity of the legal nature of objects of copyright, science and scientific works have exceptional social significance. Unlike literature and art, aimed primarily at satisfying recreational needs, science aims to obtain objective theoretical and practical information about reality. Scientific activity becomes especially significant in a postindustrial society, where knowledge becomes the most valuable resource for human activity, and access to it is greatly simplified due to the digitization of scientific information, the departure of science to the digital space.

The combination of massive storage capacity, powerful data manipulation techniques, and graphical capabilities has revolutionized both how basic research is conducted and how the resulting knowledge is preserved and disseminated in nearly all fields of science.

In this promising new research environment, scientists increasingly rely on automated knowledge discovery tools to mine and recombine vast amounts of data and literature that are flowing at rates that exceed the capacity of a single investigator to comprehend and manage [1].

In the early 2010s, the open science movement that proclaimed the need for «transparent and accessible knowledge that is shared and developed through collaborative networks» [2] began to gain in popularity.

However, supporters of this approach in practice face a serious legal barrier: copyright, which protects the interests of authors and copyright holders - mainly publishing houses. Bypassing such a barrier, copyright is violated, and scientific works are published and distributed in the open access. This phenomenon is called "scientific piracy".

2 Problem Statement

There is a trend according to which, with the advent of the digital space, more and more researchers, especially from developing countries, which include Russia, seek to obtain professional scientific knowledge through illegal downloading of scientific papers [3], that is, scientific piracy.

The current state of copyright, on the one hand, does not sufficiently protect the copyright and related rights of authors and copyright holders of scientific works, taking into account the digitalization of science, and on the

* Corresponding author: michael_naumov@mail.ru

other hand, it generates unstable economic and emotional relations between copyright holders and scientists, which in turn creates a situation of negative perception of copyright among the scientific community.

A study of the recent judicial practice of copyright infringement in the digital space, the reaction of the authorities and the scientific community to it, as well as an analysis of the consequences, will sufficiently assess the relevance and flexibility of copyright through a comparative legal analysis of the copyright system in the United States (developed country) and Russia (developing country), as well as identifying the main practical and theoretical problems in the copyright system in both countries.

3 Research Questions

In this article, we will try to answer the question of whether the current state of copyright law in area of scientific researches in developed and developing countries makes it possible to protect the interests of right holders and society in modern conditions, due to the digitalization of science. Do existing legal norms meet current conditions and is there a need for a copyright reform?

4 Purpose of the Study

The purpose of this study is to analyze the fundamentals of American and Russian legislation in the field of copyright protection for scientific works for the subsequent identification of the features of the legal regulation of this area.

5 Research Methods

The research methodology in this study consist of common scientific methods of system analyse, dialectical and formal-logical methods and law science methods such as comparative legal method, formal legal methods.

6 Findings

In connection with the development and spreading of information technologies, especially through the telecommunications network "Internet", scientific piracy has acquired a new character: it has become more widespread and more effective.

Perhaps the most famous science pirate today is Alexandra Elbakyan, the founder of the Sci-Hub.org website, which provides open and free access to many scientific articles and publications. As noted by Science magazine, Sci-Hub downloaders live on every continent except Antarctica [4].

Sci-Hub activities, as well as a number of similar programs, caused massive discontent among scientific publishers, many of which filed lawsuits with the aim of restoring violated copyrights in court. Below, we invite you to read and analyze the most famous and resonant decision of the New York Southern District Court (District Court, Court) in the case of "Elsevier Inc." v. "Sci-Hub".

Despite the fact that the final decision in the case was made on June 21, 2017, most of the court's conclusions are contained in the decision on the opinion of preliminary injunction dated 10/28/2015.

The plaintiffs in the case were several companies which belong to the Reed Elsevier holding [5] - one of the largest publishing houses in the world. The defendants in the case were the Library Genesis project (a group of developers of scientific pirate sites), Sci-hub.org and the only real representative of the website - Alexandra Elbakyan.

The essence of the claim was to take preliminary injunction to block these sites that violate Elsevier's copyright for further consideration of the case on the merits.

In its opinion, the court, referring to the case "WPIX Inc. vs. Ivi Inc." has previously indicated what conditions must be met for the application of preliminary injunction to protect copyright. In this case, the plaintiff must provide:

1) a likelihood of success on the merits,
2) irreparable harm in the absence of an injunction,
3) a balance of the hardships tipping in their favor,
4) that issuance of an injunction would not do a disservice to the public interest [6].

From the point of view of legal science, this position of the court can be called an evidentiary standard - a minimum set of factual information indicating the need to satisfy the claims of the plaintiff and make an appropriate decision. Russian procedural legislation (civil and arbitration), neither in terms of the protection of intellectual rights, nor in general, does not establish a clear standard of proof, providing the courts only with advisory and explanatory provisions specified in the decisions of the Plenums of the highest courts. Mandatory evidentiary standards are aimed at making judgments based on uniform measures, which makes them more predictable and standardized.

At the same time, these evidentiary criteria were borrowed by the District Court from the conclusions of the court decision, although in the same category of cases, but without taking into account the different legal, economic and social essence of these two cases, since the WPIX Inc. vs. Ivi Inc." related to copyright infringement for the broadcasting of television programs via an Internet platform. Actions of Ivi Inc. were of a commercially oriented nature, and also did not represent a significant public interest.

The relationship between Sci-hub.org and Elsevier clearly shows a different purpose of copyright infringement. In addition, Sci-hub.org's actions were not commercial in nature, and all funds received by the site are voluntary donations. Therefore, the application of uniform criteria for all cases of copyright infringement may lead to excessive formalism of the court proceedings and the desire of the court to resolve the issues known in advance, and not to analyze all the circumstances of the case.

As for the directly motivational part, it begins with an analysis of the evidence provided by the plaintiff, necessary to indicate the likelihood of success in essence. Thus, the representatives of Elsevier pointed out the existence of a formal violation of the exclusive rights established by 17 U.S.C. § 106. The court considered that in order to prevail on a claim for infringement of copyright, "two elements must be proven:

1)ownership of a valid copyright;

2)copying of constituent elements of the work that are original"[6].

This information was initially presented by the plaintiff to the court in the complaint and in more detail in the memorandum. They are based on a sequence of screenshots showing how a user could go to sci-hub.org, enter the necessary information (article title, DOI or URL), get a set of search results, follow the link to a copyrighted article on Science Direct via proxy server. In addition, the plaintiff also points to the post of A. Elbakyan on Twitter, indicating that "whenever an article is downloaded using this method, the Defendants save a copy of the article on their servers."

Interestingly that both American and Russian copyright law does not establish a list of admissible evidence. In accordance with the provisions of clause 55 of the Resolution of the Plenum of the Supreme Court of the Russian Federation No. 10 dated 04/23/2019, the evidence used in this case could well have been accepted by Russian courts.

So, admissible evidence is, including made and certified by the persons participating in the case, printouts of materials posted on the information and telecommunication network (screenshot), indicating the address of the Internet page from which the printout was made, as well as the exact time of its receipt. Such printouts are subject to assessment by the court when considering the case on an equal basis with other evidence [7].

It is also worth to be noting significant difference between American and Russian legislation. Thus, the District Court in its opinion directly obligates the plaintiff to prove the fact of copyright ownership and the fact of its violation. This imposition of the obligation on the plaintiff to prove his authorship and the fact of infringement of his copyright in the American judicial system was called "prima facie case".

"Prima facie" - a privileged position that serves as proof of the first or earliest appearance of the work, provided that the registration was made for a certain period (5 years) before or after its first use [8].

The use of a "prima facie case" is necessary in order to reduce the cases of unfounded claims and accusations of persons of copyright infringement. This is also determines the obligation to prove the factual circumstances.

What is generally called the burden of proof is more accurately labeled the burden of persuasion, i.e. convincing the trier of fact that the evidence produced supports the claim under the applicable evidentiary standard. Assigning the burden of proof is fundamentally important [9]. Burden of proof rules allocate the costs and risks of gathering and presenting evidence, and help filter good cases from the bad [10].

Russian legislation does not explicitly provide for the imposition of the burden of proof on any of the parties, except common procedural obligations, however, Russian Civil Code contains provisions establishing a kind of presumption of guilt of the infringer of copyright.

So, in paragraph 3 of art. 1250 states that "measures of responsibility for infringement of intellectual rights shall be applied in the presence of the fault of the infringer. <…> the absence of guilt is proved by the person who violated the intellectual rights" [11]. It also provides for the application of measures of responsibility for violation in the absence of fault.

In other words, the burden of proof in copyright infringement cases in the Russian Federation can be characterized as leaning towards the presumption of the defendant's guilt, which puts the plaintiff in a more privileged procedural position. This conclusion is confirmed by numerous judicial practice, including the Supreme Court of the Russian Federation (see clause 19 of the Review of judicial practice in cases related to the resolution of disputes over the protection of intellectual rights (Approved by the Presidium of the Supreme Court of the Russian Federation on 09/23/2015; hereinafter - Review), Determination of the Judicial Collegium for Civil Cases of the Supreme Court of the Russian Federation dated 02.03.2010 No. 49-B09-22; Determination of the Supreme Court of the Russian Federation No. 3055-ES-7224 dated 09/13/2016 in case No. A40-26249/2015).

At the same time, in accordance with paragraph 59 of the Review, an offense in the event of recovery of losses, the burden of proof of a civil offense is imposed on the plaintiff on the basis of the principle of "general delict" (or general rule) , in which the plaintiff is obliged, guided by the provisions of Articles 15, 1252 of the Civil Code of the Russian Federation, to prove the fact and the amount of damages, as well as all the elements of the corpus delicti.

Despite the fact that in 17 U.S.C. § 504 also provides for the possibility of recovering losses, it should be noted that the legal doctrines of the United States and Russia have developed somewhat different positions regarding the imposition of the burden of proof in disputes over copyright infringement.

The next issue brought up by the District Court for consideration was the determination of the possibility of causing irreparable harm in the absence of preliminary injunction.

The category of irreparable harm itself takes a special place in the US legal system. Irreparable harm is present "where, but for the grant of equitable relief, there is a substantial chance that upon final resolution of the action the parties cannot be returned to the positions they previously occupied [6].

The District Court found that irreparable harm was caused by the actions of the Defendants, since the parties did not have the opportunity to return to their original position: «It would be difficult, if not impossible, to determine how much money the Plaintiffs have lost due to the availability of thousands of their articles on the Defendant websites» [6].

Also, in support of this conclusion, the court indicates aspects of the "viral violation" where Elsevier's content could be transmitted and retransmitted by third parties who acquired it from the Defendants even after the Defendants' websites were shut down. Moreover, the irreparability of the damage caused is also confirmed by the fact that the amount of damage significantly exceeded the solvency of the infringer.

However, the most controversial part of the trial was the question of the presence or absence of harm to the public interest. In objections to the claim, A. Elbakyan notes that there are "lots of researchers especially in developing countries", who do not have access to key scientific papers owned by Elsevier and similar organizations, and who cannot afford to pay the high fees that Elsevier charges. The general opinion in research community is that research papers should be distributed for free (open access), not sold. And practices of such companies like Elsevier are unacceptable, because they limit distribution of knowledge.

In response, Elsevier argued that the public interest is conducive to the acceptance of the injunction and the satisfaction of the claim because it will «protect the delicate ecosystem which supports scientific research worldwide. The money Elsevier generates by selling access to scientific research is used to support new discoveries, to create new journals, and to maintain a «definitive and accurate record of scientific discovery». It also argues that allowing its articles to be widely distributed risks the spread of bad science - while Elsevier corrects and retracts articles whose conclusions are later found to be flawed, it has no way of doing so when the content is taken out of its control».

The question of the unfair structure of market relations arising around the sale of scientific works has been repeatedly raised by various lawyers. So, R. Watt claims that most academics live in a "publish or perish" environment, and so are willing to exert significant efforts in writing scientific works with the sole motivation of enhancing their hiring and promotions options, or perhaps not to see their options hampered relative to other academics. The companies that gather, revise, and publish this scientific output earn (in some cases) significant profits by providing the service upon which the academic community rides [12].

Nevertheless, the controversy that arose and even massive support for Sci-hub.org's position by the world scientific community and even universities [13] did not prevent the court from issuing a ruling to block the site, and then a decision to satisfy Elsevier's claim. Elsevier was awarded statutory damages for copyright infringement in the amount of $15.000.000 [5]. At the same time, the court, in its opinion, additionally noted that copyright law inherently balances two competing public interests: the rights of users and the public interest in broad accessibility of creative works, and the rights of copyright owners and the public interest in rewarding and incentivizing creative efforts (the 'owner-user balance').

The court also noted that scientific works are indeed of great public interest and it is sustained by two critical exceptions in copyright law: firstly, the «idea/expression dichotomy», secondly, the «fair use» doctrine.

Nevertheless, the court's decision today remains de facto unfulfilled, the sites Sci-hub.org and Libgen.org continued to operate using different domain names. So, in the period 2017-2018, from 300.000 to 600.000 articles were downloaded per day, which is significantly higher than the figures before 2016 [14]. In addition, in 2019, as Elbakyan said the number of visits to the site only increased and reached the mark of 1.000.000.000 views [15].

It is obvious that the current state of copyright in works of science is extremely unstable. Existing legal regulations cannot settle the existing actual legal relationship, all prohibitive measures do not reduce the cases of copyright infringement on scientific works in the Internet space. This legal conflict creates an atmosphere of legal skepticism, mistrust or disregard for copyright laws.

The phenomenon of scientific piracy itself, even with a large number of injunctions and peremptory norms of legislation, will not cease to exist and, moreover, will not even diminish in scale. In many ways, this is favored by the possibilities of modern information technologies, as well as the increasing desire of scientists to obtain new knowledge and share their own.

Moreover, it should be considered that the copyright policy has more than just its own legal purpose. For example, Michael W. Carroll suppose that U.S. copyright law provides a competitive advantage in the global race for innovation policy because it permits researchers to conduct computational analysis — text and data mining — on any materials to which they have access. Amendments to copyright law in Japan, and the European Union's recent addition of limitations on copyright to legalize some TDM research, implicitly acknowledge the competitive benefits provided by the fair use provision of U.S. copyright law [16] .

In this case, many lawyers agree that the institution of copyright needs reforms in terms of expanding the criteria for the fair use of works, which can be the first step in solving the problems associated with scientific piracy. In Russian civil law, a similar institution is established in Art. 1274 of the Civil Code of the Russian Federation. It is determines cases and limits of free use of the work in informational, scientific, educational or cultural purposes. However, it is noted that this institution is not sufficiently developed and studied [17].

Already now, many lawyers note the existence of significant problems in the copyright of developed countries. So, for example, Geiger C. and Izumenko E. note that the copyright of the European Union, and in particular the institution of "fair use" is «ill-adapted, fragmented, inflexible, often irrelevant and not keeping up with the technological developments» [18].

7 Conclusion

Thus, in modern situation, involving the massive introduction of digital technologies into science, current condition of copyright laws in developed and developing countries still has a number of significant shortcomings, which, in particular, include the problem of scientific piracy raised in the article. It should also be assumed that changes in copyright law, proposing a shift in the balance of interests in favor of the public, including the professional scientific community, will significantly help solve the accumulated problems and contradictions in this sub-branch of law.

References

1. J.H. Reichman & R.L. Okediji, Minnesota Law Review **96(4),** 1362-1480 (2012)
2. R. Vicente-Saez & C. Martinez-Fuentes, Journal of Business Research, **88,** 428-436 (2018)
3. J.M. Owen, *The scientific article in the age of digitalization* (Dordrecht, The Netherlands: Springer, 2007)
4. J. Bohannon Who's downloading pirated papers? Everyone. Retrieved December 02, 2020 (2018), URL: https://www.sciencemag.org/news/2016/04/whos-downloading-pirated-papers-everyone(date of access: 16.03.21).
5. «Elsevier Inc. v. Sci-hub» Default Judgment – #87 in Elsevier Inc. v. Sci-Hub (S.D.N.Y., 1:15-cv-04282). Retrieved December 02, 2020 (2017, June 21), URL: https://www.courtlistener.com/docket/4355308/87/elsevier-inc-v-sci-hub/(date of access: 16.03.21).
6. «Elsevier Inc. v. Sci-hub», Opinion No. 105999 in Elsevier Inc. v. Sci-Hub (S.D.N.Y., 1:15-cv-04282). Retrieved December 02, 2020, (2015), URL: https://www.courtlistener.com/docket/4355308/53/elsevier-inc-v-sci-hub/(date of access: 16.03.21).
7. Resolution of the Plenum of the Supreme Court of the Russian Federation Dated 23.04.2019 No. 10 "On the application of part four of the Civil Code of the Russian Federation". URL: http://www.consultant.ru/document/cons_doc_LAW_323470/c01923bc1d0db60508e82c505d7d29b544e91443/. (date of access: 16.03.21).
8. M.V. Samartseva, Courier of Kutafin Moscow State Law University (MSAL)) **6,** 114-118 (2017)
9. L.P. Loren, & A.R. Reese, *Lewis and Clark Law Review,* **23(2),** 622-680 (2019)
10. D. Vaver, Intellectual Property Journal **23,** 147-157 (2011)
11. Civil Code of the Russian Federation dated 30.11.1994 No. 51-FZ (ed. 31.07.2020) P.4. URL: http://www.consultant.ru/documents\cons_doc_LAW_64629/#dst0. (date of access: 16.03.21).
12. R. WattIntroduction, Review of Economic Research on Copyright Issues, **7(1),** 1-6 (2010)
13. A. Fox, & J. Brainard, University of California boycotts publishing giant Elsevier over journal costs and open access. Retrieved December 02, 2020 (2019), URL: https://www.sciencemag.org/news/2019/02/university-california-boycotts-publishing-giant-elsevier-over-journal-costs-and-open(date of access: 16.03.21).
14. D. Himmelstein, A. Romero, J. Levernier, T. Munro, S. McLaughlin, B. Tzovaras, & C. Greene, Research: Sci-Hub provides access to nearly all scholarly literature. Retrieved December 02, 2020, (2018), URL: https://elifesciences.org/articles/32822(date of access: 16.03.21).
15. Sci-Hub | Open access to scientific information Everyone sums up the year, what about Sci-Hub? Results of 2019 for Sci-Hub - a service for free reading of scientific literature. (2019), URL: https://vk.com/wall-36928352_39093 (date of access: 16.03.21).
16. M.W. Carroll UC Davis Law Review, **53**, 893-964 (2019).
17. E.V. Demyanenko, Science and education: farming and economy; entrepreneurship; law and governance, **11**, 93-96 (2017)
18. C. Geiger, & E. Izyumenko, American University International Law Review, **35(1),** 1-74 (2019)

SHS Web of Conferences **109**, 01028 (2021) https://doi.org/10.1051/shsconf/202110901028
LISID-2020

Project activity in the sphere of digital technologies

E.P. Panova*[1], *N.R.* Saenko [2], *P.S.* Volkova[3], *E.V.* Bobyreva[4] and *V.V.* Kortunov[5]

[1] Moscow Polytechnic University, Bolshaya Semyonovskaya str., 38, Moscow, 107023, Russian, panova_ep@mail.ru
[2] Moscow Polytechnic University, Bolshaya Semyonovskaya str., 38, Moscow, 107023, Russian, rilke@list.ru
[3] Krasnodar Higher Military School named after General of the army S.M. Shtemenko, Krasina Street 4, Krasnodar, 350063, Russian, polina7-7@yandex.ru
[4] Volgograd State Social and Pedagogical University, Lenin, d. 27, Volgograd, 400066, Russian, new_life@mil.ru
[5] Russian State University of Tourism and Service, Glavnaya Street 99, dp.Cherkizovo, Pushkinsky district, Moscow, 141221, Russian, kortunov@bk.ru

Abstract. The article considers the problems that arose in the situation when the world's system of education has been forced to transfer into distance learning mode because of the pandemic. For the first time mankind had got an experience when it became necessary to abandon traditional training. In such situation, it was possible to continue providing education both in schools and in the universities due to the advantages that "digitalization" of education carries, though a number of shortcomings that both teachers and students faced were also revealed. Empirical experiment allowed to find out that distance education has significant shortcomings which affect both quality of knowledge and physical and mental health of people as well as socio-cultural adaptation and socialization of foreign students. Abandonment of traditional education in favor of distant one carries significant losses in the quality of education and, as a result, decreases the ability to create a harmonious and comprehensively developed personality. Authors of the article come to the conclusion that distance education should be used as an additional form of education in comparison to basic one (traditional), that is, it can provide positive results only in the complex (traditional plus distance).

1 Introduction

In our days question of project activity is relevant in the framework of higher school education, especially in the humanitarian sphere, since this area is outside the zone of material and technical capabilities, rather it is a sphere of ideas, a sphere of theoretical knowledge making attempts to understand surrounding world. Perhaps that is why methodology of teaching foreigners Russian as a foreign language at the preparatory departments of Universities doesn't consider possibility of using project activities as a successful method of teaching. At the initial stage of teaching a foreign language project methods cannot be used because of the language barrier, as foreign students have very poor command of Russian. Though in the second term of the first year it is possible to introduce various types of project activity. The following problems arising in the process of applying project activity method at the preparatory faculty can be highlighted: complex process of social adaptation, difficulty in intercultural communication skills formation.

Nevertheless, during the second term of the first year is not only possible but just necessary for a teacher to introduce a project activity method as it helps students to overcome arising problems more successfully: including problems of sociocultural and intercultural adaptation, improving cognitive and communicative skills, increasing creative activity and the level of motivation and, as a result, improving general level of language skills. E.G. Krylov in the article "Peculiarities of teaching foreign students engineering disciplines" raises the question of using method of projects in the process of teaching foreigners, since this technology allows to increase motivation to study Russian as a foreign language. E.G. Krylov suggests to stimulate motivation using the project method (foreign students are invited to carry out projects on a voluntary basis). During such classes, students can be invited to complete a course project both in sciences and humanities, participate in the Olympiads in different subjects, to hold festivals and competitions, to participate in creative projects (theater and concert productions), make a report or a presentation at the conference, etc.

However, in the present situation prevailing in the world - self-isolation because of COVID-19 - it has become impossible to organize project activities in the system of traditional education. Therefore, an attempt was made to transfer project activities into a remote format. And as not all project activities can be carried out in a digital mode, it was decided to select several projects and conduct them on a competitive basis. Due to the fact that at present circumstances we had to abandon traditional training, organization of professional

* Corresponding author: panova_ep@mail.ru

activities at both sites (Moscow Polytechnic and Volgograd State Technical University) was arranged distantly in two ways: independently and in the on-Line class. Such experience of organizing training - when work was transferred into the remote format and traditional training had to be abandoned – was received by the educational world for the first time.

2 Problem Statement

The problem of the study is to justify the possibility/impossibility of project activity in the distant learning (e-leaning) mode, whether it is possible to form students' professional skills, critical and creative thinking, independent skills in the conditions of digitalization.

3 Research Questions

Is it a myth or reality to organize successful and effective training of foreign students in the sphere of project activities with the help of digital technologies at the initial stage of study?

Is it possible to use method of cooperation during project activities in the process of distant learning?

How productive it is to develop a person in different forms of activities in the process of distant learning performing project activities?

4 Purpose of the Study

To reveal the results of efficiency/inefficiency of student's project activity organization using digital technologies in comparison to traditional education, as well as to describe the methodology of various types of projects aimed at developing critical and creative thinking, professional skills and various aspects of human activity.

5 Research Methods

The methodological basis of the study is made up of the principles and provisions of the systemic, structural-functional and comparative-typological approaches.

An important methodological guideline in this study is the works of Russian developers of distance learning V. Samoilenco, T.V. Tretyakova, who write about the modification of modern higher education through distance learning [1-2]. Ideas about the modernization of higher education are developed in works by Ö.H. Kuzu [3]. An in-depth analysis of the transformation of higher education using digital technologies is found in the works of L. Seres et al [4]. C. Dexeus [5] writes about the revolutionary process in the higher education system, which is happening today with unprecedented speed. K. Sandkuhl spoke about the consequences of the transformation of the traditional education system [6]. Akhmetova's observations on the use of digital technologies in the sphere of inclusive education are of great interest [7]. Making various fields of activity in higher education digital is also being actively discussed in modern pedagogy [8-9].

Scientists from different countries, such as Collins, A., & Halverson, R., Thiele, AK, Mai, JA, & Post, S. believe that modern students are a new generation or "network generation", which focuses on new forms of education. Digitalization of the educational process of higher education is a natural phenomenon for modern students, they give preference to it, as they consider it more natural, adequate and goal-oriented [10-11].

M. Brown claims that digital teaching methods have been added to traditional face-to-face education, which has led to the fact that the methods are now mixed / hybrid [12].

Almaraz-Menendez, F., Maz-Machado, A., & Lopez-Esteban, C. correctly note that distance education has become more flexible, accessible [13], thus, all scientists come to the conclusion that modern education needs to be reshaped and adapted to meet the needs and requirements of a new "network" generation.

6 Findings

As part of the study, two experiments were conducted with foreign students and teaching staff of Volgograd State Polytechnic University and Moscow Polytechnic University. The first experiment was aimed at establishing effectiveness/inefficiency of education organization using digital technologies. To form a certain picture of success/failure of the process of teaching Russian as a foreign language in the conditions of distant learning (self-isolation) using digital technologies 154 students as well as 27 teachers had been chosen to participate in the experiment. The experiment made it possible to form the following opinion - to what extent foreign students managed to perceive the material in the process of distant education (DE) and to what extent distant learning using digital technologies was effective/ineffective as well as how much the quality of education using DE increased/decreased.

The questionnaire included the following groups of questions:

95 % of recipients understand material better in class; 67 % of recipients replied that they did not understand material well enough in the process of distant learning.

58 % of recipients responded that process of distant learning affects the quality of education but not much, 42 % responded that it affects the quality negatively. Students' knowledge on the subject has become worse.

The language cannot be learned distantly – answered 69 % of respondents; it is possible to learn it but not well enough – responded 41 %.

The length of classes conducted in digital mode is significantly reduced: only 40 % of students can study 2-3 hours a day, 28 % can study 4 hours a day and 12 % can be engaged in studies 5 hours a day.

Recipients replied that the main difficulty of distant learning is that at the initial stage of language learning foreign students have poor command of the Russian language and therefore they have language difficulties (language barrier) studying online. Also, distant learning does not create conditions for communication, and besides it is possible to use much more methods and techniques of teaching students in the classroom; a student manages to do more tasks and understands material better and deeper when in the classroom. Students answered that it was "better for them to study in the classroom; they had much more conversational practice events".

In addition to the above-mentioned problems, students also mentioned that online learning process leads to the decrease of their health conditions, besides a whole range of psychological difficulties arises; the reason for their complaint had also become the lack of language environment during distant learning, since it is the language environment itself that contributes to the development of communicative skills; they indicated that they had no opportunities to study at home as they often have technical failures with the Internet and problems of technological character in general. The questionnaire showed that there was another significant drawback that students mentioned and that was the lack of direct contact with a teacher. This fact causes a lot of difficulties for students in understanding and assimilating material, leads to decrease of the quality of fulfilled tasks and exercises, to the efficiency of education, to deterioration in control over student's activities and to decrease of interest in study in general. All these data affect the quality of knowledge of the Russian language, since at the initial stage of studying Russian as a foreign language foreign students don't have enough practical skills to study language all by themselves (independently).

Foreign students also noted that distant education narrows consciousness to the size of a computer screen; experience has shown that it is impossible to socialize (since there is no live communication) and to get socio-cultural adaptation, since there is no way to work in a team. It is impossible to get overall development of creative and critical thinking, since a huge number of tasks and types of creative work students get can be realized neither during classes nor during activity on the platform. So, for example, "during the process of translation [...], some difficulties connected not only with cultural differences but also with differences in principles of functioning of phonetic, grammatical, syntactic, lexical systems of the language arise" [14]; and during distant learning such forms of work become even more complicated and it is practically impossible to fulfill them.

Thus, summing up everything mentioned above, it can be concluded that abandoning traditional education and replacing it with distant one leads to a deterioration in the quality of education, since "Digital technologies only aggravate the process of separating people, make society more fragmented and individualistic. Not interacting with each other people find themselves outside the framework of socialization and become increasingly intolerant to each other. Understanding as a process of interaction in the frameworks of presently formed communication in the classroom - student-student or pupil-student - begins to collapse and that destroys teaching process in general, doing it more and more low-quality and unskilled process" [15]. In the process of distant education, it is possible to get only some basic skills; in the framework of digitalization formation of comprehensively developed and educated person becomes impossible. Distant learning, despite its advantages (transparency, budgeting, flexibility, time and space release, territorial independence, accessibility), can function only as additional, concomitant form of education being combined with traditional one.

To from an overall picture of success\failure assessment in teaching project activities using digital technologies in conditions of self-isolation an empirical study was also conducted among foreign students aged 18-28 (73 people participated in the experiment) of Moscow Polytechnic University and Volgograd State Technical University of preparatory faculties. During the experiment an attempt was made to find out in what way this form of project activities organization (using distant technologies) contributes to the development of creative skills, independence in organization of educational process, cognitive and communicative skills, critical and creative thinking and the ability to work both independently and in a team. "Project activity helps a teacher to solve a very important task: to make each student involved into active cognitive work. The project method helps students to learn how to set goals and find means and ways to achieve them. It should be noted that project activity has a certain limitation in the terms of time and resources, which, in its turn, enhances its educational potential", point out Semikina Yu. G. & Semikin D. V. [16]. Thus, project activity presents an instrumental level of organizing students' activities because this type of activity makes an accent on creation of a material objects. "In order such structure becomes possible students must make their own efforts to search, process and analyze data, plan their actions and communicate with various partners. Since such activities usually go beyond individual actions, they require team work. Further we will mainly talk about a project-oriented approach emphasizing that it necessarily includes problematization" [17].

For a long time it was believed that project activity was impossible at the initial stage of teaching\studying Russian as a foreign language. However, today, as a part of project activity at the preparatory faculty of the University a number of events are organized, such as: student conferences, Olympiads, presentations, essay competitions, creative competitions, debates, theater plays, literary events, festivals, linguistic tourism, etc. Such events expand intercultural communication, help student to adapt socially, increase students' cultural and aesthetic level, develop creative thinking, increase motivation, improve cognitive and communicative skills, teach to work in groups under teacher's guidance, form skills of self-work, etc. All these projects implemented at the preparatory faculty have a common methodological basis: 1) training within these projects is based on mutual activities; 2) activities are based on a student-centered approach; 3) final result of such activity is creation of a complete product; 4) work on the project takes place in a team (group), a teacher just directs and advises, helping students to set the goal and problem of a project and he also sets tasks.

Thus, taking into account the results of the first experiment, it is possible to speak about the barriers that can be come across in the process of introduction of digital technologies into education system. Such barriers can prevent introduction of project activities into the educational process through digitalization. So students can experience difficulties in perceiving educational material working in a distant mode (cognitive barriers), software language does not coincide with the native language of a foreign student (content barriers), students may not be ready to study distantly, and there may be different reasons for that, but the main reason is didactic [18]. In the process of education technical problems may arise, that is, both students and teachers may lack the necessary equipment (laptops, computers, etc.). In this case, financial barriers arise (costs of the latest technologies and software). An equally important problem is the problem of "digital literacy" of teachers. All these barriers must be taken into account when implementing project activities by means of digitalization.

In this respect it is necessary to mention psychological and pedagogical types of competence a student must have studying distantly. The most important types of competence are softskills competencies (such competences as ability to work in a team, to communicate, to adjust quickly, ability to manage your own psychological and emotional state, to listen, to talk and to come to an agreement) [7]. In the process of introduction project activity using digital technologies the following types of psychological and pedagogical competence were taken into account: cognitive (desire to increase psychological and pedagogical knowledge, desire to self-education), motivational (motivation for successful professional activity, existence of professionally important personal qualities), reflexive and fascilitative (situation, ability to control yourself), informative and communicative (ability to work with information, ability to build successful communication). These very competences of students are formed by the second semester of study at the preparatory stage of teaching Russian as a foreign language; that is why the decision was made to introduce design activities in the distant mode in the situation of self-isolation.

In the process of distance learning not all types of work that are used in project activity in the process of traditional education can be transferred into a distance form. The decision was made to organize project activity in the following types of work: to create a video, a presentation, a greeting card, a poster with a hashtag, a symbolic emblem with some text, to write an essay, to take a photo and come up with its name and history. Such projects can really be implemented in a digital format. Besides, it is important to note that in order to increase the level of students' motivation to study distantly, it was decided to implement the method of project activity on a competitive basis, while each project should be connected with one topic, that is, to come up with and develop such thematic blocks that would be closely connected with the history and culture of Russia.

Essay competition helps foreign students to master written speech, so in the second semester it is possible to implement such project. In this case, students learn to work with information, analyze new material, to prove their point of view, to express their thoughts in written form correctly, to develop creative and critical thinking. Reading competition, competition "my favorite national dish", competition of stories with a presentation about your country can be shown in the form of a video. Such activity presents a good methodological basis to organize project activity in a format of distant learning process. Self-production of a video on a given topic helps to develop communicative skills such as speaking, writing, reading, auditing, because before creating a project a student needs to collect material, to write a story and then to prepare an oral message to be recorded. Such video can be filled with motion in the form of multimedia, elements of text which become animated with the help of graphics, color and musical accompaniment.

Creative projects, such as a contests of postcards, contests of posters with hashtags, contest of emblems, contest of photography, drawing contest develop not only communicative skills (writing and reading), but also introduce students to the culture and history of Russia. Students not only get acquainted with the symbols and heraldry of our country and with its history, but also get to know peculiarities of our culture and etiquette. Students also practice writing skills as in these projects they need to present a text of congratulation of either a poetic or a prose character. At the stage of works checking a jury that evaluates projects according to pre-determined criteria is formed, thus achieving transparency and objectivity of control.

In order to evaluate the results of work in a digital format it is necessary to summarize objective results of the experiment. The purpose of the heuristic experiment conducted among foreign students was made in the attempt to identify the advantages and disadvantages of organizing project activities for foreign students in a digital mode. During the survey the following advantages were found: 98 % of students took part in projects and showed great interest in it, 99 % of students wrote that such type of work helped to increase their level of knowledge of Russian, their communicative and cognitive skills, that is, students noted that they got new knowledge of vocabulary and grammar, improved skills working with Internet-bases in Russian. 99 % of students replied that process of project preparation was quite difficult for them, it took a lot of time, it required deep knowledge of the language, but at the same time it expanded their ideas about aesthetic, cultural and socio-political characteristics of our country (students got an idea about national and state holidays of our country, our national cuisine, Russian cinema, cartoons, literature, they learnt socio-political and geographical realities of our country).

Perhaps that's why students answered (99 %) that work on the project allowed them to gain extensive knowledge in the field of Russian history, culture and art. They replied that participation in the projects in general helped them to understand and therefore love Russian culture more, to expand the vocabulary (they learnt a lot of words with specialized meaning) and their knowledge in grammar. They learnt Russian better, increased their level of motivation, managed to get to know more not only about Russian culture, but also of grammar and syntax of the Russian language.

Answering the task "Write 2-3 positive characteristics of distant participation in the events", students noted such advantages of distant learning as transparency and the possibility of covering great geographical and territorial spaces thanks to Internet resources (you do not need to spend time traveling, participants living in different parts of the city and country can take part in a project at any time, conduct assessments and save time sending documents). So called "erasing boundaries".

Students also replied that participation in the project activity, despite its remote nature, helped them to become more confident, to believe in their own power, helped them to fill their free time with an interesting business, and contributed to the acquisition of life-important knowledge and practical skills.

To the question: "Are there any differences between face-to-face and distant participation in such events? and What is the difference between face-to-face and distance participation?" – 98 % of recipients replied that, in the projects that take place in a real-life format there is an amazing atmosphere, on the one hand, which allows students to take their business seriously and, on the other hand, an atmosphere of a holiday is created. Process of project preparation is better and longer which leads to higher quality results. "All participants can directly contribute to the development of events, comment and evaluate final result obtained due to the participation in the project, make decisions, create a festive atmosphere of the event itself". In the process of full-time participation in the projects, students can work in teams or groups which allows to implement method of creation and further on allows students to exchange ideas achieving better results working in groups. During face-to-face participation an atmosphere of joy, unrest, competitiveness and, at the same time, peace and unity is created – people get valuable experience of empathy in creativity and joint activity.

84 % of respondents believe that participation in a face-to-face event is more difficult and require more responsibility from psychological and linguistic point of view as assessment of work on the project and assessment of final result is presented collectively, to the group, and is more objective than during correspondence learning.

89% of recipients replied that if a teacher consulted them in person, in a live format, they would have managed to prepare projects of higher quality.

In the process of interviewing recipients, the negative aspects of joint project activities using digital technologies were pointed out: weak technical support for both students and teachers, enterprises will have to install equipment and software that can be expensive for small and medium-sized workshops; it is also difficult to determine the degree of students' independent participation in the project, lack of a festive atmosphere, lack of emotions of joy, lack of healthy cooperation that are formed in the process of team work. Also, as a main drawback the recipients pointed inability to participate in the project in the living. Also, vast majority claimed that there are such shortcomings in the organization of distant learning as poor technical level of communication and poor technical preparation to distant learning both of students and universities.

When asked: «have you learne, thanks to remote participation in the competition, to communicate with other people, have you become more sociable?» - 98 per cent of recipients replied that they either had to communicate only with their teacher via Skype technologies or to solve all the arising problems all by themselves. In the process of work on the project students were deprived of mutual cooperation and mutual communication which excluded possibility of implementing socio-cultural adaptation, did not teach them intercultural communication skills. 100 % of respondents wrote that they did not find new friends participating in the project activity distantly.

7 Conclusion

In the conclusion we would like to point out that teaching foreign students Russian as a foreign language using distant technologies significantly reduced their level of knowledge and, despite the fact that introduction of project activities became possible in digital format, nevertheless, such work did not lead to effective results. The decline in the quality of education is due to the emergence of a number of barriers, primarily cognitive, communicative, financial and didactic ones. It is also necessary to mention problems foreign students have in the sphere of socialization and sociocultural adaptation. Therefore, in the future it is necessary to take into account that distant education is possible only in combination with traditional one and only in tandem with traditional education it can lead to food results.

References

1. V. Samoilenko, International Journal of Nursing Scenes, **4**, 343-344 (2017)
2. T.V. Tretyakova, E.Z. Vlasova, E.A. Barakhsanova, M.S. Prokopyev, & M.A. *Integrating Engineering Education and Humanities for Global Intercultural Perspectives. IEEHGIP* (2020)
3. Ö.H. Kuzu, Higher Education in Russia, **29(3)**, 9-23(2020)
4. L. Seres, V. Pavlicevic, & P. Tumbas, *12th International Technology, Education and Development Conference (INTED2018) Conference Proceedings*, 9491–9497 (2018)
5. C. Dexeus, *The Deepening Effects of the Digital Revolution. In E. Fayos-Solá, C. Cooper (Eds). The Future of Tourism: Innovation and Sustainability* (2019).

6. K. Sandkuhl, & H. Lehmann, In A. Rossmann, A. *Zimmermann (Eds). Digital Enterprise Computing. Bonn: Gesellschaft für Informatik*, 49–60. (2017) URL: https://dl.gi.de/handle/20.500.12116/119;jsessionid=1A142A0814932BBEF0EE7D38FA224CF9 [in Ger] (date of access: 12.02.21)
7. D.Z. Akhmetova, T.S. Artyukhina, M.R. Bikbaeva, I.A. Sakhnova, M.A. Suchkov & E.A. Zaitseva Higher education in Russia **29 (2)**, 141-150 (2019)
8. V.A. Lazarenko, P.V. Kalutskiy, N.B. Dremova, & A.I. Ovod, Higher Education in Russia **29(1),** 105-115 (2020)
9. V.A. Starodubtsev, & I.V. Ryashentsev, Higher Education in Russia **29 (4),** 63-72 (2020)
10. A. Collins, & R. Halverson, *Rethinking Education in the Age of Technology: The Digital Revolution and the Schools.* (New York: Teachers College Press. 2009)
11. A.K. Thiele, J.A. Mai, & S. Post, Journal of Physical Therapy Education 28(1) 80–93(2014)
12. M. Brown, Educause Review 50(4), 16–28 (2015)
13. F. Almaraz-Menendez, A. Maz-Machado, & C. Lopez-Esteban, International Journal of Advanced Research 4(10), 2284–2296 (2016)
14. E.P. Panova, *VolgGTU* (Volgograd, 2013)
15. E.P., Panova & N.R. Saenko, Service plus, 94-102 (2020)
16. Yu.G., Semikina & D.V Semikin, Business. Education. Law., **(3)** 399 – 403 (2019)
17. V.S. Khamidulin, Higher education in Russia **29(1),** 135-149 (2020)
18. G. Marci-Boehncke & T. Vogel *12th International Technology, Education and Development Conference* (2018)

Computer aspect of interdisciplinary research in philosophy of education and computer science

Julia Petrova [1] and *Olga* Vasichkina [*2]

[1] Faculty of Linguistics and Journalism, FSBEHE RSUE, B.Sadovaya str., 69, Rostov-on-Don, 344001, Russia, julia-pp@yandex.ru
[2] Faculty of Linguistics and Journalism, FSBEHE RSUE, B.Sadovaya str., 69, Rostov-on-Don, 344001, Russia, ms.vasichkina@mail.ru

Abstract. A large-scale global study of events both within and between disciplines provides strong evidence that leading thinkers have begun to introduce new ways of thinking to the point that most academic disciplines have undergone major paradigm changes throughout the twentieth century, according to Guidley. Increasingly, over the past forty years, we can see significant changes, if not in all, then in the main scientific disciplines. New ways of thinking have appeared in almost all disciplines. With the help of interdisciplinary approaches, a movement has emerged to integrate knowledge, moving beyond the fragmentation of knowledge related to disciplinary specialization. Computer technologies and computer concepts have penetrated almost all areas of academic research. These events can be considered as the implementation of new ways of thinking and new models of knowledge, respectively, and may be edges of the evolution of consciousness, in which computer technologies occupy a significant place. Education and communication in the age of globalization and computerization are an obvious necessity for the continued existence of society.

1 Introduction

As scientific knowledge develops across a wide range of disciplines, scientists are increasingly aware of the need to connect disciplinary areas to get fuller answers to critical questions and simplify the application of knowledge in a particular field. Nowadays there is also a growing emphasis on promoting interdisciplinary research and supporting it through a number of specific proposals. Integrative research concepts such as interdisciplinarity and transdisciplinarity are becoming more and more widely spread in research. Despite the widespread opinion that interdisciplinary cooperation is useful, there is not enough research on the problem of the philosophy of education and the philosophy of computer science, and the impact of computer telecommunications on education. According to J. Nash, the main problem in interdisciplinary research extends to "intrapersonal, interpersonal system levels"[1]. In parallel with the problems of interdisciplinary research, we suggest changes in cooperation. For the purposes of our study, collaboration is defined as "working together" with other parties to determine the achievement of mutually agreed goals [2].

2 Problem Statement

Modern science is not widely taught in educational institutions because it is considered to be complex, expensive and / or esoteric. Enlarging the ways to provide access to new intellectual tools of modern science (scientific papers, articles, monographs, research, etc.) significantly helps increase the international competitiveness of not only institutions, but also countries.

Institutions are usually organized in terms of disciplines (such as philosophy) or practices (such as education). Disciplines reveal some phenomenon in the world and make it a subject of study while practice is an integral part of the process. Thus, educational practice, for example, is part of the pedagogical process, in which higher education, as in the case of critical thinking, requires creative thinking and is an ideal place for practice, skill development and the formation of a moral personality.

3 Research Questions

Researchers of the philosophy of education, which is considered a branch of practical philosophy, address both the internal discipline of philosophy and educational practice, as well as developmental psychology, cognitive science in general, and other relevant disciplines [3]. The philosophy of education is the field of philosophy that deals with philosophical questions concerning the nature, goals, and problems of education. "...to call the philosophy of education a special branch of philosophy does not mean, however,

* Corresponding author: ms.vasichkina@mail.ru

that it is a separate branch in the sense that it can exist separately from the established branches of philosophy. ... It would be more appropriate to perceive the philosophy of education as using branches of philosophy and combining them in ways that are relevant to educational issues" [4].

The theory of constructivism in education assumes that people create their own understanding and knowledge of the world through communication, experience, and reflection on the created experience. We use the term "collaboration" interchangeably with terms such as communication, collaboration, mutual planning and integration, and assistance [5]. While most of the phenomena and problems currently being investigated are too complex to be fully described in the context of a single field of philosophy – the philosophy of education, or the philosophy of computer science, interdisciplinary research has the potential to improve communication and accelerate the implementation of discoveries in practice.

4 Purpose of the Study

The philosophy of education, as a branch of applied or practical philosophy, studies the nature and goals of education, as well as philosophical problems that arise in the theory and practice of education. This field of philosophy, since it is the practice of all human societies, its social and individual manifestations are diverse, and its influence is so profound that the broad nature of this problem affects questions of ethics, social and political philosophy, epistemology, metaphysics, philosophy of thought, language, reasoning, value and moral education, teaching and learning [6].

The term "philosophy of computer science" is not as well known as "philosophy of education". One of the original characteristics of the term "philosophy of computer science" we see in the work of Timothy R. Colburn "Philosophy and Computer Science (Explorations in Philosophy)" (2000), in which the scientist traces the philosophy of all computer sciences. The author claims that the philosophy of computer science is a "special discipline" in the same sense as the philosophy of education [7]. It certainly differs to a certain extent from the philosophy of artificial intelligence, philosophy of information, and philosophy of mind, being a separate research area of the philosophy of science, but as a branch of philosophy it also deals with the problems of specific academic disciplines, which usually include ontological, epistemological, and methodological problems, as well as questions about logic, ethics, and semantics, in which we can trace the philosophy of all computer sciences.

Computer technologies and computer concepts have penetrated almost all areas of academic research. This interdisciplinary specialization makes it possible to combine the study of computer science with other academic disciplines, combine the study of computer science with traditional academic disciplines such as physics, chemistry, sociology or biology, and/or with new fields that include a significant computational component, such as bioinformatics, cognitive science and digital art. [8].

5 Research Methods

Combining the study of computer technology and computer concepts, for example, with cognitive science, makes it possible to consider such problems as the intelligent behavior of computers and robots, to understand the human brain and human intelligence using computational models, and predict some social and ethical consequences of the intelligent machines introduction. Interdisciplinary research in this field combines philosophy with relevant courses in computer science and mathematics.

The value of interdisciplinary research is in promoting the areas of study and accelerating scientific discoveries. And since the ideas researched often exceed the limits of a single discipline or program, integration of research by means of the interdisciplinary approach prepares a workforce that solves scientific problems in an innovative way. Technology has always been at the forefront of human education. From the times of figures carved on the stone walls to the present day, when most of the students are equipped with several portable technological devices, technology continues to raise educational potentialities to a new level. When you look at where educational methods and tools have come from, and where they will take us in the future, the importance of technology in education and research becomes now more obvious than ever before. However different the history of computers in these fields may be characterized, it is clear that innovators in this field have created the most provocative and stimulating ideas in the history of interdisciplinary research.

6 Findings

The age of computers in education since the second half of the 20th century [9] has improved the quality of education in many ways. For example, computer text processing programs have made it easier to correct spelling errors, grammatical errors, and punctuation marks. Some of the latest text editors can insert text quotes into student's research papers. Similarly, the software can correct errors and check the plagiarism of the work under study. Therefore, the benefits that come with text-based computer writing software outweigh the negative impact on students' writing literacy. The Internet has opened the door to information, knowledge, and educational resources, and has expanded learning opportunities both inside and outside the educational institutions [10].

Teachers use online materials to prepare lessons, and students use them to expand their range of learning and research. Information and communication technologies and high-speed computers have revolutionized the concept of information processing, provided an environment for alternative learning and research. Computers being in use since the beginning of the last century are changing the way we learn, become a critical addition to educational perception, and thus open up more learning opportunities [11].

Currently the usage of Information and communication technologies in education brought a new era, characterized by expanded access to education, improved quality of education and by strengthening the relevance of education as a digital workplace. The Internet has changed the way and the image of education and has become an incentive to optimize teaching methods. Thanks to easy access to the Internet, distant learning is becoming increasingly popular.

The growing demand for courses and programs with more profound specialization forces the universities to differentiate their offers in the educational sphere. Currently, educational computer literacy in Russia is experiencing a new explosion caused by the pandemic of the century. The most important are all kinds of interrelated competencies needed to improve people's interaction with information, both for educational and scientific purposes.

The theory of constructivism in education assumes that people create their own understanding and knowledge of the world through experience and reflection based on the created experience. We use the term "collaboration" interchangeably with such terms as communication, collaboration, mutual planning, and integration assistance [12]. While most of the phenomena and problems currently being investigated are too complex to be fully described in the context of a single field of philosophy - the philosophy of education, or the philosophy of computer science, interdisciplinary research has the potential to improve communication and accelerate the implementation of discoveries in practice. For the purposes of this study, "collaboration" is understood as working together with other parties to determine the achievement of mutually agreed goals [2].

Different countries of the world have their own intellectual traditions and their own ways of institutionalizing philosophy in the academic universe. Society not only continues to exist through cooperation, transmission, and communication, but one can say that it exists within them. There is more than just a verbal connection between the words "community" and "communication". People live in society because they have something in common, and communication is an "example of a speech event" [13], a way in which common features, values, and interests are found and revealed. With the help of technology, opportunities for communication and collaboration expanded significantly. Traditionally, classrooms were relatively isolated, with collaboration being limited to other students in the group or in the building. The main goal in the communication process is to understand people, that is, to understand the context [14]. "...examples of multicultural and cross-cultural communication are becoming a common form of communication for many people in their daily lives" [15]. Today, computer telecommunications provide such forms of communication and collaboration that have never been dreamed about before. For example, students can read publications of scientists in blogs; send questions to scientists by email; participate in video conferences; share what they are studying with students in other classrooms, cities and countries who are investigating a similar problem; work together on group projects; use virtual scientific laboratories to conduct scientific experiments, etc. Thus, students of the 21st century have a great opportunity to form communities of researchers who share "believes, aspirations, knowledge, understanding, and common sense..." [16].

7 Conclusion

Education and communication in the age of globalization and computerization are an obvious necessity for the continued existence of society. It may seem that we are somewhat fixed on the truism. Though, the justification is that this emphasis is a means of distracting us from the excessively scholastic and formal understanding of education. One of the tasks of education is the formation of a moral personality, but this is only one task, and it is relatively perfunctory in comparison with the others. Only by recognizing the need for more fundamental and permanent teaching methods we can make sure that scholastic methods are placed in their true context. W. Jacob notes that nowadays higher educational institutions can no longer train graduates to solve all current and emerging problems basing on a single disciplinary source. Interdisciplinary research approaches are the most essential foundation for best meeting the dynamic needs of today's students of higher educational institutions [17].

The exclusivity and uniqueness of the education development in the socio-cultural space of various countries and regions remain a determining factor in the formation of national characteristics that reflect the ideas of individual peoples about national and historical values and determine the fundamental nature of national education. Simon G. notes that advances in science and information processing technologies have changed the meaning of the verb "to know". If it used to mean "store information in your memory", now this means access to information, the ability to identify it, extract it, and use it [18].

Global transformations that have affected humanity have become a criterion for the unification of education, determining international trends in its development. This new approach to education results from the influence of complex and ambiguous processes in the development of society, changes both in the nature of relations between people and the improvement of technologies, the evolution of the way of communication. "The indisputable relevance of education issues is related to the current trends in increasing the importance of the educational sphere, the formation of teaching methods that meet the realities of our days, which consist both in the transfer of knowledge and in its preservation" [19]

References

1. J.M. Nash, Am J Prevent Med. 35(2), 133–140 (2008)

2. N. Wells, R. Johnson & S. Salyer, The international journal of advanced nursing practice **12(4)**, 161-166 (1998)
3. H. Siegel, The Oxford Handbook of Philosophy of Education. Retrieved from (2009), URL:https://www.oxfordhandbooks.com/view/10.1093/oxfordhb/9780195312881.001.0001/oxfordhb-9780195312881-e-001 (date of access: 16.03.21).
4. I. Scheffler, *Reason and Teaching* (Hackett Publishing, 1989)
5. D. Drotar, Developmental & Behavioral Pediatrics, **23(3),** 175–180 (2002)
6. T. Husén, & N. Postlethwaite, Philosophy of Education **7**, 3859-3877 (1985)
7. T.R. Colburn, *Philosophy and computer science* (Armonk, Routledge, 2000)
8. Van Wormer, R., Lindquist, W. Robiner S. Finkelstein, Dimensions of critical care nursing: DCCN. **31(3),** 202-210 (2012)
9. Informational Technology and its Impact on American Education. Office of Technology Assessment, U.S. Congress, Washington, DC (1982), URL: https://www.princeton.edu/~ota/disk3/1982/8216/8216.PDF 001 (date of access: 16.03.21).
10. H.L. Lim, *Innovative Methods and Technologies for Electronic Discourse Analysis* (IGI Global, 2013)
11. Y. Petrova, & L. Olifirenko, Humanities and socio-economic sciences, **3 (112),** 18-21 (2020)
12. D. Drotar, Developmental & Behavioral Pediatrics, 23(3), 175–180 (2002)
13. Y.A. Petrova, Humanities and socio-economic sciences, **2 (87),** 88-90. (2016).
14. E.E. Nesmeyanov, & Yu.A. Petrova, *SHS Web of Conferences: Trends in the Development of Psycho-Pedagogical Education in the Conditions of Transitional Society.* **70** (2019)
15. M. Estaji, & A. Rahimi, International Journal of Society, Culture & Language (IJSCL), **6 (2),** 1-18 (2020)
16. J. Dewey, *Democracy and Education*, EBook #852/ 2015, URL: https://www.gutenberg.org/files/852/852-h/852-h.htm#link2HCH0008 (date of access: 16.03.21).
17. W. Jacob, Palgrave Communications **1,** 15001 (2015)
18. H.A. Simon, *Designing Organizations for an Information -Rich World* (Baltimore, MD: Johns Hopkins Press, 1971*)*
19. Collection of scientific papers. Theory and practice of teaching foreign languages in a non-linguistic university: traditions of innovation, prospects, (2014) URL: http://elib.fa.ru/fbook/melnichuk.pdf/download/melnichuk.pdf (date of access: 16.03.21).

Electronic resources as a means of constructing the learning environment of a children's art school

Olga Popova[1,*], Anzhela Danilenko[2], Nadezhda Kosenko[3], and Svetlana Chernikova[4]

[1] MBISE "Children's Art School of the City of Belgorod", Belgorod State National Research University, Pobedy str., 77, Belgorod, Russia, olggubareva@yandex.ru
[2] Belgorod State National Research University, Belgorod, Pobedy str., 85, Belgorod, Russia, Danilenko@inbox.ru
[3] Orel State University named after I. S. Turgenev, Orel, Komsomolskaya str., 95, Russia, kosenko@yandex.ru
[4] Orel State University named after I. S. Turgenev, Orel, Komsomolskaya str., 95, Russia, svetlanachernikovaorel@gmail.com

Abstract. The article presents a model of personalized mixed learning for the implementation of programs of general additional education in the field of fine arts. This model is presented through the metaphor of the "space of opportunities", in which the student can choose their own educational trajectory and ways to achieve educational goals. The structure of the personalized educational space is slated, the role and place of information and communication technologies in the implementation of personalized education are considered. The key elements of the model presented by us were goal-setting, criteria-based evaluation of the results of educational activities and the use of information and communication technologies. Within the framework of the proposed approach to goal-setting, the transition from simple knowledge and skills to more complex ones is carried out, as well as the synthesis of mental operations with a specific application and creative rethinking of the acquired knowledge. Criteria-based evaluation is intended to provide high-quality feedback. It is carried out on the principles of equality; personalization; clarity of criteria; evaluation of the result, not the activity of the child in the lesson; continuity. The main characteristics of the presented model are: level-based, result-oriented, interactive, and variable.

1 Introduction

One of the factors that can provide the effective development of the modern education system is its personalization. Today, countries with effective educational systems are actively developing models of mass personalized learning. In the Russian Federation, attempts to create personalized learning models aimed at meeting the educational needs of comprehensive school students and allowed them to independently determine the learning strategy are also being made [1].

2 Problem Statement

However, in the Russian Federation at the present phase, there are no models and strategies for personalized learning for the implementation of additional educational programs in the field of fine arts.

3 Research Questions

Programs of additional education, including in the field of fine arts, have significant specifics that distinguish them from the syllabuses developed for comprehensive schools. This specificity is due to the personality-oriented approach that underlies the art education system, the close interaction of the student and the teacher in the learning process, and the creative nature of artistic activity. Is it possible to develop a model of personalized distance learning for general development programs in a children's art school taking into account this specificity?

3.1 What is the structure of the personalized learning model for the implementation of additional education programs in the field of fine arts?

3.2 By what technologies can this model be implemented in institutions of additional education of children?

4 Purpose of the Study

The purpose of our research is development and approbation a model of personalized learning for the implementation of additional education programs in the field of fine arts for children's art schools.

4.1 Achieving this goal requires solving a number of research tasks:

- Definition of the basic principles of creating a personalized learning model for additional education programs in the field of fine arts;

* Corresponding author: e.malushko@volsu.ru

- Definition of the structure of the personalized educational space, its system-forming elements.

5 Research Methods

The methodological basis of the work was the research of P. N. Kirilov, N. I. Koryakina, T.Wanner, E.Palmer, K. Turvey, M. Hayler, M.Virvou [2-4]. Based on the study of their work, we identified the key factors of personalization of learning: criteria-based evaluation and mixed learning. Criteria-based evaluation is based on the comparison of the actual achievements of the child with his previous achievements with the predominance of formative and ascertaining evaluation and the level approach, as well as the presence of evaluation as an educational activity (self-assessment). Mixed learning is a technology for organizing the educational process based on combining traditional and e-learning, taking into account new didactic opportunities due to the use of information and communication technologies. The preferred form of implementation of the mixed learning model is an inverted classroom, in which there is a change of working areas: at home, the student gets acquainted with the educational material, and in the classroom, the practical training of the acquired knowledge is carried out. When developing a model of personalized learning for the development of additional educational programs in the field of fine arts, we also relied on the ideas of the constructivist approach [5-6]. The essence of this approach is the idea that knowledge can not be "transmitted" in a ready-made form, the child receives it on the basis of existing experience in the process of educational activities based on the search and solution of problems that are important to him. The constructivist approach is based on the attitude to active cognitive activity of students, to gain knowledge in the process of active goal-setting, search and solution of significant problems for them. The role of the teacher in this case is to guide and regulate this activity. The constructivist approach is associated with significant shifts in the established paradigm of education. This is, firstly, a shift in the content of education, in which the emphasis is shifted from the knowledge component ("I know what") to the activity component ("I know how"). Secondly, the transformation of the form of education, as a result of which the emphasis is shifted from learning as a process where the student is the object, to educational activities in which the student is the subject. Thirdly, there is a shift in the position of the teacher: if previously the teacher acted as a translator of knowledge and mentor, then with the constructivist approach, his role is the role of the organizer of activities.

Methodologically significant for us is the concept of skills of the 21st century, which states the need to develop not only subject, but also meta-subject skills, the key among which are the so-called "4C": communication, collaboration, creativity, critical thinking. When developing a personalized learning model for the development of additional educational programs in the field of fine arts, we also relied on a system-activity approach, in which the transition from passive learning to active learning activities is carried out and the relationship between planned results and evaluation is emphasized.

Based on the methodological basis of the research, we have developed a model of personalized learning, which is focused on creating a space of opportunities for students, in which they can independently set goals, make choices, and reflect on the results obtained.

6 Findings

The key elements of the presented model were goal-setting, which allows to structure the learning goals according to the levels of complexity of educational and cognitive tasks, criteria-based evaluation of the results of educational activities and the use of information and communication technologies.

6.1 Goal-setting

The proposed model is based on the modular principle of implementing educational programs and the principle of goal-setting levels. The system of goals of the proposed model is based on the taxonomy of R. Marzano [7] according to which three areas of knowledge are distinguished – information, mental operations, and physical operations. Based on the taxonomy of R. Marzano, we structured the learning goals into four levels. These are the levels of obtaining information, mastering information, applying the acquired knowledge, and forming meta-subject skills. Goal-setting is based on the principle of clarity of goals for the child. This principle allows us to realize the position of the student as a subject in relation to the learning goals. The goals are presented in the form of statements formulated on behalf of the student in the activity form ("I will be able to recognize", "I will be able to portray") and taking into account the SMART criteria.

Table 1. Levels of goal-setting module

Level			Level name	Specifics of educational activities	Examples of the formulation of educational goals
Meta-subject	Advanced	Level 4	4K skills	Transfer of knowledge and skills to another area	I apply the acquired knowledge in my creative activity
Subject	Basic	Level 3	Physical operations	Readiness to apply the learned elements in practice	I can create (compose) an imaginary (conditional) still-life and choose a palette of colors for it with basic colors.
		Level 2	Mental operations	Understanding and assimilation of elements	I can graphically represent the history of coloration development or describe in an essay about coloration development from ancient paintings to the application of basic colors in a painting.
		Level 1	Information	Non-mechanical assimilation of material and its reproduction	I can choose works whose coloration is based on basic colors.

Within the framework of this approach to goal-setting, the transition from simple knowledge and skills to more complex ones is carried out, as well as the synthesis of mental operations with a specific application and creative rethinking of the acquired knowledge. Structured goals are an integral part of a personalized learning model for additional general education programs in the field of fine arts. They allow us to solve a number of important tasks. Firstly, to provide a choice at the basic level and the ability to choose between the basic and advanced levels. Secondly, to provide guidelines for the educational activities of the student with less dependence on the teacher. Thirdly, to provide a methodological basis for the teacher, who will be able to offer students interchangeable tasks.

The subject content is structured into training modules lasting 4-6 hours. Learning within the framework of the module is based on the principle of an inverted class: theory is studied at home (2 hours are allocated for it), practical skills are practiced in the classroom (2 hours), for those students who have chosen an advanced level, an additional task is provided, which is performed at home and at school (1-2 hours).

Individual educational trajectory is provided through the variability of tasks. The tasks are designed in such a way that the child can choose the source of information (text, video and audio), the type of educational activity, the forms of control over the assimilation of theoretical material, and the forms of work. A student who has completed a level earlier than planned can use the free time to study an additional module on the development of softskills – flexible skills.

Tasks are designed in such a way that in the process of their implementation, flexible skills of the groups "thinking" (exploring the world, discovering new things, solving problems), "interaction with others" (acting in a team), "interaction with myself" (understanding myself, managing myself) are developed. At the heart of this approach is the "soft through hard" - the attitude that the subject content is not an end in itself, but a tool for developing social skills and experience.

Table 2. Examples of the formulation of educational goals

Level	Formulation of the learning goal	Formulation of the learning goal on behalf of the student	Flexible skills group
1.1.	To know the structure of the Delacroix color circle	I can tell you how the Delacroix color circle works.	I explore the world
1.2.	To know the main colors of the color circle and additional ones to them.	I can name and point in the image to the main colors of the color circle and additional ones to them.	I open a new one
1.3.	To know the stages of color development based on the main colors	I can graphically represent the history of coloration development or describe in an essay about coloration development from ancient paintings to the application of basic colors in a painting.	I solve problems

In accordance with the SMART goal-setting technology, the formulated goals should meet several criteria: specificity, measurability, achievability, relevance, and time-bound. Therefore, in the process of goal-setting, it is necessary to avoid generalized formulations of educational goals, such as "develop a sense of beauty" (the criterion of measurability is violated); "draw a landscape" (the criterion of specificity is violated – the volume of the material being studied and the nature of actions are unclear); "perform an exercise" (the criterion of relevance is violated – the formulation contains a teaching method, while it is unclear why you need to perform the exercise, because the educational result is not the fact of performing the exercise, but the knowledge, abilities, skills obtained).

The goals of all four levels should be connected logically, if at level 1 a simple subject result is formulated, then at level 4 it should be integrated into a more complex, meta-subject result.

6.2 Criteria-based evaluation

Criteria-based evaluation is intended to provide high-quality feedback. Within the framework of the proposed program, it is carried out using distance and e-learning technologies, which saves the teacher time for interacting with students. Level 2 tasks published on the e-learning platform are evaluated automatically or are provided with evaluation schemes (rubricators), which allows self-testing. At the end of each module, willing students present the results of their work on Level 4.0 tasks to the class and receive feedback. Feedback can also include formative evaluation — the current, frequent evaluation of the work performed, which allows the student to understand whether he has achieved the goal and quickly adjust his work, and the teacher to understand the student's educational position. This evaluation is carried out within the framework of performing creative work in practical classes.

Planned types of evaluation:

- current subject evaluation (records the percentage of tasks completed within the module);
- final subject evaluation (individual verification of the achieved subject results at the end of the development of the training module / group of modules).

A level is considered completed when the student demonstrates the grounds for achieving it by completing the corresponding tasks.

Table 3. Evaluation criteria

Evaluation criteria	1	2	3
Volume of completed tasks	x	x	x
Volume of educational material	x	x	x
Volume of work on the assimilation of educational material	x	x	x

Criteria-based evaluation is intended to provide high-quality feedback. It is carried out on the principles of equality; personalization; clarity of criteria; evaluation of the result, not the activity of the child in the lesson; continuity. Within the framework of the proposed programme, criteria-based evaluation is carried out using distance and e-learning technologies, which saves the teacher time for interaction with students.

7 Conclusion

Thus, the key elements of the model presented by us were goal-setting, criteria-based evaluation of the results of educational activities and the use of information and communication technologies. The model is based on the principle of modular learning, the implementation of which represents the movement of the student to the educational goal through the achievement of a certain result formulated in accordance with the SMART goal-setting technology (specific, measurable, achievable, relevant, time bound). The main characteristics of the presented model are: level-based, result-oriented, interactive, and variable. This model is intended for implementation in children's art schools and schools of art.

References

1. P. Kirilov, & N. Korjakina, School of opportunities. Individual trajectories of development. (2018), URL: https://edpolicy.ru/school-of-opportunity(date of access: 16.03.17)
2. K. Turvey, & M. Hayler, Teaching and Teacher Education **68,** 42-52(2017).
3. M. Virvou & E. Alepis, Computers & Education **44(1)**, 53-68 (2005)
4. T. Wanner, & E. Palmer, Computers & Education **88,** 354-369 (2015)
5. N.Özkarslı, & A. Gürdal, *Procedia - Social and Behavioral Sciences* **2 (2)**, 3965-3970
6. Q. Zhang, & Q. Kou, *Physics Procedia* **25,** 2294-2297 (2012).
7. R.J. Marzano, Designing a new taxonomy of educational objectives (Thousand Oaks, CA: Corwin Press, 2000)

SHS Web of Conferences **109**, 01031 (2021) https://doi.org/10.1051/shsconf/202110901031
LISID-2020

The influencing potential of the conflictogenic media space in the aspect of legal culture

Tatiana Rogozhnikova [1]* and *Elvina* Salikhova [2]

[1] Ufa State Aviation Technical University, 12, Karl Marx St., Ufa, 4500008, Bashkortostan, Russia, burzian@yandex.ru
[2] Ufa State Aviation Technical University, 12, Karl Marx St., Ufa, 4500008, Bashkortostan, Russia, salelah12@yandex.ru

Abstract. The authors consider the problematic field of conflictogenic media space in a linguistic-legal aspect. The goal of the article is to present the individual results of linguistic and psycholinguistic studies of polycode controversial texts. In expert practice the authors actively employ the emotional-lexical assessment of verbal and non-verbal models, the assessment of the rhythmic brain activity during the perception of the entire text, the assessment of the rhythmic coherence in the perception of individual parts of the text and their comparison, the technique for recognizing the hidden motives of the object behavior in the process of its professional activity, analysis of the individual coefficients of the text impact on the individual and mass consciousness and subconsciousness. Conflict genes are demonstrated and described in this work. These are the minimum significant units of the component of the conflictogenic media space. The specific character of the polycode text, the texture of which consists of verbal and non-verbal parts, is demonstrated. The quantitative and qualitative results of the research performed are determined by the potential of classical linguistics and related branches of knowledge for solving polycode text problems using the developed software products in the field of psycholinguistic investigations.

1 Introduction

From an expert and analytical point of view, the conflictogenic zone of the modern media space is considered by us 1) as a fragment of the socio-discursive field, which reflects the "communicative battles and fights" taking place in it, whose participants (journalists-authors of publications, including "commissioned" ones, and their heroes), being subjects of the network socio-cultural system, violate the norms of speech behavior and subsequently act as plaintiffs / defendants in litigations (for example, for the protection of honor, dignity and business reputation - art. 152 of the Civil Code of the Russian Federation or for slander and insult – art. 129, 130 of the Criminal Code of the Russian Federation); 2) as a linguo-ecological environment that has a public character, the signs of which are identified by experts in the published information - in the disputed text (see, in particular, art. 128.1 of the Criminal Code of the Russian Federation). A journalist cannot fail to reflect the dynamics of a society that is closely linked to the development of economic, socio-political relations and innovations. The media play a special role in this process and are able to have a significant impact on one or another object of their attention, on the evaluative attitude of society to this object, on its reputation, etc.

1.1 Linguistic and psycholinguistic expertise (hereinafter referred to as LE and PLE) is based in conjunction on the appropriate scientific and methodological tools and terminological subsystems. In contrast to the linguistic, psychological-linguistic and psychological types of expertise, the PLE is based on specific data from experimental linguistics, which studies the products of human speech activity based on experimental materials. Successful results of the formalization of these psycholinguistic (PL) experiments in the form of computer programs have increased the possibilities of using this scientific paradigm in judicial linguo-expert activity. Creating a visualized result of the processed data, based on statistics, on the comparison of coupled databases, enriches the expertise and eliminates the subjective interpretation of the speech product.

1.2 One of the functions of human speech, as well as any text, is the function of influence. Within the framework of the PL, methods have been developed and units of analysis have been identified, with the help of which it is possible to give an objective evaluation of the material: to measure in quantitative terms the strength of the impact of the text, to establish the quality of this impact, to decode the suggestive potential of the language phenomenon, to rank emotional and evaluative signs, to evaluate the motives and intentions of the author of the text, to calculate the "golden section" of the text as a zone of uncontrolled influence on consciousness. The expert uses special units of analysis in order to study the latent suggestive resources of the text, since the lack of linguistic manifestations of deep internal forms does not lend itself to analysis through

* Corresponding author: burzian@yandex.ru

traditional linguistic and psychological techniques and methods. The figures (see Figures 1-8) show examples of the results of working with such analytical units. Explanations of the illustrations are given below in the relevant sections of the article.

Fig. 1. The brain waves pattern activated by the Painter's comments perception in connection with the "KruShchu…" caricature publishing

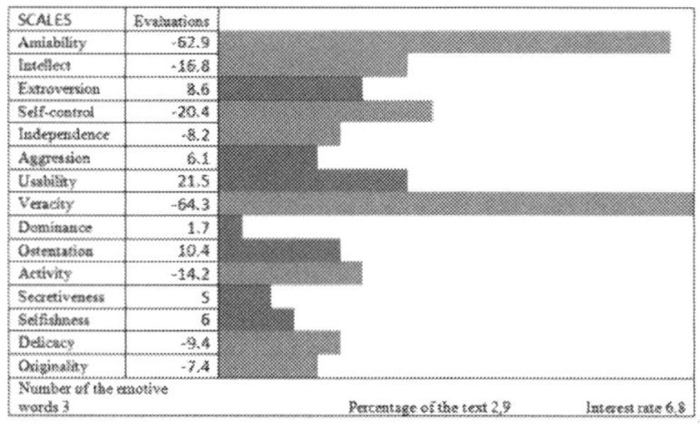

SCALES	Evaluations
Amiability	-62.9
Intellect	-16.8
Extroversion	8.6
Self-control	-20.4
Independence	-8.2
Aggression	6.1
Usability	21.5
Veracity	-64.3
Dominance	1.7
Ostentation	10.4
Activity	-14.2
Secretiveness	5
Selfishness	6
Delicacy	-9.4
Originality	-7.4
Number of the emotive words 3	Percentage of the text 2,9 Interest rate 6,8

Fig. 2. The emotional-lexical evaluations of the «Pontius Pilatus» caricature, writings and comments

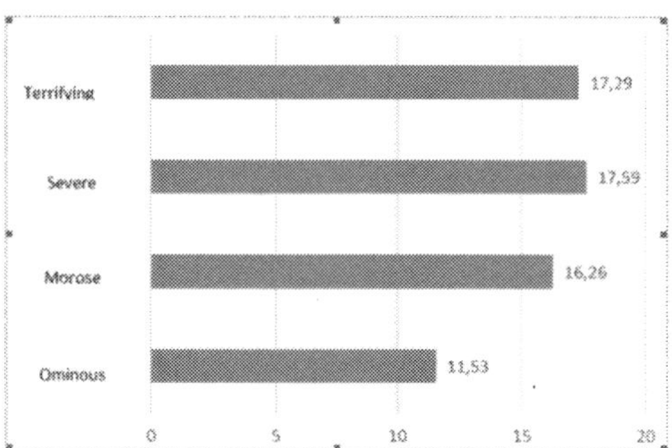

Fig. 3. The dominant features of the «Pontius Pilatus» caricature

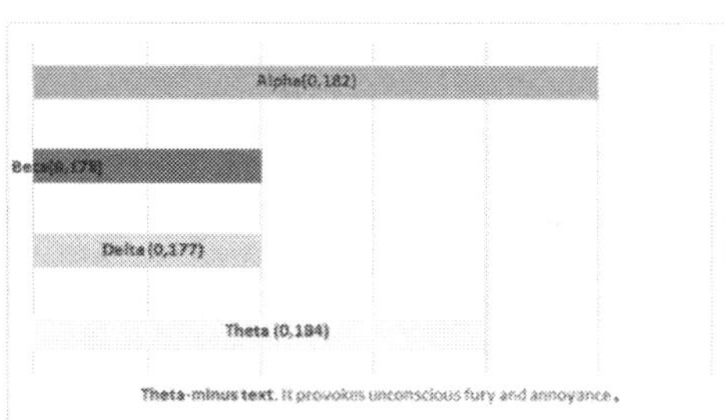

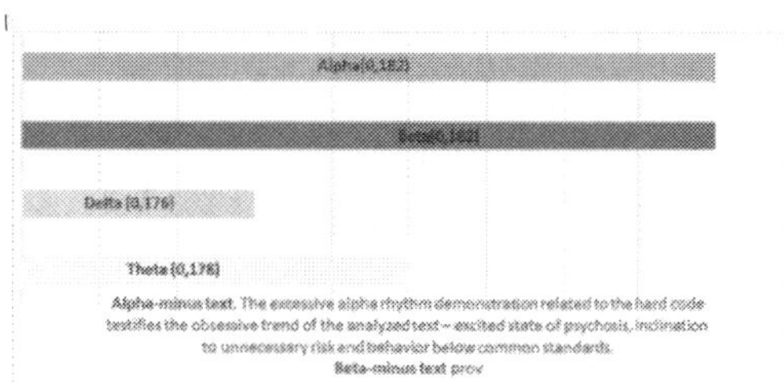

Fig. 4. The brain waves patterns activated by the perception of all the caricature writings analyzed

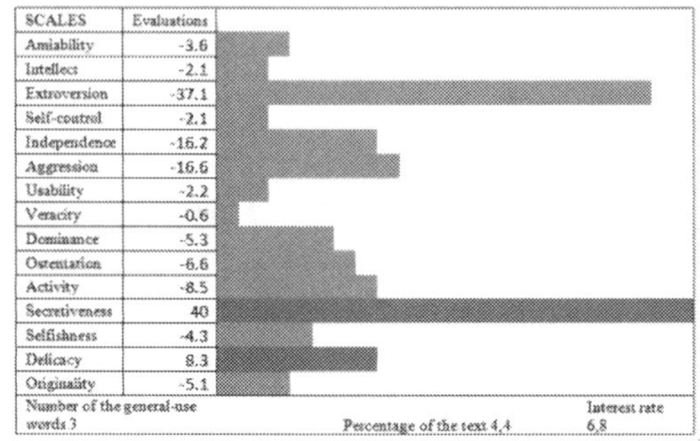

SCALES	Evaluations
Amiability	-3.6
Intellect	-2.1
Extroversion	-37.1
Self-control	-2.1
Independence	-16.2
Aggression	-16.6
Usability	-2.2
Veracity	-0.6
Dominance	-5.3
Ostentation	-6.6
Activity	-8.5
Secretiveness	40
Selfishness	-4.3
Delicacy	8.3
Originality	-5.1
Number of the general-use words 3	Percentage of the text 4,4 Interest rate 6,8

Fig. 5. The emotional-lexical evaluations of the caricature writings

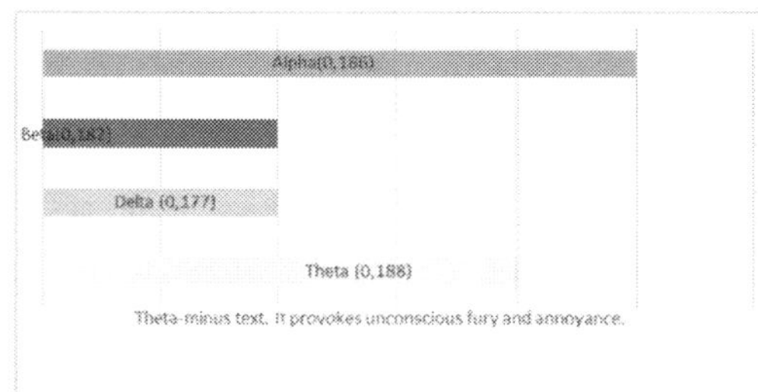

Fig. 6. The brain waves pattern activated by the perception of the «Have a lick» caricature writings

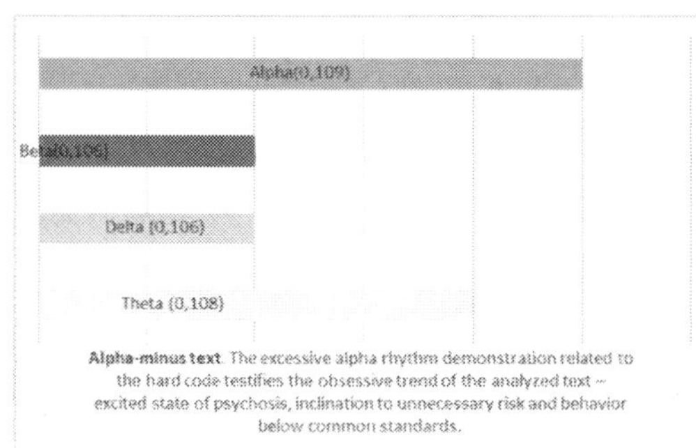

Fig. 7. The brain waves pattern activated by the «Have a lick» caricature comments perception

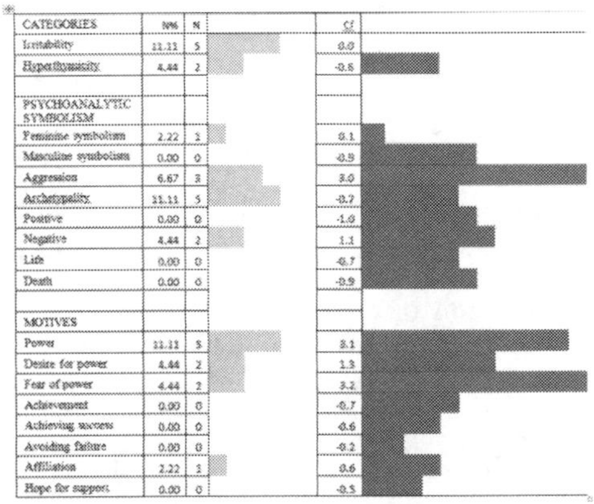

Fig. 8. The load on the categories (categorial-statistical evaluations) of the «Have a lick» caricature

2 Problem Statement

Traditionally, linguists conduct a study of the text in order to identify its semantic orientation, the expressiveness of speech units, the modality of propositions, their semantics and formal-grammatical characteristics, the specifics of the stylistic means used. Psycholinguists identify those conflictogens through which it is possible to decode the influencing potential of a language phenomenon. In order for lawyers to qualify cases of information distortion, linguists and psycholinguists should be able to differentiate communicative elements according to their purpose to provoke a conflict through separate linguistic elements – sound-letter complexes, morphemes, words and their combinations, titles, utterances, intonation, video sequences, etc. [1].

Conflictogens, representing the minimum significant units of the discursive component of the media space, act as detonators of a conflict communicative situation. The origins of the latter, as shown by the long-term expert practice of the authors of the article, indicate personal motives that encourage the author (whether a journalist, commentator of remarks in social network chats or a participant in discussions) to inform the public.

3 Research Questions

Within the framework of the publication, we propose one of the ways to understand complex LE and PLE based on the achievements of modern Russian applied linguistics and PL. As illustrations, fragments of a comprehensive expertise with the participation of the authors of the article are provided (with the preservation of confidentiality and the symbol of the subjects – the Artist, the Organization, the Head of the Organization). In a modern society characterized by information saturation, there is an increasing need to rank the information flow in such a way already at the stage of text generation, so that the addressee can effectively isolate the most significant information for himself. This is expressed in semiotically complicated (polycode, creolized) texts that include verbal and nonverbal components. In the modern media space, such a synthesis of the word and the iconic element is presented in written and oral texts of journalistic communication.

The term "creolized texts" is widely known, which is used in psycholinguistics by Yu. A. A. Sorokin and E. F. Tarasov. In their opinion, these are " texts whose texture consists of two inhomogeneous parts (verbal language (speech) and non-verbal (belonging to other sign systems than natural language)" [2]. As a working, this term is used in their works by E. E. Anisimova, M. A. Boyko, N. S. Valgina, N. M. Chudakova, M. B. Shinkarenkova and others. The researchers note that the creolized text appears as a complex text formation, in which verbal and non-verbal elements form one visual, structural, semantic and functional whole, aimed at a complex impact on the addressee. Further development of research in this area has shown that the terminological value of the concept of "creolized text" has become a subject for further discussion [3]

So, semiotically heterogeneous or complicated phenomena are treated in science as polycode (A. V. Alferov, G. V. Eiger, A. G. Sonin, V. L. Yukht), as creolized (E. E. Anisimova, Yu. A. Sorokin, E. F. Tarasov), as isoverbal (A. A. Bernatskaya), as video-verbal (O. V. Poymanova), as heterogeneous (M. A. Ishchuk, A. G. Sonin), as linguovisual (L. M. Bolshiyanova), as a kommunikat (N. N. Volskaya, V. E. Chernyavskaya), as visual (A. A. Kornienko). Under the term "polycode" text (identical in meaning in this article, the term "creolized" is used) we, following A. G. Sonin [4], T F. Petrenko [5], Z. I. Trubina [6], understand a multi-level sign formation of a synthetic nature, which represents unity in functional-structural, semantic and communicative aspects and which is built on the combination of semiotically heterogeneous components. A person who perceives video or audio information pays attention to alphabetic and other sign means and interprets them in the intentional direction set by the author of the text. The creolized text, which is the object of our research, is the text that accompanies the "static" caricatures on the pages of the social networks "Facebook " and "Instagram". In them, the Artist is the author of graphic conflictogenic works and part of the printed texts expresses his critical attitude to the situation that has developed around a well-known Organization in Bashkiria. The studied discourse environment has a focus on visual perception (text and graphic materials); spatial and temporal extension, pictorial character, continuity; a special sphere of functioning (assessed in perception as a socially significant activity).

4 Purpose of the Study

The set of visual materials with text comments of the participants-visitors of the page forms the syntactic hierarchy of the Internet page, which is built depending on the features of the perception of the sign, on the degree of unambiguity and objectivity of its interpretation. With this approach, symbols, mainly graphic and linguistic together, are organizing the text space as a whole. This function is performed by the verbal title of the material, rarely in the format of a chat in a social network, made out with the help of graphic or numeric characters. In the context of communication in a social network, both the title and the author's comment spatially frame the video sequence and accompany the text, minimizing possible interpretations of the text.

In the linguosemiotic and PL aspects, the caricature of the Head of the Organization in the form of a thimblerigger has a tough complex effect. Caricature (from caricare "exaggerate") - a satirical or humorous image in which the comic effect is created by exaggerating unpleasant features, unexpected likenesses. When analyzing this polycode text, there were no signs of a comic effect. The characteristic signs of the text were *sinister, gloomy, harsh, and intimidating*. They create negative images not only of the Head of the Organization and members of the community, but also of such structures as the court, the prosecutor's office, the Ministry of Internal Affairs, which the main character manipulates as a gambler-a thimblerigger. Linguistic synonyms of the concept - a fraudster, a swindler ("A swindler with a tailor's tool"; "a fraudster who speaks: "I twist and turn, I want to confuse everything!"" [7], the deceiver [8] - leave no doubt that the use of this image in public communication humiliates the honor and dignity of a particular person, as well as harms the business reputation of these structures, lowers their status [9], making them the subject of condemnation by society.

The caricature, as well as the inscriptions to it, have a hard complex effect on the subconscious of those who read and consider it. The sound side of the inscriptions and comments on the caricatures also adds stiffness. Variants of the pronunciation of the words *Krushchu* (from the distorted sound appearance of the verb "twist" – "1. To bring in a circular, rotational movement; to rotate, to turn. 7. Conv. To dispose of at your own discretion, to command. 8 conv. To avoid a direct answer with any tricks, trickery" [8]; the same VERSCHU from the verb "to spin" – "3. Conv. to dispose at your own discretion, to force to act at your own will" [ibid. P. 120]; KHOSHCHU from "to want" – "1. To have a desire, an intention to do something; to feel the need, the need for someone/something" [ibid. P. 1452], "to confuse" - "2. To make it unclear, difficult to understand, or complicated. 3. Conv. To confuse, to mislead. 4 Conv. To involve someone in any unpleasant business, to implicate" [ibid. P. 341]"), *Verschu, Khashchu* exaggeratedly demonstrate the accent inherent in the Turkic-speaking native speakers of the Russian language. Such examples are numerous and can be found both in other caricatures (*Gumerysch, Zaoschnik*) and in the comments to them (*molodiS, Razdascha, do konSa*). Re-checking the conclusion about the harsh complex impact of this image, accompanied by inscriptions, was carried out by us through the analysis of the Artist's comments, with which he accompanied an open discussion about the caricature on social networks. **Figure 1** shows the patterns of brain waves occurred during such a "conversation" with different commentators.

4.1 The excessive manifestation of the alpha-minus rhythm, associated with hard coding and hard impact, again indicates the manic orientation of the speech product generated by the Artist, which can lead him and those he affects to unnecessary risk and violation of the rules of behavior, including speech. A high rate of beta-minus rhythm acts as a provocateur of excitement, fear and anxiety, always introduces a stressful destabilizing component both in communication and in the situations that accompany it [10].

4.2 According to the classification of invective vocabulary in [11], the word *fraudster*, which is strongly associated with the image of a thimblerigger, belongs to the first group of "words and expressions that conceptually denote antisocial, socially (and more often legally) condemned activities". It is noted that the use of such units in the sphere of public and formalized public communication is considered an insult, that is, it is a question of such a nature of the context, "which can infringe on the honor, detract from the dignity and harm the reputation of the characterized persons, if the facts stated in the justification of the assessments are fictitious" (ibid. Pp. 209-210). The targeting of the public use of invective lowers the public status of the depicted subject, gives it the status of a socially condemned person, putting it in the role of the guilty [12]; reveals, makes publicly known the personal hostility of the addressee, ... makes the victim the subject of discussion in society, the subject of gloating. The information in the comments and descriptions of the situation around the Organization is given in the form of statements about the facts of the condemned character's activities and actions in a managerial position, which are condemned by society as illegal and immoral. Such information is aimed at humiliating the dignity of a citizen, but does not have an indecent (obscene) form, although it can significantly discredit the honor, dignity and business reputation of a particular criticized person. As E. S. Kara-Murza clarifies, "from the linguistic and legal point of view, defamatory information, first, is information about the offenses or misdemeanors (of the plaintiff), and not any negative factual information about him. Secondly, it concerns directly (the plaintiff). Thirdly, it should be expressed in the form of a statement" [13].

4.3 The interpretation of the caricature, conditionally called by us "Pontius Pilate...", is also quite transparent for the reader's perception. An artist who works in such a genre is always in a conflictogenic risk zone, because it is associated with a potential insult. The caricature, being anonymous in different epochs, has now finally lost it and has gained not only a specific goal for psychological attack, but also a certain number of supporters, on whose opinion the caricaturist relies. There was a transition to personality, which means a new type of functioning of the insult. The time has come for a march, clothed in an artistic form, on human dignity itself, regardless of the place in society of the Artist and his victim. The dialogue, which is conducted in the language of "artistic insults" [14], does not lead to the establishment of truth or reconciliation. The purpose of this kind of creativity is to arouse the determination of the opposition, morally eliminate, weaken and humiliate the enemy. The possible political bias of the work is based on the totality of the values of the social group on behalf of which the Artist acts. And here there is a question about the ethical limits of what is permissible, which explicates and makes clear the author's (Artist's) own ideas about aesthetics and morality.

In the context of the promoted attitude to the Head of the Organization on the Internet page, the interpretation is unambiguous – he carries his cross associated with guilt and death. The cynicism of caricature positions the Artist as a person who does not know the measure of what is allowed, who descends to mockery and is not able to stop at the line that cannot be crossed. This leads to the fact that the character, in accordance with the laws of associative processes, becomes a martyr, and the depicted judge (a well – known political strategist in Ufa) is a villain who has assumed the role of a prosecutor.

Figure 2 shows the results of the emotional-lexical evaluation. Recall that the special PL block of the computer program VAAL [15] takes into account not only evaluative words, but also adjectives, verbs and nominalizations used to describe and evaluate a person. As semantic scales, 15 factors are used, according to which each lexical unit is evaluated, and the emotional background is determined by the evaluations of the participants of the experiments, their expectations and experiences. Analyzing the material on the factors of emotional evaluation, the specialist has the opportunity to establish the main criteria and values for them. Emotional-lexical evaluations are calculated by calculating the average values for each of the scales. If the resulting evaluation is positive, the histogram line is red, if it is negative, it is blue. At the bottom of the histogram, data on the number of emotional-evaluative words in the text, the percentage of emotional-evaluative words from all the words of the analyzed text, as well as the normal percentage of emotional-evaluative words in the Russian language are shown.

The perception of this text material is characterized by extremely high evaluation coefficients: the emotional state of the Artist is assessed as *unfriendly* (62.9), *untruthful* (64.3); there is no *self-control* (20.4) and the *independence* of the author (8.2). If self-control is a person's inherent ability to control their own behavior, based on the will as the highest mental function responsible for conscious decisions, then the lack of independence indirectly indicates the artist's bias. You should pay attention to the indicators of *aggressiveness* (6.1) and *demonstrativeness* (10.4), which are colored in red in Figure 2. Verbal or pictorial *aggressiveness* is a form of activity aimed at insulting and / or causing harm to another; *demonstrativeness* is a trait of hysterical personalities, manifested in emotionally colored behavior through the desire to be visible and the desire to please (see the Artist's comments: *I want to continue to respect you all; I do not condemn; I enter the situation))); well, that's nice)*. This type of people is prone to deliberate exaggerations, designed for external effect and expressiveness of manifestations. Through caricature as a detailed drawing and deliberate sharpening of some traits and features, there are signs of demonstrativeness as a psychological characteristic of the author's personality.

4.4 This caricature, as well as the inscriptions to it, like all other works of the Artist, is devoid of a humorous component, which is manifested in PLE through the dominant characteristic features: *intimidating* (coefficient-17.59); *severe* (17.29); *gloomy* (16.26); *sinister* (11.53). **Figure 3** shows the signs of the polycode product "Pontius Pilate...". Caricatures with a missing humorous component become a destructive product of settling scores and have a great offensive potential.

To double-check the validity of the results of this aspect of the analysis, the experts examined the characteristic features of the inscriptions accompanying the caricature "Pontius Pilate..." (the orthography and style of the original are preserved): *Dui, Bulat Gumerysh, on his Golgof. And here's your Pontius Pilate. If, of course, you understand what I'm talking about and who I'm talking about, zaoschnik hrenov)))))*. Apparently, the author intuitively realizes that the figure does not cause smiles. This can explain the presence of five (!) signs-brackets that replace the smile after the inscription, and probably mean homeric laughter. Only, alas, the author himself, not the readers. The characteristics of the above inscription for the presence of a humorous component include: *frightening* (5.89); *dark* (5.65); *sinister* (5.61); *sad* (4.82); *gloomy* (4.21); *severe* (4.09); *heavy* (2.69); *dreary* (1.52). The results of a comprehensive analysis of the available inscriptions on the caricatures provided for analysis are shown in the following figure **(Figure 4).**

Excessive manifestation of theta-minus rhythms leads to a state of irritation and anger. It is theta waves that are ideal for non-critical acceptance of various external attitudes and beliefs that change the attitude towards others and shape the corresponding behavior. These rhythms are responsible for reducing the protective mental mechanisms that allow critical evaluation of information. These rhythms allow transformative information to penetrate deep into the subconscious (These states are used in hypnosis. See: Cherepanova [16]. It is important to emphasize that with the activation of such a pattern, critical comprehension of the perceived is lost, but emotional experiences increase. **Figure 5** shows the emotional and lexical evaluations of all the caricature inscriptions. Note that the text space is characterized by *ill will* (-3.6, blue zone), *dependence* (- 16.2), and *lack of self-control* (- 2.1).

The result, associated with the author's style, is characterized by the desire for refinement and sophistication of the works. The revealed *refinement* coefficient (+8.3, red zone) makes the Artist a recognizable author with his own individual handwriting. The calculations of the dominant characteristic signs revealed in the analysis of all the inscriptions on all the caricatures are as follows: *frightening* – 11.37, *sad* – 11.01. They are considered dominant because their coefficients go beyond the sign registered as available in the product. The analysis of all the signs revealed the following: *dark* (8.38), *dreary* (4.8), *minor* (3.47), *heavy* (1.84), *sinister* (0.16), *gloomy* (0.01).

4.5 In the final caricature series, the climactic image of the character unbuttoning his belt and taking off his trousers at the request of members of the criticized community (*The people ask, how can you refuse ...*), in order to bare his ass for them (*To lick is the right of a lawyer, Chief, do not hesitate! We are for!!!, We lick the whole chamber - as one!*) this image is intended to give a negative characterization of both the Organization as a whole, "bending" under its head, and its Head. Visual symbols and their verbal description are made in an obscene form with a gross violation of the norms accepted in society and with a violation of thematic communicative taboos that affect topics that are morally prohibited for public discussion (physiological processes, sexual actions, various kinds of perversions). **Figures 6 and 7** show the patterns of brain activity that are activated when a given polycode text is perceived.

The quality of the impact of the speech product is determined by us through the establishment of the dominant rhythm in the pattern of brain activity. With the help of the new version of the computer program *DIATON* [17], it is possible to analyze 11 main states of brain rhythms and emotions related to it. For the convenience of describing the results of the analysis, each type of brain waves is assigned the signs "+"(plus) or "–"(minus), depending on the redundancy or insufficiency of the manifestations of one or another rhythm, the neutral position is fixed, the position of full synchronization of potentials with the sign "+"(coherence of waves) and the position of hard negative impact with the sign "–" (incoherence of waves). The predominance of a certain rhythm, revealed when working with a stimulus, means that the perception of this stimulus contributes to the emergence of one or another emotional state associated with this pronounced rhythm. So, analyzing the text through the prism of the brain waves that arise during its perception, we can judge the quality of the impact on a person. The polycode text, which includes the caricature inscriptions, as seen in Figure 6, causes an increase in theta-minus rhythms, accompanied by unconscious anger and irritation.

The rhythmic activity that occurs during the analysis of comments indicates that the image leads to the generation of speech products characterized by a manic orientation. Figure 7 shows a pattern of brain activity with a predominance of alpha-minus rhythm. With a high degree of probability, it can be assumed that the image entails not only negative consequences for the reputation of the Organization and a particular person – its head, but also excites the audience states leading to possible violations of the norms of behavior.

Rechecking (using the VAAL expert system) of this conclusion through another aspect of the analysis related to the load on the categories showed the validity of the conclusion. A category is a set of tokens with a certain meaning. It is based on the frequency of use of words of the category as a whole and takes into account the indicator of the "normal" frequency of use of words by native speakers of this language. There are empirically defined norms for the use of various words. Knowing the "normal" frequency of this word in the Russian language, it is possible to calculate whether words of this category are used less often or more often by a particular person in a particular text. A category can be words that characterize a certain type of personality. Comparing with the norm the frequency of use of words of this category by a particular person, we can judge about his belonging to a certain psychotype. A category profile is a visual (in the form of a diagram) information about how words of different categories are distributed in the text.

Figure 8 shows the load indicators for the various categories; (1) the column names the category; (2) indicates the percentage of words in this category of the total text volume; (3) - the number of words of this category found in the analyzed text; (4) is a histogram of the visual representation of the quantitative characteristics of the distribution; (5) contains an estimate of the deviation of the number of words found in this category from the average language norm. This is the traditional evaluation of the z-score in content analysis, calculated by the formula: (N-E) / (standard deviation), where N is the number of words of this category found in the analyzed text, and E is the expected number of occurrences of words of this category in the text; the sixth column is a visual representation of the deviation of the number of words found in this category from the average language norm. The red color indicates a significant excess of the norm, the blue color indicates a significant deviation in the lower direction, and the gray color indicates an insignificant deviation from the norm either in the direction of its excess or in the lower direction.

In the category of "Psychoanalytic symbolism", a high level of aggressiveness was revealed. It is linked to the attraction to unnecessary risk and to the violation of norms of behavior.

5 Research Methods

We use a complex - linguistic and psycholinguistic - approach to text analysis, which implies, in addition to the traditional practice of linguoexpertology, an appeal to statistical methods with tracing the hierarchy of frequency of verbal constructions and sound-letter complexes, with counting the number of identical models, with comparing the frequency of single sounds and with comparing the normative indicators of the frequency of use of certain sounds. We used factor, feature, semantic, content analysis of the text, analysis of the suggestive potential of the verbal model. In the expert practice of the authors, emotional and lexical evaluation of verbal and nonverbal models, evaluation of the rhythmic activity of the brain in the perception of the entire text, evaluation of the rhythmic coherence in the perception of individual parts of the text and their comparison, techniques for recognizing the hidden motives of the object's behavior in his professional activity, analysis of individual coefficients of the impact of the text on individual and mass consciousness and

https://doi.org/10.1051/shsconf/202110901031

subconsciousness, evaluation of the impact of the polymodal text are also actively used.

The successful results of the formalization of these PL experiments in the form of computer programs allowed to increase the possibilities of using this scientific paradigm in judicial linguo-expert activity. The creation of a visualized result of the processed data, based on statistics, on the comparison of coupled databases, significantly enriches the expertise and eliminates the subjective interpretation of the speech product. One of the functions of human speech, as well as any written or oral text, is the phatic function, or the function of impact. Within the framework of the PL, methods have been developed and units of analysis have been identified, with the help of which it is possible to give an objective evaluation of the material: to measure in quantitative terms the strength of the impact of the text [18] to establish the quality of this impact, to decode the suggestive potential of a language phenomenon, to rank emotional and evaluative signs, to evaluate the motives and intentions of the author of the text, to calculate the "golden section" of the text as a zone of uncontrolled influence on consciousness [19].

The expert operates with special units of analysis in order to study the latent suggestive resources of the text, since the lack of linguistic manifestations of deep internal forms does not lend itself to analysis through traditional linguistic and psychological techniques and methods. On the basis of the Department of Language Communication and Psycholinguistics of UGATU, computer programs have been developed and successfully tested, with the help of which it is possible to work in five languages (Russian, English, German, Tatar and Bashkir): 1) Computer program BARIN (Automated analysis of words and text). Certificate of state registration of the computer program No. 2011618299 (M. 2011); 2) Computer program BATYR (Automated analysis of words and text). Certificate of state registration of the computer program No. 2014613238 (M. 2013); 3) Computer program SCHETOVOD (Automated text analysis). Certificate of state registration of computer program No. 2014618598 (M. 2014); 4) Computer program PULSE 2015 for processing the rhythm of prose text. Certificate of state registration of the computer program No. 2015614549 (Moscow, 2015); 5) Computer program BYURGER (Automated analysis of words and text). Certificate of state registration of the computer program No. 2016616320 (Moscow, 2016).

6 Findings

With the help of lexical, grammatical, and stylistic means, a negative picture is formed in the mind of the addressee and a corrupt and submissive image of the Organization as a whole and its Head in particular is built. The identified units of different levels of language create the semantic basis of the communicative space of the social network, which is clarified through comments, evaluations, etc. The meaning of the word includes not only the conceptual core, which allows you to distinguish this subject from a number of others, but also the figurative-sensory representation, and emotional-evaluative and stylistic meanings. The combination of these creates certain stylistic effects that give the caricature one or another tonality, which expresses the sarcastic attitude of the author to the unsub and the subject of the image.

6.1 In the context of these examples, the significance of the PL paradigm in expert researchs of conflictogenic discourse becomes self-evident. At the same time, it should be noted that the linguistic evaluation significantly complements the analysis, giving the expertise the status of a complex one. Without a competent linguistic component, the logic of describing the material is lost, since psycholinguistics acts as a different, not traditional, methodological platform designed to identify new units of analysis, describe the laws of action and impact of verbal models, as well as formalize the suggestive resources of the products of speech activity. The theoretical basis of the paradigm, laid down by the works of A. A. Leontiev, E. F. Tarasov, A. A. Zalevskaya, N. V. Chernigovskaya , I. N. Gorelova, N. O. Zolotova, S. V. Lebedeva allows you to build a flexible algorithm for analytical and experimental work of detecting and decoding the influencing resources of verbal models of different complexity.

6.2 A fundamentally important point in working on materials in PL format is an objective evaluation, implemented through quantitative indicators, through the established quality of impact, which is also decoded by ranking patterns of rhythmic activity of the brain, combinations of emotional and evaluative signs, explication of the motives and intentions of the author of the text, patterns of sound-color associative color of the verbal model, and today, the establishment of psycho-emotional states of the author of the text, its readers or viewers, if we are dealing with video text.

7 Conclusion

Since psycholinguistics can be called experimental linguistics, which studies the products of human speech activity, all the formalized data obtained, all the indicators and coefficients can be repeatedly rechecked. Moreover, within the framework of one research, we double-check our results obtained with the help of different tools developed by us, which allow us to consider one problem from different sides.

The linguist in the research proceeds from the principles of ethics of the writer / draughtsman, who in his professional and speech activities is obliged to respect the dignity and honor of the subject / object of his attention: The Artist is always able to adequately, according to the stylistic and ethical norms of the chosen genre of creativity, choose from the means provided by the language that do not discredit the subject of the description. Caricatures and their accompanying texts have a specific character, which, except for some, do not formally have an indecent form, but they are linguosemiotically qualified as being outside the norms of literary language and speech behavior in general, which are considered by society as unacceptable in public use.

Taking into account the "conflictogenic risks", experts analyze the speech act from the point of view of the explicit and hidden content of the text, describe the factors that generate the statement and the reaction to it. Knowing the patterns of speech communication, experts-philologists characterize a specific communicative conflict, and lawyers give the results of the research a legal interpretation. Being an objective study, LE and PLE expertises are not just a complex procedure used to identify speech crimes in the journalistic sphere of communication, but an applied direction that reveals new properties of the conflictogenic media space.

References

1. E.A. Salikhova, Political Linguistics **3 (81)**, 198-208 (2020)
2. Yu.A. Sorokin, E.F. Tarasov, *Optimization of speech impact*. (Moscow: High school,1990)
3. E.G. Nikitina, *The language of regional television sports reporting and the specifics of its impac*. (Izhevsk: Udmurt State University, 2014)
4. A.G. Sonin, D.S. Michurin, Questions of Psycholinguistics **2 (16)**, 164-174 (2012)
5. T. F. Petrenko, *University readings* (Pyatigorsk: PSLU. 2015)
6. Z.I. Trubina, Crede Experto: Transport, Society, Education, Language **3**, 163-174 (2019).
7. Dictionary of Russian words Dictinary.ru. URL: https://www.dictinary.ru/определение%20слова/напёрсточник.html (date of access: 16.03.20).
8. Large explanatory dictionary of the Russian language (BTSRYaz, Saint Petersburg: Norint, 2007)
9. I.A. Sternin, L.G. Antonova, D.L. Karpov, M.V. Shamanova, *Identification of signs of humiliation of honor, dignity, belittling of business reputation and insults in the linguistic expertise of the text* (Yaroslavl: Yaroslavl State University, 2013)
10. T. Rogoznikova, *Current issues of linguistics and didactics: The interdisciplinary approach in humanities (CILDIAHSS-2018).* URL: https://doi.org/10.1051/shsconf/ 20185001146. (date of access: 16.03.20).
11. *Culture of Russian speech: An encyclopedic dictionary-reference* (Moscow: Flinta-Nauka, 2007)
12. *The price of a word* (Moscow: Gallery, 2002)
13. E.S. Kara-Murza, Political Linguistics **1(27)**, 47-71 (2009)
14. A. Ustinov, Domestic notes **6(63)** (2014), URL: http://www.strana-oz.ru/2014/6/oskorblenie-kak-hudozhestvennyy-zhest. (date of access: 16.03.20).
15. V.I. Shalak, M.N. Dymshits, *Software product. The psycholinguistic expert system* (VAAL. Moscow, 2005)
16. I.Yu. Cherepanova, *Verbal suggestion: theory, methodology, and socio-linguistic experiment* (Perm: Perm State University, 1996)
17. *Program for the expertise of suggestion texts DIATON, version of SLOVODEL Laboratory "Vedium". Certificate of state registration for the computer program.* No. 2008611081. (Moscow, 2008)
18. T.M. Rogozhnikova, (2017) *Proceedings of the 7th International Scientific and Practical Conference «Current issues of linguistics and didactics: The interdisciplinary approach in humanities» Advances in Social Science, Education and Humanities Research (ASSEHR)* **97** (2017)
19. T.M. Rogozhnikova, E. A. Salikhova, Eurasian Advocacy **1 (37)**, 26-38 (2019)

SHS Web of Conferences **109**, 01032 (2021) https://doi.org/10.1051/shsconf/202110901032
LISID-2020

Legal protection of photographs in the digital age: Russian regulation

Marina Rozhkova[1,2*], *Olga* Isaeva[3,4]

[1] The Institute of Legislation and Comparative Law under the Government of the Russian Federation; Immanuel Kant Baltic Federal University
[2] Cheremushkinskaya street, bldg. 34, 117218, Moscow, Russian Federation; 14 A. Nevskogo ul., Kaliningrad, 236041, Russian Federation, rozhkova-ma@mail.ru
[3] The Institute of Legislation and Comparative Law under the Government of the Russian Federation
[4] Cheremushkinskaya street, bldg. 34, 117218, Moscow, Russian Federation, ol.v.isaeva@ya.ru

Abstract. Appeared in 1826, the photography has been much developed so far. Being mostly used for pragmatic purposes at the dawn of its existence, today, the photography is not only art, but also serves to depict a person at a certain point in time and even for personal identification. The photographs are used as evidence and as technical means of reproduction of other works of authorship. However, although the photographs have always been tacitly divided into informative and aesthetic ones, and various legal regimes of photography have already been enshrined in the Russian legislation, it is still not really clear how all those regimes relate to each other. Trying to answer that question the authors reveal problems and gaps existing in the current Russian regulation of photographs and make some proposals that could be useful for both the scientists when furthering the discussion on the topic and the legislator when developing the existing rules.

1 Introduction

The world's first photographic images are believed to have been taken in 1822 by Joseph Nicéphore Niépce. Having become fascinated with the chemical fixing of light images and trying to improve the printing method of lithography, the French inventor succeeded in reproducing drawings on pewter plates and lithographic stones by the aid of the sun. The photographic process was called heliography (in French, héliographie), or literally "sun writing". The most famous heliograph of Nicéphore Niépce called "View from the Window at Le Gras", which appeared in 1826, is thought to be the first life photo. Although the heliography was found hardly suitable for life photography, it was widely used by bookmakers when making inserts with pictorial reproductions.

Later, aiming to proceed with experiments on fixing of light images, Nicéphore Niépce became associated with Louis Jacques Mandé Daguerre, who, after Niépce's death, established the art of photography on a practical basis. Daguerre developed the daguerreotype (in French, daguerréotype), a new means of the image reproduction based on the light sensitivity of silver iodide. January 7, 1839, the day when the French Academy of Sciences announced the invention of the daguerreotype, is considered as the day of the photography invention. On June 14, 1839, having provided Daguerre with a lifetime pension, the French Government bought the rights to the process and gave it freely to the world, except England where the process was patented (Patent No. 8194: 'A New or Improved Method for Obtaining the Spontaneous Reproduction of All Images Received in the Focus of the Camera Obscura') [1]. It is worth mentioning that the daguerreotype was not used for a long time: in the second half of the 19th century, it was replaced by other photographic technologies that were easier and cheaper in their use.

It is also worthy of note that at the beginning of its existence and almost until the end of the 19th century, neither the photography was considered as art, nor the creators were considered as authors through whose creative effort the works had been created. The photography was used exclusively for pragmatic purposes: as a means of fixation for the purposes of research, as a technique for documenting events and different items, as an auxiliary (training) tool for artists, etc. Thus, the photography became of particular importance for law enforcement purposes and for the development of forensics which led to the creation of photo identification.

Herewith, the photography was found to be commercially useful. Being an affordable alternative to canvas created by artists, photographs became collectible items: many companies produced small-format photos of celebrities, landscapes from distant places, architectural monuments, and genre scenes. That was when the first photographic albums appeared, when group and

* Corresponding author: rozhkova-ma@mail.ru

individual portrait photographs became extremely popular, and when advertising photo cards were created.

At the same time, photographs were not exhibited as works of art. They were exhibited either along with lithographs, or as demonstrations of photographic documentation in science sections, or separately, being presented as examples of an independent field of activity meant to be neither technical nor artistic [2]. It was only at the end of the 19th century that the photography started seriously pretending to be a result of creative activity. That time, the creators of artistic photographs began to use special shooting and printing techniques, e.g.: to apply filters and special lenses, to process the picture print using chemical compounds in order to obtain a more expressive artistic effect, to create compositions using photomontage, etc. As a result, in 1908, the photography received protection under the Bern Convention.

2 Problem Statement

The photographs have always been divided into information ("report" ones that fix certain information) and aesthetic photos (reflecting the artistic perception of the author, the photographer). Due to that fact, it is not surprising that some publications, if not court decisions, intuitively disagree with the statutory thesis stating that every photograph shall be deemed a result of intellectual activity. In this context, the position of minimum standards for the protection of works of authorship, which has been developed in the German law, contain valuable insights.

Among the conditions excluding the creativity of a work the German theory of creativity specifies the extent to which the author is able to create an original work of authorship and the mechanical or random nature of the author's actions when creating the item resembling a work of authorship. To the extent that the form of a work is determined by factors, which restrict significantly the author's ability to express their personality in their work, the latter cannot be deemed an original work of authorship even though it is inherently new. In this context, the lack of opportunities for the author to create an original work of authorship precludes the creative nature of their work. In other words, the German copyright does not protect every intellectual item resembling a work of authorship by its outward signs; the German copyright protects only the items of a certain quality allowing to ascertain a close connection with the personality of the author [3].

To wit, according to the German doctrine, the photographs of a "report" (informational, technical) nature – e.g. photographs of a conference participants or those of a board of directors, lookalike photos of tourist attractions, portrait ID photographs, paparazzi photos depicting private lives of stars, etc. – won't be deemed intellectual property items and won't therefore acquire the corresponding legal protection. However, the photographs reflecting the personal touch of the author will undoubtedly be recognized as the copyrighted works. Although it is difficult to define clear criteria for drawing a line between aesthetic and informative photos, these types of photographs are actually very different that becomes particularly obvious at authorial photo exhibitions.

The approach developed in the German law is fully consistent with Directive 2006/116/EC of the European Parliament and of the Council of 12 December 2006 on the term of protection of copyright and certain related rights stating that "[a] photographic work within the meaning of the Berne Convention is to be considered original if it is the author's own intellectual creation reflecting his personality…" (recital 16 in the preamble to Directive 2006/116/EC). Moreover, as the European Court of Justice explained, the author of a photographic work is able to express his "personality" in several ways and at various points of its production: when choosing the subject's pose and the lighting (in the preparation phase), or the angle of view and the atmosphere created (when taking a photograph), or choosing a developing technique he wishes to adopt that includes by means of a software [4]. In other words, according to the European approach, a photograph shall be not only aesthetic to be copyrightable but it also shall reflect the personality of its author.

The division of photographs into informative and aesthetic, a significant number of special applications and software allowing to edit photos, the problems related to the protection of privacy in the digital age along with some other issues create the need for developing different approaches in granting legal protection to photographs. So, how do the various legal regimes of photography, which have already been enshrined in the legislation, relate to each other?

3 Research Questions

The Russian legislation establishes several legal regimes that may apply to photographs depending on the "nature" of the images.

3.1. Legal protection of a photograph depicting a person's image at a certain point in time

According to Article 152.1 of the Civil Code of the Russian Federation (hereinafter referred to as the Russian Civil Code) the notion of the "citizen's depiction" covers, inter alia, photographs, videos, and other works of visual art depicting an individual. On its surface the mentioned article seems to be aimed at resolving the issues arising from the use of copyright objects depicting a person. However, the content of the specified legal regime along with the placement of the provisions at issue into a chapter of the Russian Civil Code taking no credit to copyright shall be deemed indicative of its completely different purpose.

3.2. Legal protection of photographs considered to be personal data

The issue of classifying citizens' photographs as personal data and biometrics has been raised for discussion both in doctrine and in legal precedents more than once. However, neither unanimity of opinion nor unambiguous understanding of the clarifications given by the competent executive bodies has been reached. This situation shows that there are some uncertainties in the legal regime at issue, and there is no therefore uniform practice in the application of such rules.

3.3. Legal protection of photographs considered to be works of authorship

The copyright protection of photographs is seemingly the most general and common legal regime. However, the absence in the Russian legislation of clear criteria for the copyrightability of photographic works not only gives rise to professional discussions on whether it is possible to recognize photographs lacking originality as copyright items, but also leads to the incoherence of the judicial practice. As a result, inherently uncopyrightable photographs tend to be granted copyright protection, while the copyrightability of original photographic works may be called into question.

3.4. Legal protection of photographs created by means of applications or software

Amid the expanding capacity of graphic editors and the increasing use of artificial intelligence software, the question arises as to whether the photographic images shall be copyrightable when their originality has been achieved through post-processing rather than through photography itself. Can we even speak of the originality and any kind of legal protection if all the "creativity" belongs to the software?

3.5. Legal nature of photographs created without any participation of a legal subject

Generally, the photographs created without any participation of a legal subject have no cultural value and are not therefore deemed copyrightable. At the same time, pictures taken at random may not only be informative (e.g. surveillance photos) but also spectacular (e.g. a snowboarder's shot from the perspective of a camera falling in the snow or a photo "created" by animals or plants). Thus, the legal regime of such photos is worth being thought and talked about.

4 Purpose of the Study

In the digital age, the division of photographs into informative (fixing certain information, "report" ones) and aesthetic (creative, reflecting the artistic perception of the author – photographer) is of particular importance. The purpose hereof was to reveal various legal regimes reflecting the specific nature of photographs, that perform both an informational and aesthetic function, and to compare those regimes with each other.

5 Research Methods

The research was being carried out through analysing the Russian and the foreign doctrine as well as law enforcement and judicial practices.

6 Findings

As it has already been mentioned above, the Russian legislation does not explicitly distinguish between aesthetic and informative photographs. However, the Russian legislation does grant legal protection to photographs in their various capacities prompting suggestions that such a distinction might be tacit.

6.1. Legal protection of a photograph depicting a person's image at a certain point in time

The current Civil Code of the Russian Federation contains Article 152.1. titled "Protection of the citizen's depiction".

Before making comments on the said Article, it's worth mentioning that a similar rule existed in the prior legislation. Article 514 of the Civil Code of the Russian Soviet Federative Socialist Republic dated 1964 (the RSFSR Civil Code) stated the following:

The publication, reproduction and sharing of works of visual art bearing the depiction of an individual shall be admissible only subject to their consent or, after the individual's death, subject to the consent of their children and the live spouse. Such consent is not required in cases when the said is done in the state or public interests, or when the depicted individual sat for the depiction for a payment.

That Article was included into Section IV titled "Copyright" and was supposed to deal with the issues specified in its title: "Protection of interests of the citizen depicted on the works of visual art". That means the said Article settled a problem arising when a work (a copyrighted item) was to be reproduced and shared.

Herewith, Article 152.1 of the Russian Civil Code has nothing to do with the copyright, but it is essentially focused on three objectives at a time.

First and foremost, the Article under consideration sets out the rules for the publication and further use of a citizen's images (photographs, videos, and other works of visual art) depicting an individual at a certain point in time. Since Article 152.1 of the Russian Civil Code expressly states that a depicted person has the right to authorize another person to use their image, we can say that a citizen is granted a legal possibility to dispose of the right to their image whether for free or against payment. In other words, the point at issue is the introduction of the right to a citizen's depiction into commercial turnover.

Second, whereas Article 152.1 of the Russian Civil Code is placed into the chapter on intangible goods, the Article may be considered as being aimed at limiting the intrusion on a person's privacy. The inviolability of private life of a citizen may be violated, inter alia, by means of publication of their photographs even if such depictions do not discredit honor and dignity of the depicted person but do report on their private (personal) or family life. This further justifies the set requirement to obtain the consent of the depicted person to use their image and, where appropriate, the consent of other persons explicitly named by the law.

Finally, the provisions of Article 152.1 of the Russian Civil Code establish a legal framework for the exercise of the right to appearance (physical appearance), which the doctrine traditionally considers only within the scope of inalienable rights as being vested in that person by virtue of birth [5]. The widespread practice of photo shooting (when a model gets paid for being depicted) makes the exercise of the right to appearance possible. For the purpose of the foregoing, it is noteworthy that the model may dispose of the right to the resulting photographic image, i.e. give their consent to its publication or any other use. The point is that the law does not provide for a rule allowing the model to revoke the given consent while the jurisprudence recognises the legality of using the images without any prior consent of the model provided that the fact of sitting for payment is proven by relevant evidence [6].

Curiously, foreign legal systems provide for different approaches to the latter issue. For instance, being a model and a TV presenter, an actress was judged by the French Supreme Court (Cour de Cassation) to have no right to preclude the publication of her nude photo, which had been taken by a famous photographer, in an art book, because when she had consented to be photographed, she had also consented to the further publication and exhibition of the photo for any purpose (no. 06–10.305, 2007). At the same time, according to Article 97 of the Italian Copyright Act (Law No. 633 of April 22, 1941), the consent of the portrayed person shall not necessarily be obtained when the portrait is being reproduced for scientific, didactic, or cultural reasons, but neither display nor commercial distribution is permitted when it would prejudice the honour, reputation or dignity of the portrayed person.

6.2. Legal protection of photographs considered to be personal data

When Russia acceded to the Council of Europe Convention for the Protection of Individuals with regard to Automatic Processing of Personal Data (ETS No. 108) dated 1981 (hereinafter referred to as Convention 108), which aimed at solving the problems caused by the use of new technologies and ensuring fundamental human rights when the information is being automatically processed, it led to the adoption of the Federal Law of July 27, 2006, No. 152-FZ on Personal Data (hereinafter referred to as the Law on Personal Data). The said law reflects most provision of Convention 108 and, as it is stated in its Article 2, aims at providing for the protection of the rights and liberties of person and citizen in the processing of their personal data, including protection of the rights to the inviolability of private life, personal and family secrets.

Article 11(1) of the Law on Personal Data sets out the concept of biometric personal data, which are defined as "[the] information which characterizes the physiological and biological features of a single individual whereby he/she can be identified (biometric personal data) and is used by an operator to identify a personal data subject". In light of this definition, lawyers generally classify photographs as biometric personal data [7], [8]. However, they do not take into account that not every photo of a citizen is suitable for biometrics.

A review of technical standards in force affords grounds to state that biometrical personal data would cover the photograph which contains distinctive biometric features allowing to recognise the depicted person. In other words, a photographic image of a citizen could be deemed biometric personal data provided that it meets certain requirements (related to lighting, head position, camera location, image resolution, colour saturation, etc. as specified in GOST R ISO/IEC 19794-5-2013 [9]) and is processed in order to identify the personal data subject. The photographs collected with no identification purposes and used only to confirm certain actions shall not be considered as biometrics.

Accordingly, it is necessary to mention the Federal Service for Supervision of Communications, Information Technology, and Mass Media (Roskomnadzor) explicitly stating in its Clarifications [10] that biometric personal data cover, inter alia, a colour digital photographic image of a person in a foreign passport, photographs of employees or visitors of state and municipal bodies and enterprises (organisations) contained in their access control system (ACS). At the same time, it is noted that photographs of the employees in their personal files shall not be deemed personal data of the employees.

This raises the following question: shall a photographic image of a person, which does not fall under the category of biometric personal data, be deemed personal data? The provisions of Article 3(1) of the Law on Personal Data defining personal data as "any information directly or indirectly concerning an individual who is defined or is being defined (personal data subject)" seem to answer in the affirmative. However, this view is shared only by some lawyers [11], [12]. Speaking of photographs, others classify them into biometric personal data and a citizen's depictions to be protected under the abovementioned Article 152.1 of the Russian Civil Code [13].

In our opinion, the answer to this question should be based on the general premise of the Law on Personal Data specifying that any information, including photographs, could be deemed personal data if it is processed in order to identify a person. Such an approach corresponds to the European legislation [14]. Therefore, a photo of a citizen, which is not to be deemed biometric personal data under Article 3(1) of the Law on Personal Data, may be deemed personal data, because it contains subjective information on the person enabling their identification (recognition), and sometimes, it also contains other additional data, e.g. date and time of the shooting, the camera settings, GPS location, etc.

However, in cases when the photograph of a citizen is processed for other purposes having nothing to do with their identification, we shall talk about a normal use of a photograph falling under the scope of the abovementioned Article 152.1 of the Russian Civil Code.

6.3. Legal protection of photographs considered to be works of authorship

According to Article 1259(1) of the Russian Civil Code, photographic works and works created by means analogous to photography shall be deemed copyrighted items. However, the Russian Civil Code specifies only two requirements for works (including photographic ones) to be copyrightable: a work shall be created by some creative activity of its author (Article 1257) and be expressed in any objective form (Article 1259(3)).

Thus, the Russian legislation only sets out requirements to the creation process (the work creation activity shall be creative) and establishes no requirement (apart from objective form) to the result of such activity. This fact has been recognized by the Court for Intellectual Property Rights stating in its Ruling No.C01-664/2017 [15] the following:

The current legislation does not establish special conditions that would be necessary to recognise photographic works to be copyrightable; therefore, the author (photographer) shall enjoy the copyright to any work (photograph) by virtue of the very fact of its creation and regardless of its artistic value and importance.

Herewith, the Supreme Court of the Russian Federation has further developed that idea explaining that: "…the mere lack of novelty, uniqueness and (or) originality of the intellectual product cannot point to the fact that such a product is not a result of creative activity and is not therefore a copyrighted item" [16].

The approach, according to which any photograph shall be deemed copyrightable regardless of its originality and artistic value, can hardly be considered correct. Pursuing that approach would mean that a legal protection to be granted to a "work of authorship" shall be granted to any photograph even deprived of originality and individuality: e.g. lookalike photos of tourist attractions taken at the same angle [17], [18], photos of products offered for sale [19], [20], portrait ID photographs, "report" photos taken during any meeting or even occurrence [21], etc.

As it has already been discussed above, the European legislation is developing in a different way following the approach, according to which a photograph would be deemed a work of authorship provided that it is original and reflects the personality of its author. In our opinion, such an approach shall be adopted by the Russian legislation: the originality should become a requirement for the copyrightability of works. As for photographs, a photo should be deemed original if it is possible to characterize it as reflecting the personal touch of the author and being, in any case, distinctive from others. The said may be achieved through a specific style or a shooting technique or angle, by presence of a distinctive element on the picture, by choosing a time for shooting and the composition of elements, etc.

In furtherance of the foregoing, a due regard should be given to the question of whether the photograph of a work of authorship (e.g. a photo of a picture or of a sculpture in a gallery) shall be deemed a derivative work (an alteration of another work) [22] or a reproduction of that work by means of photography [23]. In our opinion, a simple fixation of a work of authorship by any means, including both videorecording and photography, shall be deemed a reproduction of that work, which is a way to use the work (Article 1270(2) of the Russian Civil Code) and is therefore subject to the consent of its author. In such cases, a photograph is usually no more than a simple technical means for reproduction and may not be generally deemed a copyrightable item.

At the same time, when someone else's work of authorship has been photographed by accident or is just a background, such a photograph can easily be deemed an original work of a derivative nature, which can be created only subject to the consent of the author (right holder) of the work used for its creation (Article 1260(3) of the Russian Civil Code).

6.4. Legal protection of photographs created by means of applications or software

Within the context of the originality of photographs, the question arises as to whether a photographic image might be considered a work of authorship if its efficiency has been achieved through a competent use of the predefined functions of Photoshop or another graphic editor, rather than through the creativity of its author. This question becomes particularly relevant in response to the ever-increasing use of programmes taking advantages of artificial intelligence.

On the one hand, today, the artificial intelligence is involved in both the process of taking photos by means of smartphones and the process of editing photos. Technologies in general, and artificial intelligence in particular make it possible to edit photographs without much intellectual effort: it is enough to upload a photo and to indicate what needs to be changed (the style of the photo, the composition, the white balance, etc.), to use automatic correction, or to choose a style of a famous painter (as Prizma suggests for instance) in order to get a completely different photo. However, it is difficult to believe that the originality of the image achieved through such editing may be considered as reflecting the personal touch of its author.

On the other hand, some authors can leave their personal touch and express their individuality, their own style and a particular vision of the object being photographed only by using Photoshop or another editor. The mixing of "ordinary" images with unexpected fantasy details, the author's colour correction, an unusual collage of several images, special effects, etc. may "change the reality": the resulting image does usually reflect the ideas that are simply impossible to be brought to life without certain functions of graphic editors.

Herewith, such editing often requires far more time and intellectual resources than the creation of the original (initial) image itself.

In such a case, it seems to be very reasonable to speak about creating a work of authorship: the author creates a new original image, that reflects their personality, by means of a graphic editor. It is necessary to pay attention to the following point: if such a photographic work has been created basing on another original photo, there are all grounds to point to the creation of a derivative work (Article 1260(1) of the Russian Civil Code) that binds the author of the derivative work to get the consent of the author of the original (initial) work to use the original copyrighted item.

6.5. Legal nature of photographs created without any participation of a legal subject

The photographs created without any participation of a legal subject can be conveniently classified into those created automatically and created by accident due to some external (natural) factors.

The first group includes photographs taken by red-light-running cameras or in automatic mode, e.g. by outdoor cameras [24]. The photographs obtained in this way cannot be subject to copyright protection not only due to the lack of creativity but also due to the absence of the legal subject. However, such photographs are commonly used as evidence when it's necessary to establish whether an act has been committed, including in court proceedings [25], [26], [27].

The second group may include photographs "created" by animals. Since only an individual can be a legal subject, a photograph taken by an animal will not be copyrightable. In other words, neither the owner of the camera, nor the owner of the animal will have the copyright to such a photo: the photo shall customarily be considered as a public domain item.

For instance, in a famous case on the copyright to photos taken by a monkey [28], although David Slater (the photographer to whom the camera belonged) tried to justify his copyright by the fact that it had been him who provided conditions under which the creation of the images at issue became possible, the copyright protection was denied [29].

It is also noteworthy that the U.S. Copyright Office has supplemented the Compendium of U.S. Copyright Office Practices with an explicit statement in section 313.2 that the Office would not register works produced by nature, animals, or plants. The first cited example is "a photograph taken by a monkey" [30].

7 Conclusion

The study allows concluding that the current Russian legislation provides several regimes for legal protection of photographic images establishing no clear boundaries for each of the regimes. Due to the lack of official explanations by representatives of legal profession, the photographs, which do not seem to be copyrightable, are granted copyright protection, and an ordinary image of a citizen may be covered by two legal regimes at a time. We hope that the problems and the gaps in legal regulation we have outlined, and the proposals we have made would form the basis for further scientific discussion on the topic considered herein.

References

1. S. Edwards, 'Beard Patentee': Daguerreotype Property and Authorship. Oxford Art Journal 36.3, 369-394 (2013) doi:10.1093/oxartj/kct030
2. V. Levashov, *Lectures on the History of Photography* (IE Gusev, Moscow, 2014)
3. A.V. Kashanin, Civil Law Review **4**, 23-62 (2007)
4. Eva-Maria Painer v. Standard Verlags GmbH and Others, EU:C:2011:798 (ECJ, 2011)
5. M.A. Rozhkova, *Civil Code of the Russian Federation. Article-by-article Commentary on Chapters 6-12* (Statut, Moscow, 2014)
6. Case No. 33-46886, Appellate decision (Moscow City Court, 24.11.2014), URL: https://sudact.ru/regular/doc/GXJJ6aPUNpdR/ (date of access: 11.12.2020)
7. A.V. Dolmatov, E.A. Dolmatov, Vestnik of Saint Petersburg Juridical Academy **3**, 53-60 (2018)
8. M.S. Krivogin, Journal of the Higher School of Economics **2**, 80-89 (2017) doi: 10.17323/2072-8166.2017.2.80.89
9. *Information technologies. Biometrics. Biometric data interchange formats. Part 5. Face image data. (2015). GOST R ISO/IEC 19794-5-2013 from January 1, 2015* (Standartinform, Moscow, 2015)
10. *Federal Service for Supervision of Communications, Information Technology, and Mass Media. Clarifications regarding the consideration of photo, video, fingerprint and other information as biometric personal data and the peculiarities of their processing (September 2, 2013)* (Moscow, 2013)
11. M.S. Krivogin, Proceedings of Voronezh State University. Series: Law **1**, 108-117 (2018)
12. M.G. Moshkovich, Glavkniga **21**, 74-78 (2018)
13. D.R. Galiullina, Archive. History. Present **15**, 264-268 (2015)
14. Determining what is personal data. Data Protection Act, URL: https://ico.org.uk/media/for-organisations/documents/1554/determining-what-is-personal-data.pdf (date of access: 11.12.2020)
15. Case No. A57-29754/2016, Ruling No. C01-664/2017 (Court for Intellectual Property Rights, 30.10.2017), URL: https://kad.arbitr.ru/Card/1eca1cbb-8902-41d4-b955-03b6b42d8f72 (date of access: 11.12.2020)
16. Resolution No. 10 On Application of Part Four of the Civil Code of the Russian Federation (Plenum

of the Supreme Court of the Russian Federation, 23.04.2019)

17. Case No. A40-239418/2016, Ruling No. C01-812/2017 (Court for Intellectual Property Rights, 30.10.2017), URL: https://kad.arbitr.ru/Card/57f25041-f56b-4d94-9d42-29ffc866d13a (date of access: 11.12.2020)
18. Case No. A60-47992/2017, Ruling No. 309-ЭС19-4080 (Supreme Court of the Russian Federation, 19.04.2019), URL: https://sudact.ru/vsrf/doc/liCZ7K9swuQr/ (date of access: 11.12.2020)
19. Case No. 33-881, Appellate Decision (Moscow City Court, 28.01.2019), URL: https://www.mos-gorsud.ru/mgs/services/cases/appeal-civil/details/e4264e40-5c3c-4a90-83bc-9f58f35fabde?caseNumber=33-881/2019 (date of access: 11.12.2020)
20. Case No. A50-29320/2015, Ruling No. 309-ЭС17-4997 (Supreme Court of the Russian Federation, 26.06.2017), URL: https://sudact.ru/vsrf/doc/Dk3pDXuMNWmZ/ (date of access: 11.12.2020)
21. Case No. A49-6557/2015, Ruling No. 306-ЭС16-11177 (Supreme Court of the Russian Federation, 19.08.2016), URL: https://sudact.ru/vsrf/doc/eDuTe8iTtPNv/ (date of access: 11.12.2020)
22. A. Dovgalyuk, Zakon.ru, (2015) URL: https://zakon.ru/blog/2015/05/08/fotografiya_kak_obekt_avtorskix_prav__chast_3 (date of access: 11.12.2020)
23. A. Dovgalyuk, Zakon.ru, (2015) URL: https://zakon.ru/blog/2015/5/7/fotografiya_kak_obekt_avtorskix_prav__chast_2 (date of access: 11.12.2020)
24. E.T. Sidorov, D.A. Temnyakov, Bulletin of the Economic Security **1**, 230-238 (2016)
25. Case No. 46-АД 18-15, Ruling (Supreme Court of the Russian Federation, 10.09.2018), URL: https://sudact.ru/vsrf/doc/1oiO7GAbMrHG/ (date of access: 11.12.2020)
26. Case No. 57-АД19-42, Ruling (Supreme Court of the Russian Federation, 09.08.2019), URL: https://sudact.ru/vsrf/doc/RWJcweP00md4/ (date of access: 11.12.2020)
27. Case No. 78-АД20-1, Ruling (Supreme Court of the Russian Federation, 18.03.2020), URL: https://sudact.ru/vsrf/doc/ka42Yxy7eXH8/ (date of access: 11.12.2020)
28. Naruto v. Slater, 888 F.3d 418 (2018)
29. A. Guadamuz, Internet Policy Review **5** (2016), doi: 10.14763/2016.1.398
30. I. Bardov, Which photographs shall be considered as copyrighted items? (2017), URL: https://bardov.legal/qanda/kakie-fotografii-yavlyayutsya-obektami-avtorskogo-prava (date of access: 11.12.2020)

Current issues of using digital technologies for environmental protection: legal aspect

Anatoliy Ryzhenkov[1*], and *Lidiya* Burinova[2]

[1] Kalmyk State University, ul. Pushkina, 11, Elista, Russian Federation, e-mail: 4077778@list.ru
[2] Kalmyk State University, ul. Pushkina, 11, Elista, Russian Federation, e-mail:burinova@mail.ru

Abstract. The global changes of recent years in the world around us indicate that the line between the physical and digital worlds becomes blurred. More and more sectors of social life move online, and digital technologies have an increasingly strong influence on the industrial, agricultural, scientific, medical, educational, and other processes. Digitalization technologies have a particular influence on the evolution of traditional ideas about the methods of environmental protection. Though most international legal acts do not place a direct emphasis on this, many environmental problems caused by the globalization era can hardly be resolved without using digital technologies (for example, in terms of overcoming the consequences of global climate change). Further development of digital technologies will reduce the consumption of hydrocarbons and greenhouse gas emissions, help to mitigate the consequences of the emergence of e-waste. The use of environmental digital technologies within the framework of global and national environmental monitoring, improvement of the safety of the operation of industrial enterprises, in agriculture, as well within the framework of enhancement of the comfort of living in the urban environment, is quite promising. However, in Russian environmental legislation, there is now no direct mention of the possibilities of using digital technologies in the field of environmental protection, which should be corrected by supplementing the law with a special chapter.

1 Introduction

The global changes of recent years in the world around us indicate that the line between the physical and digital worlds becomes blurred. More and more sectors of social life move online, and digital technologies have an increasingly strong influence on the industrial, agricultural, scientific, medical, educational, and other processes.

Digitalization technologies have a particular influence on the evolution of traditional ideas about the methods of environmental protection. Though most international legal acts do not place a direct emphasis on this, many environmental problems caused by the globalization era can hardly be resolved without using digital technologies (for example, in terms of overcoming the consequences of global climate change).

The close interrelation between digitalization and the Sustainable Development Goals (SDGs) formulated in the UN Summit Outcome Document "Transforming our World: the 2030 Agenda for Sustainable Development" in 2015 appears not less important since it is not possible to ensure the balance of economic, environmental and social interests in the globalization era without using digital technologies.

The coronavirus pandemic has had a particular influence on the discussion of the role of digitalization for environmental protection. First, it paralyzed the operation of many industrial enterprises. As a result, a number of countries registered reduction of emissions of harmful substances, including CO_2, NO_2, and the blue sky returned to some megalopolises as a consequence of this [1]. Second, the problem of mass infectious medical waste has emerged, for which many countries had not been ready. Third, there has been an increased demand for digital technologies, in particular, for those which make it possible to remotely monitor technological processes, including emissions into the environment. Digitalization has become especially important for monitoring of the state of the environment in general (and not only emissions of a given enterprise), which ensures remote assessment of environmental threats to the areas of particular municipalities and regions and adoption of prompt measures to eliminate them.

2 Problem Statement

In this situation, we observe a greater role of the studies of the possibilities of law to regulate social relations associated with digitalization of certain areas of social life to reduce the negative impact of society on ecological systems, both natural and changed by humans.

3 Research Questions

* Corresponding author: 4077778@list.ru

Elimination of the consequences of the coronavirus pandemic will be of a long-term nature, and it will definitely have an effect on the dynamics of achievement of 17 Sustainable Development Goals (SDGs) that was before 2020. The importance of achievement of the goal of ensuring public health will inevitably divert the attention of both governments and the public away from other goals, which seemed so important to everyone not long time ago (fighting against climate change, ensuring sustainability of seas and oceans, etc.). However, neglect of some goals in the interests of other ones can hardly be maintained even in a middle-term perspective since they are closely interrelated. In our view, in order to identify the dynamics of their interaction and work out a new strategy for the achievement of all SDGs in the period of the pandemic and after it ends, we need new technologies, which are provided by digitalization. However, before we start analyzing the environmental and legal aspects of digitalization, it is necessary to answer a more general question of what should be understood by the term "digitalization".

For example, some authors note that digitalization is "not only the use of digital technologies and the creation of new possibilities, including the exercise of interests in gaining profit, it is a process of transition to digital business, which requires detailed regulation by law" [2]. Others think that "digitalization can be considered as a process of application of digital technologies in various fields, while digital transformation implies specific results arising from their use" [3].

Therefore, the emphasis is placed on technical and/or economic aspects. For our part, we think that it is a much wider phenomenon. Digitalization is a complex of economic, management and social processes associated with the application and widespread expansion of digital, computer, information, electronic, network (telecommunication) technologies, and AI systems in modern life [4]. Further we will consider digitalization exactly in this sense.

At the moment, there is no generally accepted classification of the main areas of digitalization of legal protection of the environment either in international legal acts or doctrinal studies, though its certain elements have already been examined in detail.

We think that digitalization of environmental protection can be mentioned in two aspects: digitalization of "classic" areas of legal protection of the environment and those which emerged not long time ago and were caused by the shift of countries to the sixth technological paradigm.

Within the first aspect, we suggest distinguishing digitalization of state environmental management (especially in terms of environmental supervision and licensing), digitalization of calculation of the payment for the negative impact on the environment or estimation of the volume of emissions, discharges, etc. A particular place in this group is occupied by digital solutions in the field of protection of environmental human rights related to the use of digital technologies to file complaints, applications and suggestions on the portals of state authorities and local government bodies that perform environmental protection functions, as well as consideration of citizens' opinions while discussing projects of various solutions of authorities [5].

Within the second group, we propose to distinguish the following elements:

1) Digitalization of production processes in agriculture that improves its sustainability. This area can both include several sectors related to digitalization of the very production processes (use of drones for spraying of fields, robotic automation of agricultural machinery, etc.) and be of a purely informational nature (use of mobile phones to deliver agricultural products, access to information about the soil condition and air humidity that is necessary for farmers, advice on fertilizer application, etc.) [6]. Artificial intelligence technologies can be used in the future [7]. Digitalization becomes especially important with respect to the development of the market of organic goods [8]. Its significance could consist in the development of a "digital passport" for certain kinds of agricultural products that would enable consumers to track the place and time of production of the goods, their chemical composition, use of fertilizers, etc. Particular steps in this direction have already been made. For example, Russia has created the Unified State Register of Organic Producers, which is maintained in electronic form. It seems equally promising to develop an open electronic database of unscrupulous producers of food that have been held liable for production and circulation of foodstuffs which do not comply with the established quality and safety requirements [9]. Hence it follows that the main areas of digitalization of agriculture and legal incentives for their introduction should be reflected in the national legislation on agriculture.

2) Digitalization and the "smart urban environment". Without having the possibility to consider all aspects of the "smart urban environment", let us focus on only one of them – the digital logistics of transport flows. Within this area, we distinguish two environmentally important sectors. First, the management of urban transport flows can be improved due to digital technologies, which will reduce the time spent in traffic jams, the fuel consumed in vain (which will decrease the volumes of oil production and processing), and the emissions from cars into the air (which will reduce urban pollution). Second, the development of digital technologies will eventually enhance the logistics of cargo transportation, which will improve their efficiency, reduce empty runs and result in the same positive effect as in the case stipulated above.

The improvement of the "green logistics" will lower pollutant emissions by as much as 3.6 billion tons by 2025. This can be facilitated by introducing new digital platforms (Uber Freight, Convoy or Transfix) that help to increase the overall efficiency of a supply chain by cutting out unnecessary intermediaries, sharing transport capacity, and sharing warehouse space. Moreover, the internet of things makes it possible to identify optimized patterns and potential breakdowns in the supply chain, which also contributes to logistics players' corporate social responsibility initiatives. In the maritime logistics industry, using big data to optimize vessel speeds and paths reduces waiting times in ports and, therefore, CO2 emissions. Another way to resolve logistical issues

 https://doi.org/10.1051/shsconf/202110901033

would be development of government incentives and subsidies [10].

3) The energy sector is the most promising area for environmental digitalization; it can increase the efficiency and cost-effectiveness of the processes of production, transmission, and distribution of energy.

Reduction of electricity consumption, further integration of electricity produced by means of renewable energy sources into the system of national and international electricity markets will lead to decrease in the use of hydrocarbons in the energy sector as well as emissions of harmful substances related thereto. In addition, there is a close relationship between electricity generation as well as industry and the transport sector (manufacture and use of electric vehicles). There are opportunities for digitalization also in industry, where the use of these technologies strengthens control over wastewater discharge, improves its physical and chemical treatment, which is already happening gradually.

4) It is possible to improve the rationality of the use of certain types of natural resources with the help of digital technologies. For example, due to them, it is possible to use forest resources more rationally, reduce water losses by creating online networks and using sensors that transmit information about leakages and consumption of water.

5) Digitalization and climate. Climate change and digital technologies are undoubtedly two of the most defining features of our civilization. Digitalization of production can have several areas. For example, digitalization of the manufacture of electric cars (and electric planes and ships over time) will make it possible to reduce dramatically emissions and discharges and to improve the climate. Introduction of hydrogen engines in production will play here a particular role, which will lead to the end of the oil era and reduction of greenhouse gas emissions. Meanwhile, the economic growth has not yet been separated from the growth of emissions and the use of natural resources. The historical trend is that for every 1 percent increase in global GDP, CO2-equivalent emissions increase by about 0.5% and resource intensity by 0.4%. The current business practices contribute to the increase in the global gap between natural resources supply and demand up to 8 billion tons by 2030, which will result in a loss of 4.5 trillion dollars in the economic growth by 2030. To settle this issue, a legally binding climate agreement was concluded at the UN Conference in December 2015. The agreement is aimed at preventing the average temperature on the planet from rising above 2˚C relative to the pre-industrial era, and if possible, to reduce it to 1.5˚C. It appears that, to make this activity effective, a coordinating body should be established under the UN to assist all interested countries in introducing digital innovations in the field of climate. Within this area, we should emphasize that digitalization of certain sectors of the economy (transport, industry, agriculture, etc.) can create synergies by strengthening the results in each sector, which is, in our view, the aim of the 2015 Paris Climate Agreement.

6) Digitalization and environmental monitoring. Big data and artificial intelligence algorithms can be used for environmental monitoring, including that one which ensures conservation of endangered land species, as well as for early warning of natural disasters such as earthquakes, forest fires, floods, and droughts. It is especially important to receive these data about the condition of specially protected natural areas (reserves, natural monuments, etc.) [11]. Tools for data collection and satellite survey can help to ensure ocean sustainability by preventing overfishing and monitoring of the level of health and pollution of marine ecosystems and ocean habitats. Intelligent networks can help track trends in energy consumption and ultimately reduce greenhouse gas emissions.

Space environmental monitoring and improvement of informational interaction between different countries are of particular importance in this case [12]. For example, in Russia, the state space monitoring system includes several centers located in the European, Siberian and Far-Eastern parts of the country. The Earth remote-sensing data received from domestic and foreign satellites are matched with a network of ground-based observation sites that provide high reliability and calibration of the remote sensing data [13].

7) Blockchain digital technology can revolutionize the environmental protection efforts since such technologies can ensure decentralized and sustainable management of resources, including water and energy consumption. Moreover, blockchain technology could help to save marine biodiversity preventing such practices as overfishing by means of traceable and transparent supply chains. Finally, no discussion of new technologies would be full without the internet of things. Though the term "the internet of the environment" can be more suitable in this case, the sensors connected to the internet can help to control and prevent deforestation, which is associated with 15 percent of global greenhouse gas emissions. They can be used also to fight against poaching by keeping track of endangered animals, monitoring their location, and introducing intelligent security systems.

8) Digitalization and scientific studies. Due to digital technologies (including mobile applications) even unskilled citizens can make an important contribution to processing of empirical data, for example, those related to observation of species. Digital technologies can unlock the potential of already collected data, for example, citizens can help with the digitization of natural history collections. Search engine data can now be used not only to trace such common processes as pollen release and mosquito outbreaks but also to forecast more important biological events [14].

9) Environmental digitalization already has numerous consequences, which are not always clearly perceived by us. For example, during the period of the coronavirus, the transfer of school and university education to digital form caused a decrease in the demand for paper, which will lead to a decrease in felling of trees. In a similar manner, the transfer of postal communication to digital form (electronic mail) led, in addition to the decrease in the demand for paper, to a decrease in the postal transportation, in the demand for fuel, and reduction of emissions into the air. Moreover, the transfer of business

https://doi.org/10.1051/shsconf/202110901033

negotiations (or scientific conferences) to digital form caused reduction of trips and fuel consumption, which will inevitably have an environmental effect.

Along with the abovesaid objective prerequisites for the introduction of digital technologies in environmental protection activities, we should also note a subjective factor existing in this process. The matter is that the prospective consumers of digital environmental services should, at least, know about their availability, and, in addition, rely on them. Even if we speak of not interaction with authorities by means of digital platforms but only of digital technologies that enable citizens to control water and electricity consumption via mobile applications, even in this case consumers should believe that their devices and data are protected, and their personal information will not be stolen.

It should be noted that along with undoubted advantages (even with allowance for the subjective factor of their perception by society), digital technologies pose a number of inevitable threats to the state of the environment. First, we can say that the Internet causes hidden (invisible) pollution. This manifests itself in the huge electricity consumption, in terms of which there is an impact on the environment.

Second, the use of digital devices themselves (computers, telephones, etc.) leads to the emergence of such a phenomenon as "digital e-waste", which is a serious environmental problem. According to the available data, about 50 million tons of e-waste is produced every year, and only 20 percent of it is recycled. The other 80% of e-waste is usually buried underground. Though the vast majority of e-waste is produced by developed countries, its most part is handled (i.e., dismantled and recycled in so-called e-waste dumps) in developing countries.

Though utilization of e-waste may seem a more reasonable option, it has its own disadvantages as well. Contact with harmful toxic materials such as lead, chromium, and cadmium can cause serious health problems, including respiratory diseases and lung cancer, to which working children are especially vulnerable. Despite the piles of e-waste, the demand for new high-tech gadgets continues to grow and aggravate the environmental consequences. Almost all digital technologies imply the presence of metals and minerals, known as rare earth elements, which are necessary for their functioning and are only sometimes recycled from e-waste. A smartphone contains 16 various rare earth metals. Even when metals are not rare, the process of their extraction causes serious environmental problems since it causes significant damage to the soil, pollutes water with chemicals used in the extraction process, and requires expensive cleaning operations.

Third, rising sea levels could cause damage to submarine cables (particularly in coastal areas in the US) and ultimately disrupt the functioning of the Internet in the coming 15 years. Similarly, the depletion of the ozone layer that in turn contributes to the increase in the intensity of ultraviolet rays from the sun will most likely hamper the electromagnetic frequency along which Wi-Fi travels [15].

4 Purpose of the Study

The purpose of the study is to consider the main areas of environmental protection digitalization for further change of environmental legislation.

5 Research Methods

Such research methods as comparative legal analysis, synthesis, formal logical and other methods of scientific knowledge were used within the framework of the conducted study.

6 Conclusion

Therefore, there are advantages and disadvantages of environmental digitalization, and there are no reasons to idealize this factor of modern life. Further development of digital technologies will reduce the consumption of hydrocarbons and greenhouse gas emissions, help to mitigate the consequences of the emergence of e-waste. The use of environmental digital technologies within the framework of global and national environmental monitoring, improvement of the safety of the operation of industrial enterprises, in agriculture, as well within the framework of enhancement of the comfort of living in the urban environment, is quite promising. However, in Russian environmental legislation, there is now no direct mention of the possibilities of using digital technologies in the field of environmental protection, which should be corrected by supplementing the law with a special chapter.

References

1. COVID-19 Implications & Responses, Digital Transformation & Industrial Recovery, United Nations. Industrial Development Organization, (2020)
2. V.V. Cherevko, D.V. Shepelev, Actual problems of modern science and practice **2**, 135 (2019)
3. S.M. Zubarev, Actual problems of Russian law **6**, 26 (2020)
4. Y.V. Gaivoronskaya, O.I. Miroshnichenko, A.Y. Mamishev, Indiscreet charm of digitalization. The legal paradigm **18-4**, 43 (2019)
5. A.P. Anisimov, B.B. Inzhieva, A.Y. Ryzhenkov, *Ubiquitous Computing and the Internet of Things: Prerequisites for the Development of ICT. Studies in Computational Intelligence* **826** 325-332 (Springer Nature Switzerland AG, 2019)
6. The Networked Farm, Digital Revolution: More Efficient Agriculture, More Reliable Harvests, (2020), URL: https://www.research.bayer.com/en/digital-farming.aspx (date of access: 11.11.2020)
7. C. Giesler, Digitisation in agriculture - from precision farming to farming 4.0, (2020) URL:

https://www.biooekonomie-bw.de/en/articles/dossiers/digitisation-in-agriculture-from-precision-farming-to-farming-40 (date of access: 30.11.2020)

8. A.S. Shcherbakova, The North and the market: shaping the economic order **1**, 92-102 (2017)
9. A.Yu. Mokhov, Scientific Bulletin of the Russian Academy of VF. Series "Jurisprudence" **1**, 94 (2019)
10. J. Dallemand, Is digital transformation the key to environmental sustainability in logistics? (2020), URL: https://blog.datumize.com/digital-transformation-and-environmental-sustainability-in-logistics (date of access: 30.11.2020)
11. A. Anisimov, K. Lidzheeva, A. Ryzhenkov, Problemy Ekorozwoju **12-2**, 47-58 (2017)
12. S.V. Veretekhina, Theory and practice of project education **3**, 69-74 (2019)
13. S.N. Bobylev, Geography, Environment, Sustainability **11-1**, 149 (2018)
14. K. Arts, R. van der Wal, W.M. Adams, Ambio **44-4**, 664 (2015)
15. Digital and Environment, (2020), URL: https://dig.watch/trends/digital-and-environment (data of access: 30.11.2020)

Current issues of digital legal education in the context of the Covid-19 pandemic

Anatoliy Ryzhenkov[1]*, and *Buinta* Inzhieva[2]

[1] Kalmyk State University, ul. Pushkina, 11, Elista, Russian Federation, e-mail: 4077778@list.ru
[2] Kalmyk State University, ul. Pushkina, 11, Elista, Russian Federation, e-mail: inbuinta@mail.ru

Abstract. The main areas of digitalization of higher education, including legal one, have been actively discussed by representatives of authorities and the educational community over the past few years. The COVID-19 pandemic summed up these discussions in a certain way, making teachers and students shift to the practice of introducing digital technologies into the educational process without proper preparation and prior training for them. The article discusses the problems of digitalization of the higher education system of the Russian Federation faced by law faculties after the beginning of the pandemic in the spring of 2020. The authors state the positive results of the transition to digital technologies as well as the existing problems, many of which are of a systemic nature and are still to be resolved. At the same time, they point out that the requirements for the competencies of law graduates change in the context of the mass digitalization of all sectors of social life. Along with the knowledge, abilities and skills that are commonly recognized for lawyers, they must have the competencies to interact with authorities that have also shifted to a digital mode for communication with citizens and legal entities.

1 Introduction

The main areas of digitalization of higher education, including legal one, have been actively discussed by representatives of authorities and the educational community over the past few years. The COVID-19 pandemic summed up these discussions in a certain way, making teachers and students shift to the practice of introducing digital technologies into the educational process without proper preparation and prior training for them.

2 Problem Statement

Therefore, digital higher legal education has now become a reality, which makes it possible to assess both the preliminary theoretical concept and the problems with its implementation in order to adjust the initially set goals and develop a new strategy for the development of digital education. This will become relevant in the period after the pandemic and in the return of the educational process of higher education institutions to the pre-COVID mode.

3 Research Questions

To analyze the goals and objectives of digital education, first, it is necessary to determine the terminology. A. Marei considers digitalization "as a change of the paradigm of communication and interaction with each other and society" [1]. Other authors believe that digitalization is "not only the conversion of information into a digital form but also a comprehensive solution of an infrastructural, managerial, behavioral, cultural nature" [2].

According to L.V. Kovtunenko, "the digital educational environment is a system of effective and comfortable provision of information and communication services, the need for which is due to modern trends in the modernization of education" [3]. V.S. Moskalyuk points out that the progress in the development of the Internet and mobile technologies are the key pillars of digitalization. Digitalization of education "causes the transformation of the labor market by instilling new skills and competencies through the radical transformation of the educational process and the role of teachers. Digitalization fundamentally changes the obsolete methodological framework of the school and also gives access to information not only in the form of text but also in audiovisual form, which requires its permanent search and selection of the most adequate content and its analysis" [4]. The process of transition to digital technologies under consideration can be designated by the term "digital transformation of education" – that is, a qualitative change in "both the educational process itself and the educational activities on the basis of the development of breakthrough information (digital) technologies" [5]. Therefore, digitalization of education is one of the strategic areas in the system of education management, the purpose of which is the transition to a new model of provision of educational services with the use of possibilities of the Internet and new methodological

* Corresponding author: 4077778@list.ru

techniques aimed at developing new abilities and skills of students in processing and generalizing information flows.

This will ultimately lead to the situation where completely new professions directly related to digital technologies will be in demand in the future: virtual lawyers, developers of educational trajectories, experts in cyber prosthetics, moderators of platforms for communication with state authorities, recycling technologists, cross-cultural communication managers [6].

The necessary legal framework for the transition of higher education institutions to digital technologies had been formed even before the beginning of the pandemic. For example, according to paragraph 2 of Article 12.1 of Federal Law No. 273-FZ of December 29, 2012 (revised on July 31, 2020) "On Education in the Russian Federation", various learning technologies, including distance learning technologies and e-learning, are used in the implementation of educational programs. The latter is understood as the organization of educational activities with the use of the information contained in the databases and applied in the implementation of educational programs, information technologies and technical means ensuring its processing, as well as information and telecommunication networks ensuring the transmission of the said information through communication lines and interaction between students and teaching staff. Distance learning technologies are understood as the learning technologies implemented mainly with the use of information and telecommunication networks in indirect interaction (at a distance) between students and teaching staff (Art. 16 of the Law).

The National Program "Digital Economy of the Russian Federation", which includes the Federal Project "Human Resources for Digital Economy", in particular, suggests that 10 million people in Russia will have completed training in online programs for the development of digital literacy by the end of its implementation period (December 31, 2024) and from 50,000 to 120,000 people should be admitted annually to training in higher education programs in the field of information technology. A range of activities for education digitalization was planned at the level of constituent entities of the Russian Federation. Among these activities, Order of the Ministry of Digital Development, Communications and Mass Media of the Russian Federation No. 428 of August 1, 2018 "On Approval of Explanations (Methodological Recommendations) on the Development of Regional Projects within the Framework of Federal Projects of the National Program 'Digital Economy of the Russian Federation'" mentions introduction of the target model of the digital educational environment, including renewal of the material and technical framework, computer facilities, software and presentation equipment in educational organizations; development of regional information services and resources in the field of education; use of information services and resources of the federal information and service platform of the digital educational environment; introduction of modern digital technologies in the administrative and management activities of educational organizations, planning of lessons and maintenance of schedules, electronic gradebooks and diaries, student movement registers, compilation of reports (analytics and reports based on big data, electronic reports, cloud accounting), etc.

The leaders of other countries had high hopes for digitalization too. For example, the German federal government considered it as a way to ensure the transfer of knowledge and innovations in science, however, it also expected citizens to improve their digital literacy in order to be able to fully participate in education and life of society [7].

This significant attention to the transition to digital technologies in the higher education system is explained by the fact that the modern state and society, both in the Russian Federation and in other countries, are at the stage of transition to a new technological order, which is an unprecedented level of technological equipment of a modern person. Even before the pandemic, it was obvious that the classroom-lesson system, which developed during the Middle Ages, had become obsolete. Even two hundred years ago, teachers served as almost the only source of knowledge for students, since the library system was underdeveloped, and there were no other forms of educational communication. However, in the 21st century, with the spread of the Internet, everyone has unlimited access to educational resources, whether they be textbooks, practical exercises, monographs, or other sources of information that can be used in the educational process. In these conditions, the role of teachers had to change, as well as the lecture-seminar form, since digital technologies make it possible to radically change the previous approaches, diversify and saturate the educational process with new methodological techniques and views. However, before the pandemic, large-scale digital innovations were used in practice only in a few universities in a limited range of educational courses. In addition, it is fair to note that digital technologies were widely used as an addition to the traditional teaching methods in many higher education institutions (interactive whiteboards, presentations, access to electronic libraries, online consultations, use of the Legal Reference Systems "Garant" and "Consultant Plus" in the educational process, etc.). Much was done for the development of network education with the use of the possibilities of the Internet [8].

Two promising areas of legal online education development had formed even before the pandemic: massive open online courses from such providers as Coursera, Udacity, and online study programs offered by traditional universities as an alternative to extramural studies. Open educational resources are "various types of publicly available educational and scientific materials with 'open licenses' that give the possibility to everybody to use them without any difficulty – in particular, to copy, modify, create new resources on their basis. The leading universities in the world create and spread digital courses of lectures, electronic textbooks, study guides, training modules, audio and video materials, tests, computer software, etc., thereby expanding the availability of educational services and promoting the improvement of their quality" [9]. However, despite this trend, the shift of the function of teachers from the "only carrier of knowledge" transferring it to students that do not have other sources of it to coordinators and organizers of students' independent work, in order to form their practical competencies and skills for self-study, had not been realized completely by the educational community and had not received any support before the beginning of the pandemic. The world trend of individualizing educational trajectories, rapidly increasing possibilities for students to choose educational programs and modules and expanding their access to information had not been implemented in the Russian Federation either.

We should note that within the said trends of the "pre-COVID" era representatives of authorities and society already clearly realized that in the digital era any country wishing to maintain its position in the global division of labor had to change the structure of public investment, increasing the contribution to the human rather than physical capital [10]. It was also obvious that the role of advanced training programs became greater in the digital era. Not only young people but also all other age groups will be trained within their framework, and this requires the development of new and the adjustment of the available academic programs.

Therefore, at the time of the beginning of the pandemic, there were two formed topics for disputes on the role of digitalization in the education of lawyers – first, this is the issue of the methods and technologies of teaching legal disciplines in the digital era, and, second, the issue of the specific competencies of graduates which will be in demand in the conditions of the formation of digital economy in the country as well as the digitalization of all other areas of social life. After the beginning of the pandemic, the following problems were identified within the framework of the two abovementioned issues.

1) In various higher education institutions of the country, professors and students had to shift as soon as practically possible to the distance learning mode, with the use of the platforms of ZOOM, MOODLE, TEAMS, etc., and not all higher education institutions organized high-quality training of professors and students for work in the mode completely new for them. While students, being accustomed to modern gadgets, in general easily shifted to the new learning mode, this breakdown of the usual teaching modes caused serious technical problems for many professors (especially elderly ones).

2) In the course of the settlement of the problems of transition to the distance learning mode, the question of forming new competencies of law graduates related to the change in the needs of the labor market, which now requires experts able to communicate, for example, with authorities in a digital mode, remained open [11].

At the same time, there is now no single list of these new digital competencies of lawyers. This is why some authors believe that young experts must "have skills in electronic document management, understand the technology of entering into smart contracts, use digital signatures, have skills in cybersecurity, be able to find evidence on the Internet, possess the knowledge of digital forensic technology, etc." [12].

Other scholars note that possible options for the penetration of digital technologies in the area of professional activity of lawyers, include state and municipal purchases in electronic form; submission of applications for state registration of legal entities and individual entrepreneurs, acquisition of rights to immovable property items via the public services portal; maintenance of various registers including those related to the activity of notaries in electronic form; publication of information related to the consideration of cases, etc. on the websites of courts; work with legal systems through the use of the Internet; protection of intellectual property rights related to the use of network space, etc. [13].

In our view, it is hardly possible to draw up an exhaustive list of such areas of the application of digital technologies within the framework of the legal profession. However, it is obvious that it is necessary to develop such new competencies of future lawyers on the basis of negotiations between the educational community and employers' representatives. At the same time, the search for these new competencies should not be limited only to the study of digital technologies, since in the 21st century employers increasingly often need employees with a range of skills related to several branches of knowledge (jurisprudence, economics, political science, mathematics, management, etc.) rather than narrow experts. Therefore, it appears that digitalization of certain aspects of legal activities (for example, development of computer software that enables citizens to draw up a civil law contract by themselves) will lead not to the disappearance of the legal profession but to change in the set of competencies of these experts and the imposition of new job duties on them.

3) Innovations in digital legal education should not be reduced only to technical innovations associated with a new mode of organization of the educational process but should also cause a number of legal, organizational and structural changes in higher education institutions. For example, it is necessary to change the regulatory framework governing the development of digital educational and methodological resources at universities, including in the field of the protection of copyrights of professors publishing their lectures and other materials on digital platforms.

This circumstance is often pointed out in scientific literature. In particular, a number of authors note that, in the context of the rapid development of digital economy, there is a need to create coherent, global and comprehensive legal guarantees, including reliable guarantees of legal protection governing the use of digital technologies in order to minimize the risks of digitalization and legitimize new assets, both tangible and intangible. International organizations and states actively develop strategies to change laws on the use of modern digital technologies. However, the main problems consist in the fact that, on the one hand, the suggested strategies are sectoral and relate to only some digitalization aspects and, on the other hand, the solutions are often aimed at implementing the political agenda to the detriment of the promising agreed global legal strategy [14].

4) In overcoming the consequences of the pandemic, the main role was assigned to the constituent entities of the Russian Federation, which, with consideration of the

epidemiological situation, could vary the list of measures for reduction of the threat from COVID-19, in particular, make decisions on the period of transition of universities to the distance learning mode. In a number of regions, this policy of the governors should be recognized as generally successful, however, the rapid introduction of prohibitions and restrictions took a significant toll on the economy and the development of higher education in other constituent entities of the Russian Federation.

5) The following promising areas of digitization of the educational environment within the framework of law faculties can be distinguished now (November 2020): development of electronic (online) courses of individual authors, expansion of online education and the possibility for students from different cities to complete study (advanced training) at university; development and introduction of new methods of online examinations and tests (in this area, it is still necessary to decide on the technology of development of educational computer games, when examinations are conducted not as a conversation on zoom or teams online platforms or not even in the form of common tests but in the form of computer games with a certain legal plot involving certain legal problems solved by students and making it possible to achieve victory in the game and earn a credit); development of new digital textbooks, which are not just text, they also open access to multimedia content – video, interactive presentations, etc.; development of new approaches to communication and interaction between students and professors with the use of the possibility of digital learning and individualization of this communication.

6) Digitalization of education in the context of the pandemic, in addition to the obvious positive effects related to the protection of the health of students and professors, as well as the use of new opportunities and technologies, has many non-obvious effects. For example, digitalization of education in any country leads to reduction in paper consumption, which means achievement of an important nature protection goal related to less deforestation.

7) Higher education institutions of Russia training lawyers and developing distance learning systems face particular and quite typical difficulties now. The most significant problems include unreadiness of the information and educational environment of higher education institutions for support of the complete process of e-learning; lack of teaching methods for the electronic environment, a mandatory system of advanced training in the field of electronic technologies; insufficient provision of e-learning with educational and methodological materials; unreadiness of professors to implement e-learning; lack of awareness of its potential and the need to use it; lack of a sufficient number of highly qualified programmers to solve multi-factor tasks of informatization of higher education institutions [15].

It appears that to solve many of the mentioned problems related to the introduction of digital technologies in higher education institutions, to improve their efficiency it is necessary to build a dialogue between all interested parties, including students, professors, parents, and representatives of employers so that the final result would be voluntarily recognized by all of them.

4 Purpose of the Study

The purpose of the study is to consider the main areas of digitalization of Russian higher legal education, to identify the influence of the coronavirus pandemic on the mandatory introduction of digital standards in the educational process of higher legal education institutions and law faculties.

5 Research Methods

Such research methods as comparative legal analysis, synthesis, formal logical and other methods of scientific knowledge were used within the framework of the conducted study.

6 Conclusion

Therefore, digitalization of education in higher education institutions is an issue of concern to all interested parties in the field of education, including professors, students, their parents, business, as well as state authorities. The skills in using digital learning technologies become increasingly relevant both in Russia and in all other countries. Along with the transition to distance learning technologies, a particular problem of law faculties is still training of future experts capable of coping with problems and finding solutions with the use of digital technologies as a means to overcome new challenges caused by digitalization of many sectors of social life. The use of the possibilities of digital technologies can make teaching and learning more flexible as well as lead to the expansion of possibilities of students and the improvement of learning skills throughout life.

Meanwhile, both in Russia and many other countries, the use of digital technologies gives rise to a number of problems related to the lack of digital skills of professors and students as well as their unwillingness to shift to new learning technologies. The settlement of this problem requires thoughtful awareness-raising activities and the development of special advanced training courses. This will require additional incentives and funding, for which the Russian Federation is not yet ready.

References

1. A. Marei, *Digitalization as a paradigm shift* (2020), URL: https://www.bcg.com/ru-ru/about/bcg-review/digitalization.aspx/ (date of access: 31.10.2020)
2. E.L. Vartanova, A.V. Vyrkovsky, M.I. Makseenko, S.S. Smirnov, *The industry of Russian media: digital future: academic monograph* (Mediamir, Moscow, 2017)
3. L.V. Kovtunenko, *Actual problems of activity of divisions of UIS: collection of materials of the all-*

Russian scientific and practical conference, 178 (Voronezh, 2019)

4. V.S. Moskalyuk, Science and education today **10**, 15 (2019)
5. V.N. Yuzhakov, A.A. Efremov, Russian law: education, practice, science **6**, 19-20 (2018)
6. Atlas of new professions (2020), URL: http://www.skolkovo.ru/public/media/documents/research/sedec/SKOLKOVO_SEDeC_Atlas.pdf (date of access: 30.10.2020).
7. M. Bond, V.I. Marín, C. Dolch, S. Bedenlier, O. Zawacki-Richter, International Journal of Educational Technology in Higher Education **15:48**, 4 (2018)
8. A.P. Anisimov, A.J. Ryzhenkov, A.A. Serebryakova, Espacios **39-25**, 32.
9. L.L. Sabirova, Bulletin of the Mari State University **12-2**, 75 (2018)
10. G.F. Shafranov-Kutsev, Bulletin of the Tyumen state University. Socio-economic and legal research **3-4**, 11 (2017)
11. D.M. Osina, G.P. Tolstopyatenko, A.A. Malinovsky, *Engineering Economics: Decisions and Solutions from Eurasian Perspective*, 392-398 (2020)
12. T.V. Epifanova, A.N. Pozdnyshov, Science and education: economy and Economics; entrepreneurship; law and management **5**, 25 (2019)
13. E.V. Elnikova, Appendix to the journal "Business law" **3**, 55-60 (2019)
14. E.L. Sidorenko, P. von Arx, Digital Law Journal **1**, 26-27 (2020)
15. T.I. Krasnova, In the world of scientific discovery **11**, 18 (2014)

The phenomenon of digitalization of the legal business education

Andrey Sadkov[1], *Natalia* Kvitsinia[2]*, and *Viktoria* Sandalova[3]

[1] Volgograd State University, University Avenue, 100, Volgograd, Russia, e-mail: sadkov@volsu.ru, ORCID 0000-0002-6148-0576
[2] Volgograd State University, University Avenue, 100, Volgograd, Russia e-mail: gimchp@volsu.ru, ORCID 0000-0002-6538-9796
[3] Volgograd State University, University Avenue, 100, Volgograd, Russia, e-mail: sandalova@volsu.ru, ORCID 0000-0001- 8600-4974

Abstract. This article attempts to consider the phenomenon of digitalization of education in the construction of a digital economy, to identify the most significant problems, risks, as well as the vector of modification of the educational process from the standpoint of legal and ethical regulation. Results: according to the results of the study, the authors made proposals to improve state standards in the field of law in terms of updating competencies in the context of digitalization of the economy, identified promising educational innovations, proposed ways to modernize the educational process. Conclusion: the system of legal business education is in a state of transformation, requires updating of state standards in the field of training "jurisprudence", updating the curriculum, content and typology of seminars, as well as the introduction of educational innovations in the educational process.

1 Introduction

Digitalization of education is a topic that is currently under discussion. Special portals dedicated to digital education (for example: http://www.digital-edu.ru/; http://window.edu.ru) have become very popular, bloggers have become more active, subject magazines have begun to appear. Audience activity is due to a number of objective and subjective factors. Let's consider some of them.

First of all, lawyers were forced to go into problems of digitalization due to the state programs both in the field of economics and in the field of education itself [1]. Among these program documents, a special place is taken by the program "Digital Economy of the Russian Federation" (hereinafter referred to as the Digital Economy Program) approved by the Government of the Russian Federation No. 1632-p dated July 28, 2017, adopted to implement the Information Society Development Strategy in the Russian Federation for 2017–2030 years (approved by Decree of the President of the Russian Federation dated May 9, 2017 No. 203). One of the main directions of the Digital Economy Program, which attracts attention, is the "personnel and education" direction, which envisages the creation of key conditions for the training of personnel in the digital economy and the improvement of the educational system itself.

The development of the Concept is determined by the growing contradictions between the objective needs of the digital economy and the established tradition of training lawyers in the higher education system. The necessity of its appearance is dictated by the importance of the task of creating a fundamentally new mechanism of training legal personnel for the needs of the digital economy, the desire to ensure the employment of graduates in conditions of tough competition in the legal services market.

Russian practice shows that in the past two years there has been a tendency to actively introduce digital technologies into the Russian economy, as evidenced by the program "Digital Economy of the Russian Federation". According to experts, digital transformation is a new reality, in which an important place is given to digital data, which determines the development of production. The ability to work with Big Data brings the country to a high competitive level. New technological prospects directly affect legal professions, which are gradually changing both in Russia and abroad.

Secondly, the modernization of education is also necessary because a new generation of digitalnatives has grown up, whose representatives demonstrate a tendency to use new technologies in their daily lives. At the same time, research by IT specialists reveals the use of appropriate technologies not only in the professional sphere, but also for socialization and communication. Consequently, the digitalization of education will make it more adapted to the modern audience, which will lead to an increase in the competitiveness of the university in the education market, the creation of additional value and the involvement of students.

In addition, we shall notice, that today such parameters as the level of technical equipment of student workplaces in libraries, the availability of free access for students and teachers to information resources that ensures the educational process and research work, open

* Corresponding author: gimchp@volsu.ru

wireless network, wireless network coverage in educational and administrative premises of the university, access speed in the Internet wireless network – are important for the state and public professional accreditation of the educational organization.

The noted and other arguments indisputably testify to the need of digitalization of legal business education, which implies the solution of other tasks associated with this goal [2]. So, in order to equip future lawyers with "advanced" knowledge, it is necessary to develop personal educational pathways for students, to form the basic competences of a lawyer, to modify the certification system that minimizes the ability to reproduce ready-made knowledge. It is important to ensure the education of a harmoniously developed and socially responsible person.

The Association of Solicitors in England and Wales (The Law Society) has published a report on the future of the legal profession. Experts believe that in the next five years, all participants in the legal market are expected to see global changes associated with the emergence of robotic substitutes for lawyers providing low-cost legal services. Competition will affect not only the solution of simple tasks related to conciliation procedures in the field of economic disputes, but also will affect complex legal issues. The tendency to change the situation not in favor of practicing lawyers is already developing. The software that can read and correctly interpret legal texts, helping in seconds to find solutions to complex legal issues has been developed.

Programs have been created, such as Leverton, which recognizes 18 languages and is capable of self-learning. With each new document, it builds an algorithm of the original legal solution, identifies errors, prepares reports on the terms of the agreement. Leverton is quickly learning to understand the complex legal language of contracts. Artificial intelligence "Watson" from IBM is capable of self-learning and development. In a few seconds, it makes an analysis of tens of thousands of legal documents, regulatory and judicial acts and prepares an original legal opinion. This software answers legal questions, forms claims and contracts, based on current practice.

British experts believe that thanks to technologies, including big data, online justice, the availability of justice will increase in the next five years and the cost of legal services will decrease, which threatens the loss of legal services markets for law firms specializing in these topics.

On the Internet, bot programs are widely used to assist in filling out various legal documents [3]. The bot is able to form claims for different life situations. An Australian law firm has presented an online consultation service with a lawyer. The robot program is able to make a speech for the court hearing and to answer questions during a conversation with a user.

In February 2018, the first robot lawyer specializing in consumer protection began operating in Russia. The robot is able to understand human speech, recognize abbreviated oral text, respond competently and provide high-quality legal advice.

The Russian law enforcement practice already uses the service of robots, which processes claims of individuals in Sberbank, which allowed to reduce about 3 thousand lawyers in this department. Mobile applications are being actively introduced to enable users to fill in and submit applications to appeal against parking fines. Similar programs for divorce proceedings work.

2 Problem Statement

The current regulation of the educational process makes it difficult to personalize education, introduce personal educational pathways into the educational process, allow students to acquire the necessary knowledge and skills in the field of digital technologies. In the conditions of existing educational standards and educational control procedures, building personal educational pathways for a significant number of trainees becomes very difficult. In accordance with the trends in the modern economic sphere and the tasks in the field of digitalization of society, the creation of a new generation of lawyers requires a change in the educational paradigm that determines the need for:

- developing new educational standards and training programs for law students aimed at the development of general digital literacy, forming digital competence, allowing law graduates to freely integrate into the digital economy [4];
- providing opportunities for educational institutions of higher education, providing training for lawyers, to develop and implement in the educational process the basic educational programs for students, taking into account the specific needs of the digital economy;
- making a competitive selection of educational organizations conducting legal training in order to empower them to develop competence models of lawyers capable of functioning in the digital environment, in this regard, cooperation with educational institutions of economic and technical profiles is possible, which will ensure interdisciplinary development of the necessary knowledge and skills;
- developing professional standards for legal professions, taking into account the digital competence of candidates for lawyers capable to act effectively in the digital economy, in this task, the main organizing role can be taken by associations and unions of lawyers with the coordinating role of the Association of Russian Lawyers [5];
- training of a new generation of teachers who are able to develop and solve the problems of legal support of the digital economy, to create educational content, including its mobile version, relevant to the market of digital services, using digital tools in the

electronic textbooks and manuals with video and audio support;
- creating a program of teaching staff retraining in law schools, adapted to innovative introduction of modern digital technologies in the educational process;
- updating of educational technologies with the introduction of personal learning pathways into the educational process, allowing students to choose formal and informal ways of forming basic digital competencies, to introduce new forms of certification of graduates with the involvement of employers who have experience in trading on digital platforms with digital products and startups;
- aligning the material and technical base as well as information and educational environment of legal educational institutions with the requirements of the society digitalization;
- developing relevant educational content that meets the needs of the educational market, along with the preparation of electronic copies of text information (digitized electronic textbooks) and the creation of a new generation of programmable textbooks (bots, robots-textbooks), tied to the basic educational program, operating in an interactive mode, able to update the legal framework, judicial practice, to form audio-visual information, update it as it becomes obsolete [6];
- changing in the assessment of the quality of training legal educational institutions graduates, allowing to identify the level of their digital competence, the ability to independently integrate and function in the digital economy;
- creating a system of motivation of students of law schools in the development of digital competencies for this: a) to introduce appropriate certification standards of the level of competencies in the digital economy for graduates of law schools, to use interested employers who can remotely participate in their certification with subsequent employment; b) to implement the mechanism of scholarships (grants) for students who have successfully mastered the educational programs of the digital economy, using the achievements of the leaders of the Russian market - Sberbank, Severstal, Norilsk Nickel, Rosgosstrakh and others with experience of continuous training of their employees [7];
- introducing a system of benefits for law firms engaged in practical training of law students in digital competencies;
- preparing proper level of knowledge and skills in digital technology of students in their training process and at the stage of their final exams.

3 Research Questions

The active introduction of digital technologies leads to the reorganization of the educational process, which will be largely based on the use of artificial intelligence technologies [8]. However, law enforcement, which involves a competency-based approach, is essentially a creative function. Despite the planned digital transformation of education, the relevant ministries have not defined clear objectives, especially for the Humanities, including law. Lawyers in higher education are trained according to standards designed for consumers of legal personnel of the old technological structure.

It seems that the defining factor of the new business education model should be operational relations with the labor market, which will prevent the threat that training will no longer be in demand. The existing legal framework makes it possible to establish such relations. The fact is that the standards involve students to undergo practical training, while the practical skills acquired during it are determined by the main educational university programs. It seems that such a practice should be systemically structured and involve a gradual immersion in the profession of a business lawyer. The practice of replacing the collection of materials for writing a master's thesis or obtaining teaching skills, participating in the university's admission campaign, etc., is also unacceptable. In addition, it makes sense to follow the standards of laboratory courses both outside universities and direct employers that integrate students into the profession as much as possible.

Since the standards require that universities update their educational programs annually, taking into account the development of economics and technology, it is necessary to include in the variable part of the educational program the possibility of studying the legal aspects of the digital economy as a practical course that deals with legal support of high-tech projects.

It is necessary to reverse the existing low academic mobility of students and teachers in Russia – a situation that is rare, and sometimes directly prohibited in most leading scientific countries. The university should strive for the alternation of languages, which will contribute to the deepening of students ' understanding and broaden their horizons, as well as the enrichment of cultural experience and outlook.

4 Purpose of the Study

Digitalization of education is possible due to the fact that several general-purpose technologies were developed and implemented simultaneously [9], which created ideal conditions for development: mobile devices, cloud computing, big data analytics platform, artificial intelligence, biometric technology, and most importantly – the Internet.

In this aspect, the phrase "digitalization of education" should be used to describe a transformation that goes further than simply replacing an analog or physical resource with a digital or informational one. The simplest example: textbooks do not just turn into e-books, but represent a set of interactive experiences.

It is quite possible that digitalization will take its place in a number of real discoveries, such as, for example, printing, which, being essentially a purely technical invention, became a catalyst for human progress, which determined the development of civilizations of the second half of the past millennium.

5 Research Methods

The scientific article is based on the use of both general scientific and private scientific methods. General scientific methods include: dialectical method, logical methods (analysis, synthesis, induction and deduction). Special methods include comparative and formal legal methods. When analyzing the current legislation, the following interpretation methods are consistently used: literal, systematic, historical, doctrinal, comparative legal. The methodological basis of the article is the interrelated methods, means and methods of cognition of social phenomena. The article uses such general scientific methods and methods of cognition as induction and deduction, analysis and synthesis, analogy, comparison, etc. A special attention was paid, in particular, to the method of comparative law, system and functional approaches, as well as the formal legal method and forecasting method.

6 Findings

6.1 Progressive educational tools

In general, today, as DigitalMcKinsey experts rightly point out, methods used in Russian education, training formats, educational programs, approaches to interacting with potential employers require adaptation to the needs of the digital economy so that the educational system can continue to train competitive personnel [10].

The positive aspects that have already been introduced include online courses and even remote mega-universities, which provide an opportunity to obtain higher education in the most favorable conditions for the student (at the place of residence, etc.). Thus, according to the State Program "Development of Education", it is planned to increase the number of online courses to 2500 by 2019, and by 2020 to 3500 [11].

Some Russian universities already work both within the framework of Western educational platforms (for example, the Higher School of Economics places its courses on Coursera), and with the use of domestic software products. The most famous of them is the National Open Education Platform (https://openedu.ru/), where, on September 1, 2018, there were 310 online courses in various areas of training. At the same time, in the discipline 40.00.00 Jurisprudence there were 21 online courses, of which it seems that only the web development of Tax Law (St. Petersburg State University) and Legal Support for Startups (Moscow State University V. Lomonosov) refer to business education. Thus, the shortage and need for online courses that provide distance learning of business law disciplines are obvious.

But the flow of other educational innovations may be truly inexhaustible: learning through social media (such as Twitter and Facebook), virtual simulators, mobile gaming applications, educational chat bots. Crowd-learning (appeal to the public) is seen as a promising modification, where students, amateurs and experts exchange ideas, generate and discuss content, solve problems, vote for the best solutions. A classic example of the work of the public is Wikipedia, which is created and constantly updated by joint efforts. Possible areas of crowdsourcing in business education include collecting and selecting empirical material, creating opportunities for exchanging opinions and discussing works online and providing opinions and data for use in projects and research.

In addition, the use of blockchain technology provides ample opportunity for education in general and business education in particular. With its help, it is possible to provide a transition from the centralized storage of students' academic achievements to a more democratic model, within which such records are kept by a wide range of participants in the system. The blockchain technology allows any participant to add new entries into a single chain of digital events, for example, the number of points, to store educational documents, written papers and examination papers, tests and the results. Students can also receive reputation points for performing intellectual tasks, such as checking business projects or scientific articles of other students, or pass on several reputation points to encourage the creation of an idea to another student or a teacher. Opening new opportunities for trading with a reputational currency, the use of blockchain technology creates a relationship to business training as to the object of sale [12].

6.2 Risks of digital education

The lawyer of the digital age is a person who is able to work not only in the real but also in the virtual space. The electronic environment, integrating into the social world, changes the habitual human habitat, transferring it to cyberspace. Digitalization and virtualization, which have become part of the digital revolution, create many previously non-existent ideological, moral, ethical problems, partly generated by the information environment, which plays the role of social communication and, at the same time, levels out the accepted moral ideals not only in a single professional environment, but also in society as a whole.

It can be stated that the network space has captured a certain part of society, in this regard, the traditional morality and culture have changed significantly, self-consciousness based on other virtual values has appeared

[13]. The network culture of interpersonal relations has affected the spheres of spirituality, religion and education. In this regard, the issue of ethical and moral aspects of training legal personnel has become more urgent.

The virtual environment has transformed the value-semantic basis of legal education and the information needs of students. Universal access to knowledge has given rise to the prospect of personalized learning. Under the influence of the information environment, a new type of students is formed, independently determining their educational trajectory, using the potential of self-education and self-development, connecting together their studies and work on the basis of network educational technologies and academic freedom in choosing content, forms, methods, place and time of training.

The learning process goes to remote sites. Between the student and the teacher there is a "technical intermediary" in the form of an electronic program, electronic platform, bot, robot or other product of the digital revolution. In cyberspace, there is a new morality in the relationship between a learner and a teacher. Being in a virtual space where electronic data circulate and there is no morality, a learner loses social orientation. Training a lawyer in such conditions is fraught with a change in his perception of objective reality, a reassessment of traditional values, which ultimately generates new ethics of relations not only between the student and the teacher, but also between the student and the potential employer, who can see the candidate for the position only in a virtual aspect.

Managing the educational process remotely, the teacher sometimes does not know who he is dealing with. His role as a mentor is lost and goes into a different quality. The teacher has a new technological function of the communicator, uniting students into groups, and groups are formed not by them, but by the interest of students in the academic discipline. Remote technology changes the social status of the teacher. Academic degree and title, which were previously indicators of the level of teacher training, come to the background. The teacher is in demand if he is interesting to the student.

The virtual environment transformers the status of the learner [14]. He becomes a consumer of information that is identified in the digital space as a virtual object, not a person. Accordingly, the ethics of relations change, in this case, not a person communicates with a person, but a digital image with an object. The student does not feel emotions towards the teacher, this in turn changes the attitude to the subject being studied, as in the traditional system of education the interest in the subject is largely formed by the personality of the teacher.

Students have to independently assess educational content, make a decision about the reliability of the information received, which turns into an effective form of influence on socio-cultural and moral regulatory mechanisms. The student chooses the content, which does not always coincide with the teacher's worldview. The use of the information necessary for students forms the system of activity motives and is the basis for the development of professional competencies.

The transition to the depersonalization of the student is one of the side effects of the digital revolution, which will lead to a change in unformed in psychologically unformed student. In this regard, along with professional competencies at the lawyer's training stage, it is necessary to form the competence of "information and digital culture", an integral part of which is digital literacy. It is the orientation in the digital space that makes it possible to understand how digital reality works, how a person interacts with digital technologies, what moral qualities, as socially significant in the professional activity of a graduate of a legal educational institution, need to be developed in a digital environment.

The digitalization of society and the economy, leads to a change of priorities in the choice of professions, the decline of social demand for education in the Humanities in general and Law in particular. The number of first-year students studying in Moscow in the field of digital technologies in 2018 increased to 19.6 thousand people. This is 11% more than in 2017. The number of vacancies for IT specialists in Moscow exceeded 20 thousand - this is 25% more than a year earlier [15].

The current practice makes it possible to assume that the demand for professions in the information technologies and digital sphere will dramatically exceed supply in the near future; their prestige in society will automatically rise. There will be a gap between the level of knowledge and skills of people who received technical education and people graduated from humanitarian and law schools. The elite within the professional community, skilled in information technologies (knowledge-class) and limiting their circle from the inflow of differently-minded will be formed. There will be a new morality based on professional affiliation.

It is possible to avoid this scenario if integrate academic disciplines (modules) providing digital competences into educational training programs for lawyers and align taking into account the intellectual parameters of graduates of higher education institutions specialized in IT with graduates of the humanitarian and legal educational institutions and, thus, increase their sustainability in the labor market. At the same time, the prestige of legal professions will remain. As early as 2019, large companies in the non-technology sector will launch their own machine learning laboratories, which will allow people with no technical education to be included in the training.

The digital network space largely depends on the ethics of relationships between people, machines and nature. Human desire to accelerate the process of digitalization of all socio-economic parts of life, including education, can work in the opposite direction and generate voluntary social isolation of people.

If the traditional form of education consolidated the student community, socializing all students, the digital, distance learning, on the contrary, isolates students from each other. In cases of training lawyers, this is a negative factor, since one of the qualities of lawyers is the ability to communicate in society. The development of communicative competences in real and virtual spaces is one of the tasks of legal education. Therefore, there is an urgent need to create a virtual, digital ethics as a field of applied ethics, including moral and philosophical reflection and moral evaluation of virtual communication, to prepare the theoretical basis of the norms and principles governing the behavior of people in the digital environment, and, most importantly, to create legal mechanisms to ensure compliance with these norms and principles.

In the world there are precedents of restrictions in the use of robotic programs to communicate with people. Thus, in California, a law prohibiting the use of bots in various online services (for example, chatbots), if the person (the owner of the bot) tries to mislead the interlocutor about the artificial nature of the bot. The law requires the owner of the program to disclose information that would be clearly formulated and located in a well-viewed place on the web-page. In addition, the computer program should be designed in such a way as to automatically inform the interlocutor that the bot interacts with him, and not a living person. The document comes into force on July 1, 2019.

Live communication between the student and the teacher, as well as contacts in the virtual environment should be balanced and determined by the curriculum as well as lectures and seminars, while the share of real and virtual time can be designed with due regard for the formation of competencies. The intention to reduce training only to the acquisition of skills and abilities hides the danger of forming utilitarian thinking and, as a consequence, a distorted worldview of students.

7 Conclusion

According to the results of the study, in the context of digitalization of the economy the proposals to improve state standards in the field of law in terms of updating competencies have been made, promising educational innovations have been identified, and ways to modernize the educational process have been proposed. In connection with the concerns expressed, it is also important to prevent the impoverishment of intellectual content in the pursuit of advanced digital form.

The solution of the problems identified in the article and other strategically important for Russia predetermines the need for scientific understanding of the phenomenon of digitalization of legal business education. It is obvious that taking into account the objectivity of the process of digitalization, as well as the huge potential of the introduction of technologies into the practice of law, legal education is particularly important, which should provide fundamental interdisciplinary training of professional personnel. This article has only outlined the most common vector of understanding of this global problem.

References

1. V.B. Vexov, *Proceedings from "International and national trends and perspectives in forensic science"* (Nizhny Novgorod, 2020)
2. Ministry of Public Education, *Order "On Approval of the Target Model of Digital Educational Environment" (December 2, 2019, No 649)* (Moscow, 2019)
3. V.A. Sadkov, M.V. Strokina, N.I. Baturina, N.V. Kvichiniy, *Competitive Russia: Foresight Model of Economic and Legal Development in the Digital Age: Proceedings of the International Scientific Conference in Memory of Oleg Inshakov (1952-2018)*, 545-552 (Springer Nature, Cham, 2020) https://doi.org/10.1007/978-3-030-45913-0_64.
4. V.B. Vexov, *Topical problems of practical orientation of the educational process in universities of the system of the Ministry of Internal Affairs of Russia*, 35-40 (Volgograd, 2010)
5. G. Sidorov, Digital University: the use of digital technologies in modern educational institutions (2017), URL: https://www.itweek.ru/idea/article/detail.php?ID=192831 (date of access: 12.12.2020)
6. T.V. Deryugina, V.F. Borisova, Legal Concept **17 (4)**, 162-166 (2018)
7. O. P. Kazachenok, A.V. Lavrentyeva, *Ubiquitous Computing and the Internet of Things: Prerequisites for the Development of ICT*, 333-340 (Springer Science + Business Media, Cham, 2019)
8. O.P. Kazachenok, *Competitive Russia: Foresight Model of Economic and Legal Development in the Digital Age: Proceedings of the International Scientific Conference in Memory of Oleg Inshakov (1952-2018)*, 407-413 (Springer Nature, Cham, 2020)
9. D.E. Prokudin, *Information technologies in education and their role in the formation of man-made culture (Dissertation of doctor of philosophical sciences)* (St. Petersburg, St. Petersburg State University, 2012)
10. Digital University, URL: https ://2018.edcr unch.ru/tracks/digital-university (date of access: 12.12.2020)
11. Government of the Russian Federation, Digital Economy Program (December 24, 2018, No 16) (Moscow, 2018)
12. M.V. Goncharova, O.V. Nikulina, I.V. Baltutite, A.N. Sadkov, *Competitive Russia: Foresight Model of Economic and Legal Development in the Digital Age: Proceedings of the International Scientific Conference in Memory of Oleg Inshakov (1952-2018)*, 561-568 (Springer Nature, Cham, 2020)
13. D.A. Tokarev, V.A. Usanova, N. Kagalnitskova, V.A. Sandalova, *Ubiquitous Computing and the Internet of Things: Prerequisites for the Development of ICT*, 215-222 (Springer Science + Business Media, Cham, 2019)
14. T.V. Deryugina, *Legal Framework for Bioeconomics and Biosafety: Monograph*, 385-393 (NY: Prospekt, Moscow, 2020).

15. T.N Bochkareva, A.R. Mubarakchina, Digital education in the Russia: realities and perspectives. Humanities (2019), URL: https://cyberleninka.ru/article/n/tsifrovoe-obrazovanie-v-rossiyskoy-federatsii-realii-i-perspektivy (date of access: 12.12.2020)

SHS Web of Conferences 109, 01036 (2021) https://doi.org/10.1051/shsconf/202110901036
LISID-2020

Modern financial and legal relations trends and economy digitalization

Igor Semukhin[1], *Svetlana* Kolovaylo [2*] and *Natalia* Kravchenko [3]

[1] Crimean branch of the Federal State Budget-Funded Educational Institution of Higher Education "The Russian State University of Justice", Simferopol, Russia, semmich@mail.ru
[2] Crimean branch of the Federal State Budget-Funded Educational Institution of Higher Education "The Russian State University of Justice", Simferopol, Russia, bolero_8@mail.ru
[3] Crimean branch of the Federal State Budget-Funded Educational Institution of Higher Education "The Russian State University of Justice", Simferopol, Russia, nataliakravchenko85@gmail.com

Abstract. The article considers the trends in the development of financial and legal relations, which are caused by changes taking place in economic and social relations through the introduction of digital technologies. In conducting the study the authors analyzed the totality of currency, banking relationships, insurance relationships, budgetary legal relations and financial market relations. As a result of the study, the authors concluded that it is necessary to improve the mechanism of legal regulation of the use of such financial instruments as blockchain, cryptocurrency, etc., as well as the supervisory functions of the Bank of Russia credit bureaus in order to increase confidence in financial market participants.The authors also justify the need to introduce specific taxes (taxation on payments made for the purchase of goods and services over the Internet, or a tax on turnover from commercial activities on the world wide web) in order to prevent digital businesses from tax avoidance. An implication of the analyses is identifying tools that ensure a transparency of the budget system. Control approaches themselves have also been transformed. Thus, risk-based approaches have been used, as well as ratings of control objects by the level of reliability. It also requires further analysis of the regulatory tools that can be applied by public authorities, changes in the object and tax base, the procedure for collecting taxes, and possible tax benefits. The results can be used in further research to develop an effective legal mechanism for regulating financial and legal relations in modern economic condition.

1 Introduction

The current stage of the development of public relations is characterized by the active introduction of digital technologies. It should be noted that the distribution of coronavirus infection COVID-19 caused by SARS-CoV-2 in 2020, determined a more rapid transformation of relations, manifested in the rising of remote activities, increasing the volume of electronic sales of goods and services including government services to accelerate the digital transformation of enterprises. Thus, since 2008, the number of Internet users has grown 2.5 times and reached 4.1 billion in 2020, the number of Internet users and people who use online shopping services has doubled, the volume of retail trade on the Internet has almost quadrupled (3.8 trillion US dollars), experts predict an increase in the provision of financial services through financial services twice [1-2]. These transformations, which are specific to the current stage of development of public relations, have predetermined changes in the composition and structure of financial and legal relations, which necessitates the analysis of the content of these legal relations in order to develop a legal mechanism for their regulation in the future.

2 Problem Statement

It should be noted that the process of introducing digital technologies in all spheres of public life, along with improving the quality and availability of services, increasing the productivity and competitiveness of enterprises, reducing production costs, determines the emergence of such risks as unauthorized access to information, mass unemployment, digital inequality, etc. Thus, according to the Ministry of internal Affairs of the

* Corresponding author: bolero_8@mail.ru

Russian Federation, in 2019, the number of crimes committed using information technologies increased by 60-70% compared to the previous period [4]. In accordance with the passport of the Federal project Regulatory regulation of the digital environment, it is planned to change the current legislation, including those related to the financial activities of the state [4]. Thus, a necessary condition for the introduction of digital technologies is to determine the legal status and procedure for the turnover of digital financial assets, establish conditions for VAT taxation, export operations of works and services, regulate the taxation of participants in crowdfunding platforms, and ensure the legal settlement of legal relations related to the development of the digital economy and the activities of investment platforms.

3 Research Questions

In order to offset the negative consequences of the introduction of information technologies, it is necessary to analyze trends in financial and legal relations caused by the introduction of digital technologies, such as:

3.1 Analysis of trends in the development of insurance, banking, currency relations in the field of financial legislation.

3.2 Analysis of trends in the development of tax and budget legal relations in the field of state loans.

4 Purpose of the Study

Taking into account the transformations inherent in the modern stage of development of public relations, which are manifested in the active introduction of information technologies, it is relevant to analyze the changes in financial and legal relations during the transition to the digital economy.

5 Research Methods

The goal was achieved through the use of general scientific methods, such as the analysis of literature and legal documents on the subject of research, the study and generalization of information, abstraction, induction and deduction. The method of analogy and modeling was used. When studying the legislation, a comparative legal method was used.

6 Findings

Defining the essence of financial legal relations, it should be noted that in all doctrinal approaches to determining the content of the term presented in the scientific literature, the authors are United in indicating the scope of these legal relations. Taking into account the variety of financial legal relations, various approaches to their content and classification have been formed in the scientific literature. So, generalizing approaches to classification depending on the sphere of financial activity or the content of the financial and legal norm, it is possible to present such types of financial relations as budget, tax, insurance, state credit, currency, banking, financial market regulation, financial control [5-6]. According to the authors this approach to the classification of financial legal relations gives the most complete picture of the totality of legal relations that are the subject of financial law.

In order to conduct this research, it is also necessary to define the content of the term "digitalization of the economy". As a result of the analysis of scientific approaches, the authors came to the conclusion that this category characterizes the process of changing social relations through the introduction of information and telecommunications technologies that improve the efficiency of economic activity. It should be noted that the most significant changes have been made in currency and banking legal relations, as well as in the field of financial market regulation, which is due to the emergence of a completely new type of relations related to the use of electronic money (cryptocurrency). According to scientists, the global financial crisis of 2008 is a prerequisite for the active introduction of digital technologies in banking legal relations. Thus, the failure of the banking system and the low turnover rate of financial assets caused the stagnation of the entire world economy. These trends predetermined the introduction of digital network technologies and tools, which led to a significant increase in competition in the financial sector, including through financial and technological innovation, as well as the formation of a parallel interactive virtual space that no longer obeyed traditional laws and management methods. The main advantage of new financial technologies should be considered the possibility of cloud data storage, which fully corresponds to the rapidly changing context of financial transactions from mobile devices, at any time of the day and from anywhere in the world. The legislation of the Russian Federation are not clearly regimented and only indirectly defines the concept of cloud technology in the "strategy of information society development in Russian Federation to 2017 - 2030" as "... an automated way to provision computing power, including software in remote access mode via the Internet at the client's request or... the technological model, which is any physical resource from the virtual model." [7].

However, Russian legislation cannot yet provide an adequate response to the challenges of modern times in terms of legal regulation of rapidly developing financial instruments on the digital market. This is due to the specifics of financial regulatory technologies, their high integration into digital categories, such as cryptocurrency, blockchain, and others, as well as the lack of a rule of law regulating the circulation of the above-mentioned financial instruments. The lawmaker is late, and the digital financial market is developing too fast. In this context, it is fair to conduct a legal assessment of the Federal law of 31.07.2020 N 259-FZ "on digital financial assets, digital currency and on amendments to certain legislative acts of the Russian Federation " [8], which will enter into force on January 01, 2021.

Thus, at the legislative level, the concept of "digital financial assets" (hereinafter referred to as DFA) is defined, which are digital rights secured by monetary claims. Accounting and handling of the DFA are carried out only by changing the information code, in other words, changing the distributed registry. At the same time, DFA will be recorded in the system in which they are issued, and they will be modified at the issuer's direction, which will allow the Bank of Russia to determine DFA that can only be purchased by qualified investors, classifying them according to certain digital characteristics. Transactions with the DFA will be concluded through a specialized exchange operator.

The above-mentioned regulatory legal act defines the content of the term "cryptocurrency", which will now be replaced by the term "digital currency", as a digital code, a set of data that is not a monetary unit, but which can be accepted as an investment or payment, considered as property in the new versions of the law on bankruptcy and enforcement proceedings. It should be noted that, despite the legal regulation of the use of digital currencies, the law prohibits not only accepting payment for goods and services in digital currency, but also spreading information that digital currency can be used as a method of payment for goods or services.

The transformation of banking legal relations in the context of digitalization of the economy is manifested in the use of RegTech tools that are used for customer identification, detection of suspicious activity and fraud prevention, reporting, and compliance control. The use of these tools makes it possible to increase the effectiveness of financial control over the activities of financial market participants by automating part of administrative procedures, digitalizing the interaction of participants, and improving the quality of reporting information.

Financial and legal relations in the insurance sector have also been transformed. Thus, according to experts, the main trend of development of insurance market in 2019 was to automate their provision [9], allowing, on the one hand, more quickly and to provide quality insurance services to settle insurance claims, and identified the need to develop and study tools to monitor the organization of the insurance business. For example, the changes made to the Federal law "On the organization of insurance business in the Russian Federation" on April 24, 2020, made it possible to exchange information within the legal framework through technical means, mobile devices, tablets, etc., and the individual policyholder also has the opportunity to provide information to the insurer through a financial platform, the operation of which is planned from January 2021, which will "promote competition in the financial market, increase convenience, transparency and efficiency of receiving services by clients in the "single window" mode, elimination of barriers to access to clients' financial products" [10].

According to the amendments the operator of the financial platform is neither a broker nor a dealer, which eliminates the ambiguity in the regulatory settlement of these legal relations and duplication of requirements imposed by the Central Bank of the Russian Federation for the implementation of its activities. The emergence of a new subject of insurance legal relations – the operator of the financial platform, determines the need to determine its place in financial legal relations, as well as their content. This entity, together with the implementation of financial services activities, will perform financial monitoring functions. It should be noted that in order to ensure the effective functioning of financial service operators, it is necessary to change the current legislation, in terms of the adoption of special laws on the registrar of financial transactions, on the financial aggregator, as well as legislation on products and the responsibility of bot developers.

These transformations of banking and insurance legal relations, as well as the regulation of the securities market, allow us to conclude that the functions performed by the Central Bank of the Russian Federation as a subject of financial legal relations are significantly transformed. Thus, it is responsible for solving such tasks as improving the Supervisory powers of the Bank of Russia and credit bureaus in order to increase confidence in financial market participants.

Tax relations have also been transformed. The modern system of taxation is based on principles that were laid down in the works of A. Smith and have not changed much over the centuries, such as fairness, convenience, efficiency or economy. These principles remain relevant, and such principles as benefits and solvency are the Foundation of the tax system of every country in the world. At the same time, trends in the development of this system of public legal relations are characterized by changes in the priorities of socio-economic development, sources of income, transformation of the tax systems, organization of tax collection, by shifting the focus from one object of taxation to another, expanding tax bases or reducing benefits. These transformations are due to the impact of technological progress and globalization. Taking into account these changes, it is necessary to determine the legal consequences for the modern tax system, analyze the regulatory tools that can be applied by public authorities, change the object and base, the procedure for collecting taxes, and tax benefits.

It should be noted that in a short time, digital assets have become an integral part of the life of modern enterprises. The digital transformation of industrial enterprises is uneven - some companies are already using digital technologies at all stages of the product lifecycle (from development to maintenance), while others are only studying the benefits and costs of such implementation. However, almost all businesses use Internet space to sell their products and services [11]. According to the research of PwC, the development trends of business entities in the next five years will be to reduce costs by 3.6% and increased income by 2.9% due to the digitization of products and services and develop new offerings digital services [12]. These transformations will contribute to the growth of tax revenues due to revenue growth.

However, it should be noted the risks of reducing budget revenues by reducing the tax base, due to the widespread use of digital goods and services. These trends are due to the fact that digital goods reduce the tax base due to the lower cost of digitized goods such as, for example, audiobooks. Another factor that determines the reduction of the tax base may be the complexity of determining the subject of taxation, since there are so-called income without citizenship.

The impact of digitalization on the tax system, as well as methods for solving problems in the context of the transformation of socio-economic relations is presented in table 1.

Table 1. Cross-analysis of the impact of digitalization on the tax system

Factors that determine changes	Transformations		Problem solving methods
	Economy	Taxation	
Digitalization	increase in the volume of purchase and sale of digital services and digitized goods	reduction of tax receipts for consumption of traditional (non-digital) goods and services	introduction of a tax on consumption of digital goods and services or expansion of the base of existing consumption taxes
	the problem of personal and corporate data protection	potential growth of "cyber crimes" and "cyber threats"	introduction of a tax on the collection and use of personal data for Big Data owners
	growth in the volume of transnational transactions related to technological mergers and acquisitions, as well as income " without citizenship»	blurring (leveling) the tax base when taxing profits, by entering into agreements with citizens of other countries without their physical presence in these countries	harmonization of national legal norms with international tax legislation by improving existing transfer pricing models for digital goods and services
The use of blockchain technologies	transparency of operations, free access to information about transactions	possibility of fundamental changes in the tax administration system based on automatic calculation of tax liabilities and their debiting from company accounts	development of an action plan, tools and methods for implementing blockchain technologies for automated tax collection, unification of tax administration and staffing
Robotics, automation, M2M (machine-to-machine interaction)	reduced number of low-skilled industrial jobs, lack of STEM specialists, increased unemployment and income inequality	the need to compensate the budget for the loss of social taxes that are currently paid by employees in production	determination of the feasibility and possibilities of introducing new compensatory forms of taxes (robot tax, universal basic dividend, etc.), introduction of a tax credit for loans for STEM education and retraining of smart enterprises

These changes in socio-economic development caused by the introduction of information technologies have determined the need to adapt international tax systems. After judicial investigations into tax evasion by digital giants such as Amazon and Google and finding them innocent, the European Union has begun work to protect national tax systems and minimize the risks of non-payment of tax obligations by digital companies and platforms. One of the proposals is to introduce specific taxes (a tax on payments made for the purchase of goods and services over the Internet, or a tax on turnover from commercial activities on the world wide web) in order to prevent digital businesses from avoiding responsibility [13].

At the moment, international practice is characterized by the introduction of additional provisions in the tax codes on taxation of foreign supplies of digital services and goods. An example of taxation of imported / exported digital goods and services is the indirect tax of 10% introduced in Australia from 2019. Tax rates for digital goods and services vary from country to country. So, in New Zealand, the corresponding tax rate is 15%, in Japan-8% [14]. In the European Union, the supplier of electronic goods and services pays VAT in the buyer's country. The Russian Federation has introduced the so-called Google tax, which is essentially a value-added tax on digitized goods and electronic services sold via the Internet by foreign it companies [15].

Unauthorized use of information is another negative consequence of the digitalization of public relations, which can be leveled through, among other things, tax instruments. According to a study by the analytical company "Javelin Strategy", in 2019, the largest number of cases of online fraud were registered, resulting in $ 16 billion in damage [16]. In addition, digitalization, the use of digital platforms and blockchain, leads to the formation of data arrays, known as Big Data, where large amounts of personal and corporate information are accumulated. The latter is collected and used by large and small electronic systems that form catalogs of habits, tastes, recent user requests, personal data, Bank details, and so on (Poritskiy, Oliveira & Almeida, 2019).

It should be noted that one of the directions of legal regulation of digitalization processes in the Russian Federation could be the adoption of the law "on digital/electronic economy". The adoption of this normative legal act would consolidate the conceptual framework and to determine basic categories such as digital economy, electronic business, electronic Commerce, electronic contract, electronic services, eGovernment, competent authorities, electronic communications, online store, consumers of e-services, etc., and also subject composition relations, their rights and obligations in the field of digital economy, mechanisms for the protection of their rights and legitimate interests, the liability for violations in this sphere and consequences of abuse of rights, features of conflict resolution.

The adoption of this normative legal act would correspond not only to the basics of the development of the information society in Russia (to improve the information legislation in terms of its codification), but also would give participants of digital economy, which is developing very rapidly and replacing traditional forms of business, more opportunities to exercise their rights and interests and at the same time protect them in case of violation.

The introduction of digital technologies in the budget sphere ensures the transparency in the formation, distribution and use of public funds by providing citizens with an opportunity to evaluate the activities of budget resources management and exercise public control. One of the tools to ensure transparency of the budget system is the state integrated information system for public Finance management "Electronic budget". Control approaches themselves have also been transformed. Thus, risk-based approaches, as well as ratings of objects of control by the level of reliability, were used when determining the plan of control measures implemented by the Federal Treasury. When regulating budget legal relations, it should be noted that it is necessary to make changes to the current legislation in terms of introducing special rules that would regulate the information support of the budget process.

7 Conclusion

Thus, as a result of the research, the authors came to the conclusion about a significant change in public relations, which is the subject of financial law, under the influence of the rapid introduction of digital technologies.

So, it requires improvement of the legal mechanism of regulation of financial instruments such as cryptocurrency, blockchain etc. Changes in insurance and banking legal relations, as well as financial market regulation, determine the need to improve the Supervisory powers of the Bank of Russia and credit bureaus in order to increase confidence in participants in the financial services market and ensure the stability of its functioning.

It should be noted that there is a characteristic change in the priorities of socio-economic development, sources of income, transformation of the tax systems, organization of tax collection, by shifting the focus from one object of taxation to another, expanding tax bases or reducing benefits. Taking into account these changes, it is necessary to further analyze the legal consequences for the modern tax system, regulatory instruments that can be applied by public authorities, changes in the object and base, the procedure for collecting taxes, and tax benefits. Another characteristic feature is the introduction of specific taxes (a tax on payments made for the purchase of goods and services over the Internet, or a tax on turnover from commercial activities on the world wide web) in order to prevent digital businesses from avoiding responsibility.

Also, as a result of the analysis, the authors justify the need to adopt the law "on digital / electronic economy", which would consolidate the conceptual apparatus and define such basic categories as the digital economy, electronic business, electronic Commerce, electronic contract, electronic services, electronic government, electronic authorized bodies, electronic information exchange, online store, consumers of electronic services, etc. , as well as the subject structure of relations, their rights and obligations in the digital economy, mechanisms for protecting their rights and legitimate interests, responsibility for violations in this area and the consequences of abuse of rights, features of conflict resolution.

Digitalization of budget legal relations is manifested in the use of tools that allow ensuring transparency of the budget system, such as the state integrated information system for public Finance management "Electronic budget". Control approaches themselves have also been transformed. Thus, risk-based approaches, as well as ratings of objects of control by the level of reliability, were used in determining the plan of control measures implemented by the Federal Treasury. When regulating budget legal relations, it should be noted that it is necessary to make changes to the current legislation in terms of introducing special rules that would regulate the information support of the budget process.

References

1 Accounts Chamber of the Russian Federation. Digital technologies and cybersecurity in the context of the spread of COVID-19 (2020), URL: https://ach.gov.ru/upload/pdf/Covid-19-digital.pdf (date of access:14.04.21)

2 Central Bank of the Russian Federation. The main directions of development of financial technologies for the period 2018-2020 (2018) URL: //cbr.ru/Content/Document/File/84852/ON_FinTex_2017.pdf (date of access:14.04.21)

3 Ministry of internal Affairs of Russia. Brief description of the state of crime in the Russian Federation for January-December 2019 (2019), URL: https://мвд.рф/reports/item/19412450/ (date of access:14.04.21)

4 Government Commission on digital development and the use of information technologies to improve the quality of life and business environment. Passport of the Federal project "Regulatory regulation of the digital environment", 9. (2019), URL: //legalacts.ru/doc/pasport-federalnogo-proekta-normativnoe-regulirovanie-tsifrovoi-sredy-utv-prezidiumom/(date of access:14.04.21)

5 A.A. Musatkina, & E.V. Chuklova, *Financial Law: Textbook.* (Moscow, Russia: INFRA-M, 2015)

6 A.I. Zemlin, O.M. Zemlina, & N.P. Olkhovskaya *Finasncial law of the Russian Federation: textbook for universities* (Moscow, Russia / Yurayt Publishing House, 2020)

7 President of the Russian Federation. Decree of 5.9.2017 N 203 "On the Strategy for Information Society Development in the Russian Federation for 2017 – 2030" (May 15, 2017 N 20, item 2901) (Moscow, Russia: Collected Legislation of the Russian Federation, 2017)

8 Government of the Russian Federation. Federal Law of July 31, 2020 N 259-FZ "On digital financial assets, digital currency and on amendments to certain legislative acts of the Russian Federation" (August 03, 2020 No. 31). Moscow, Russia: Collected Legislation of the Russian Federation, 2020)

9 State Duma of The Russian Federation. The main trends in the development of the digital economy in the financial sector. Legal aspects of regulation and practical application. (Moscow, Russia: State Duma Publication, 2019)

10 Government of the Russian Federation. Federal Law of July 31, 2020 N 259-FZ "On digital financial assets, digital currency and on amendments to certain legislative acts of the Russian Federation" (August 03, 2020 No. 31). (Moscow, Russia: Collected Legislation of the Russian Federation, *2020).*

11 J. Luoma-Kyyny, Data sovereignty and soft infrastructures are key enablers for the next phase of the European data economy (2020), URL: //www.sitra.fi/en/news/data-sovereignty-and-soft-infrastructures-are-key-enablers-for-the-next-phase-of-the-european-data-economy/ (date of access:14.04.21)

12 The future of financial services. PricewaterhouseCoopers (2020) URL: //www.pwc.com/gx/en/industries/financial-services/publications/future-of-financial-services.html (date of access:14.04.21)

13 After Amazon, EU considers next steps on new digital taxes. (2017) URL: //tech.eu/features/17463/amazon-eu-new-digital-tax-laws (date of access:14.04.21)

14 *Taxes on Survey Monkey Purchases.* (2017). Retrieved from https://help.surveymonkey.com/articles/ru/kb/Taxes/
15 Trakman, L., Walters, R., & Zeller, B. (2019). Tort and data protection law: Are there any lessons to be learnt? *European Data Protection Law Review,* 5(4), 500-519.
16 Buzzard, J. (2020). Identity Fraud in the Digital Age Retrieved from https://www.javelinstrategy.com/coverage-area/identity-fraud-digital-age

Civil law aspects of territorial public self-government activity during the economy's digitalization

Sergei Sharonov[1], and *Nikolai* Korovin[2*]

[1] Moscow University of the Ministry of Internal Affairs of Russia, 117437, 12, Akademika Volgina str., Moscow, Russian Federation, Sharonov345@mail.ru
[2] Department of Business Law, Arbitration and Civil proceedings, Volgograd State University, 400062, 100, Universitetsky ave., Volgograd, Russian Federation, Nikolay.vlz79@gmail.com

Abstract. The article deals with civil law aspects of territorial public self-government (hereinafter - TPSG) in terms of economy's digitization. The authors show the private law specificity of territorial public self-government, give the classification and characterize civil aspects of activity of bodies of territorial public self-government. The problems of the work of the structures of territorial public self-government in the indicated conditions are identified and the amendments to regulatory legal acts are suggested that contribute to the solution of the mentioned problems.

1 Introduction

Chapter 8 of the Constitution of the Russian Federation is devoted to local self-government, in turn the bodies of territorial public self-government are a constituent part of national self-government. According to Article 27 of the Federal Law issued on October 6, 2003 No. 131-FZ "On General Principles of Local Self-Government in the Russian Federation" (hereinafter referred to as the Federal Law issued on October 6, 2003 No. 131-FZ), bodies of territorial public self-government appearing by means of self-organization of citizens for the implementation of their own initiatives concerning local issues at their place of residence in a certain territory. According to Article 27 Paragraph 2.5 of the mentioned Law and Article 50 paragraph 2.3 of the Civil Code of the Russian Federation, territorial public self-government can be registered as a legal entity, in particular as a non-profit organization, namely as a public organization (hereinafter – a structure of TPSG).

The structure of territorial public self-government has the right to work for the satisfaction social and domestic needs of citizens, including the actions connected with the improvement of the corresponding territory. As a legal entity, the body of territorial public self-government (TPSG) can enter into legal relations with other participants of civil law contracts, acting as a buyer of goods, a customer in service sector, etc. A number of scientific publications are devoted to the research of territorial public self-government (TPSG), but insufficient attention has been paid to the analysis of civil law aspects of territorial public self-government (TPSG).

The relevance of the topic is also due to the acceleration of digital transformation of the real sector of the economy [1]. Currently, in accordance with the national program "Digital Economy of the Russian Federation", approved by the minutes of the meeting of the Presidium of the Council for Strategic Development and National Projects under the President of the Russian Federation No. 7 issued on June 4, 2019, a number of tasks are being implemented, including the task of introduction of civil law transactions by means of use of digital technologies.

As a result of the implementation of this program, amendments were passed to the civil legislation in terms of supplements to the list of civil rights objects with digital rights objects [2]. The national program contributes to the creation of a stable and secure information and telecommunication infrastructure for high-speed transmission, processing and storage of large amounts of data, accessible to all organizations.

The tools of such infrastructure available for the organization of bodies of territorial public self-government (TPSG require a detailed analysis) in the current scientific, technical and legal realities.

2 Problem Statement

The implementation of the national program "Digital Economy of the Russian Federation" is realized by means of the introduction and improvement of digital services and information systems, for example, such as: "Common web site of state and municipal services", Information system " My Mediator ", etc. The establishment of bodies of territorial public self-government along with legal entities of various organizational and legal forms has the opportunity of using digital services. At the same time, there is specificity of legal regulation of the establishment and further activities of such a legal entity as body of

* Corresponding author: Nikolay.vlz79@gmail.com

territorial public self-government due to its heterogeneous legal nature. In the conditions of rapidly developing digital technologies, there is a need for their application at all stages of the functioning of bodies of territorial public self-government, including the moment of actions aimed at the registration of such a legal entity.

The analysis of existing legislation and the need for the use of digital technologies in the context of the research topic allows us speaking about the existence of unresolved problems that hinder the achievement of the goals of the national project. The government should ensure the acceleration of changes in legislation, taking into account scientific and technical progress [3]. The right to use available digital technologies by bodies of territorial public self-government, taking into account the specificity of the implementation of their work, should be reflected in the relevant regulatory legal acts.

3 Research Questions

In the study of civil law aspects of the activity of structures of territorial public self-government during the digitalization of the economy, it is necessary to take into account their legal nature and the peculiarities of legal regulation of the relations arising from this activity, since such relations are governed by the norms of both private and public law.

4 Purpose of the Study

The purpose of the article is the identification of the private law nature of territorial public self-government. In the context of this goal, it is necessary to solve the following tasks. First, it is necessary to identify the civil law aspects of the work of territorial public self-government. Second, it is indispensible to classify the identified aspects. Thirdly, it is needed to characterize these aspects. Fourthly, it is necessary to show the impact of digitalization of the economy on the activity of bodies of territorial public self-government which is being analyzed, to identify problems of legal regulation and solve them.

In order to solve the first and the following tasks, let us use the accumulated knowledge and experience in the area under analysis, and we mean here the theses of scientists and scientific publications. In the scientific community, the concept is quite widespread, according to which the territorial public self-government, registered as a legal entity, has a dual nature, while the legal capacity of territorial public self-government can be realized through municipal and civil legal relations. According to E.G. Komissarova the definition of the legal nature allows us giving legal description of a legal phenomenon and identifying its "legal essence" [4].

S. V. Malyugin in his article devoted to the study of the category of "legal nature", expresses the opinion that in law enforcement activity this category is applicable "to the actual and legal composition of legal relationship", and in scientific activity it reflects "the essential features of various legal phenomena", allows "to understand their place and value in the legal field "[5].

A. V. Fedotov connects an approach based on the constitutional rights of citizens (the right to assemble and conduct meetings at the place of residence and the right to association) with the understanding of the nature of territorial public self-government, and the defines territorial public self-government as a form of legal support for the implementation of these rights. The scientist connects the nature of public interest with the expression of the will of the majority of citizens living in the territory [6].

E. V. Razdyakonova in her article devoted the legal status of territorial public self-government, mentions that "the disclosure of the legal nature of territorial public self-government as a key element that is simultaneously included into the system of public and private law, predetermines the inadmissibility of the exclusive assignment of territorial public self-government to one of them" [7]. According to A. A. Larichev the establishment of structures of territorial public self-government as an element of self-government and the registration of the bodies of territorial public self-government as a legal entity "are the phenomena of different legal nature"[8].

O. I. Bazhenova notes that the public-law nature of territorial public self-government (the resulting organization and management features, the scope and content of rights and obligations) should be taken into account in the development and adoption of decisions by the bodies of territorial public self-government affecting civil law transactions. The author criticizes the civil approach to the category of a legal entity (body of territorial public self-government) and believes that it does not ensure the autonomy of civil law [9]. When revealing the legal nature of territorial public self-government, the scientist notes that territorial public self-government "is an element of the local self-government system" and is based on "territorial principles." According to the author, the main principles and content of the competence of territorial public self-government (including TPSG bodies) are predetermined by its legal nature [10].

In turn, in the context of the goal set in the article, the author finds it necessary to analyze the establishment of territorial public self-government in terms of civil law. The author identifies the private legal nature of territorial public self-government by means of the examination of the essential characteristics of this legal phenomenon and its correlation with civil law.

In the absence of the need to have any economic activity, the body of territorial public self-government can be registered (established from the moment of the registration of the Charter) by local governments as one of the elements of the local government system and not to have the status of a legal entity [11]. These bodies of territorial public self-government have the status of a subject of municipal legal relations and on their own behalf cannot work for the improvement of the territory, other economic activities are meant to meet the social and domestic needs of citizens living in the relevant territory. E. A. Bratseva in her research on territorial public self-government notes that bodies of territorial public self-government without the rights of a legal entity cannot participate in civil legal relations and cannot be considered as their subject [12]. K.M. Khudoley and D.M. Khudoley note that in contrast to the territorial public self-government without the registration of a legal entity, the establishment of a body territorial public self-government gives a higher volume capacity, as the latter is has the right to have economic activity [13].

Within the mentioned purpose of the article, it is the territorial public self-government registered as a legal entity that we are interested in from the scientific point of view. Judging from the analysis of the works in the field of territorial public self-government, it can be concluded that the need for the possession of a sufficient amount of legal capacity to satisfy all interests (including social and other interests) determines the establishment of a body of territorial public self-government. From the moment of state registration, the body of territorial public self-government acquires the legal capacity of a legal entity and in accordance with Article 2 Paragraph 2.1 of the Civil Code of the Russian Federation may be a participant of relations regulated by the corresponding provisions of the Civil Code of the Russian Federation. It should be noted that the digital service "Common portal of government and municipal services (functions)" allows you to submit the documents for the registration of a legal entity without visiting the relevant registration body.

The emergence of scientific research dealing with this type of legal entity as a public organization of territorial public self-government and the introduction of their results into legislative use connected with the legal status of territorial public self-government is not covered in this article. The establishment of a body of territorial public self-government in accordance with the existing power and authority, satisfying the legitimate interests of citizens living in the territory of the body of territorial public self-government, can be a party to civil law relations under construction contracts, leases, provision of services (works), supply of goods, etc., concluded with other participants of civil law transactions [14].

The first aspect of civil law activity of territorial public self-government can be connected with the existence of the need to enter into the legal relationships mentioned above, what is impossible without the registration of a legal entity. The second aspect of the civil legal activity of territorial public self-government encompasses the legal relations arising because of this activity. The third aspect of the activity under analysis is private-law properties of the content of legal relations arising from the work of bodies of territorial public self-government. Let us classify the mentioned aspects as: 1) organizational elements; 2) subject-object elements; 3) behavioral elements.

The first aspect of the activity under analysis of a body of territorial public self-government is characterized by the emergence of legal personality and it is associated with the moment of registration of the specified legal entity and with the opportunity of participation in civil legal relations and with the ability of being a subject of legal relations regulated by civil legislation.

The second aspect of the civil legal activity of the bodies of territorial public self-government is characterized by the structure of the legal relationship under consideration, which includes: subjects (bodies of territorial public self-government and business entities); objects (material goods (property, things, results of work and rendering of services)), through which the social and everyday needs of citizens living in the territory of the TPSG are satisfied); rules of law (provisions of the Civil Code of the Russian Federation governing the relations in which these entities enter).

The third aspect is characterized by the ability of the body of territorial public self-government to acquire the rights and obligations fixed in civil law and bear legal responsibility for their implementation in civil law relations. The author of this article, earlier, came to the conclusion that "the measure of possible and the measure of the necessary behavior of establishment of territorial public self-government depend on the specific civil law contract, to which the body of territorial public self-governemnt is a party, and they are determined by the subject matter and other essential terms of the contract" [14].

The structure of territorial self-government (its authorized persons), prior to the emergence of contractual relations, has the right to exercise care and caution when choosing a reliable counterparty (supplier of goods, service provider, etc.). The information about goods or services can be found on the Internet at the corresponding websites of manufacturers or service providers, etc.

You can find out the degree of conscientiousness of a potential counterparty on the information resource "List of arbitration cases" by means of the search for information on the participation of a specific legal entity or individual entrepreneur as a defendant in legal suits held by commercial courts of the constituent entities of the Russian Federation. The claims proving the repeated non-fulfillment of contractual obligations will be the reason for the refusal to such a counterparty. An additional source of information may be the "Databank of enforcement proceedings" service, with which you can check the availability of enforcement proceedings in the Federal Bailiff Service of a potential counterparty, i.e. an individual entrepreneur or legal entity.

In accordance with Paragraph 1 of Article 7 of the Federal Law issued on August 8, 2001 No. 129-FZ "On State Registration of Legal Entities and Individual Entrepreneurs", a service has been developed ("Provision of Information from the Unified State Register of Legal Entities / Unified State Register of Individual Entrepreneurs"), which provides the information about a specific legal entity or an individual entrepreneur for free. On the basis of the provided information, it is possible to establish the powers of persons planning to sign a civil contract with a structure of territorial self-government.

Modern digital technologies make it possible to negotiate with potential contractors about contracts by means of various video conference programs, exchange of financial and economic documents via the e-mail, sign payment and other documents with an electronic signature.

If, in the process of contractual relations, unresolved claims arise, for example, the structure of territorial self-government has claims against the service provider under the contract, then you can submit a claim and documents electronically to the Arbitration Court by means of "My Arbitrator" system, and you can participate in the court session via a video conference.

The entire set of modern digital services contributes to an effective work of a territorial self-government organization. At the same time, there is an unresolved problem associated with the current sanitary and epidemiological situation caused by the threat of the spread of a new corona virus infection (Covid-19). So, according to Part 4 of Article 27 of Federal Law No. 131-FZ issued on October 6, 2003, bodies of territorial self-government are elected at meetings or conferences of citizens living in the respective territory. In accordance with Paragraphs 5 and 6 of Part 7 of Article 27 of the Law mentioned above, at conferences and meetings of citizens, the income and expenditure estimates of structures of territorial self-government are approved, statements of territorial self-government bodies are presented and approved, and other issues of activities of territorial public self-government are studied. The provisions on territorial self-government and other types of regulatory legal acts (for example, the Charter of a municipality) regulating the procedure for the establishment, functioning and termination of work of territorial self-government) in municipalities of various constituent entities of the Russian Federation provides for a procedure according to which the territorial self-government body is authorized to act in the interests of citizens during a certain period. For example, in accordance with Article 11 Paragraph 3.2 of the Regulations on Territorial Public Self-Government in Volgograd (approved by the decision of the Volgograd City Duma of February 21, 2007 N 41/1010), the Charter of a territorial self-government body establishes: procedure for the formation, termination of powers, rights and obligations, term of office of territorial public self-government bodies. In practice, in most cases, in the territorial self-government organization such as the Council of territorial self-government is elected and then functions. It consists of three or more people, headed by the Chairman of territorial self-government body elected from the members of territorial self-government Council. The Chairman of the Council acts without a power of attorney on behalf of the structure of territorial self-government, represents it in its relations with the population, judicial and other bodies, organizations of all forms of ownership, signs financial and other documents. As a rule, the members of the Council and the Chairman of the territorial self-government are elected for a term of 4 years.

After the expiration of the specified period or when need arises, there is a necessity to hold a meeting (conference) of citizens living in the area of territorial self-government for the re-election of persons mentioned above and to study the reports on the activity of bodies of territorial self-government.

In Volgograd Region, in accordance with Clause 18 of the Decree of the Governor of Volgograd Region issued on March 15, 2020 No. 179 "On the introduction of a high readiness regime for the functioning of command and control bodies, forces and means of the territorial subsystem of Volgograd Region of the unified state system for the prevention and elimination of emergencies" (amended by October 30, 2020), heads of public organizations and associations operating in Volgograd Region were advised to freeze mass events with the presence of citizens, for the exception of activities carried out in agreement with the operation headquarters implementing measures which help to prevent the emergence and spread of the new coronavirus infection (COVID -19) in Volgograd region.

According to clause 3.3 of this Resolution, citizens who have reached the age of 65 are recommended to self isolate at the place of residence. It is not difficult to assume that even upon receipt of an agreement from the operational headquarters for such a mass event as a meeting (conference) of citizens living in the territory of territorial self-government, the citizens themselves, due to the risk of the spread of coronavirus infection, may not ensure their attendance, depriving the meeting (conference) the required quorum for its eligibility. Without the procedure for the reelection of the Council and the chairman of the body of territorial self-government, there is a danger of freezing the activities of a body of territorial self-government due to the absence of a person authorized to sign civil contracts and take other necessary actions. It seems that the solution of this problem can be the development of a digital service as part of the national program "Digital Economy of the Russian Federation", which allows holding meetings (conferences) of citizens remotely, via video conferencing, and appropriate amendments to the current legislation.

For example, the Government of the Russian Federation has developed a draft of the Federal Law "On Amendments to the Federal Law "On Joint Stock Companies" in terms of creation the opportunity for holding general meetings of shareholders in the form of a video conference for the discussion of issues on the agenda and taking decisions on issues, put to a vote, using information and communication technologies without specifying the venue". In connection with the mentioned above, the author suggests amending the Federal Law issued on October 6, 2003 N 131-FZ and supplementing Clause 2 of Article 27 of the Law with a new paragraph as follows: "Meetings and conferences of citizens can be held by joint remote presence using information and communication technologies without specifying the venue". The suggested solution will help not only the existing organizations of territorial self-government, but also will simplify the procedure of the establishment of new bodies of territorial public self-government.

5 Research Methods

The methodological basis of the research is dialectical method of cognition, method of analysis, synthesis and other methods of scientific knowledge.

6 Findings

Summing up the article, the following conclusions can be drawn:

Firstly, territorial public self-government is an essential element of constitutional and municipal law, but the analysis shows that the activities of bodies of territorial self-government have distinct civil law aspects.

Secondly, the essential characteristics of the organization of territorial self-government include legal personality (the ability to be a subject of civil law) and the content of a legal relationship with the participation of the subject under analysis.

Thirdly, the legal relations that develop during the work of a body of territorial public self-government are subject to regulation by civil legislation and intermediate the participation of the territorial self-government structure in civil transactions.

Fourth, the set of the identified civil law aspects of the activity of territorial public self-government proves that the activity of the structures of territorial public self-government belong to civil law.

Fifth, the activities of the structures of territorial public self-government at all stages of its functioning can be carried out most efficiently thanks to the "tools" of digitalization of the economy. The problems of the implementation of territorial public self-government in the current difficult epidemiological conditions require sophisticated measures.

7 Conclusion

Thus, on the basis of the results obtained, it can be argued reasonably that the signs of such a legal phenomenon as the activity of an organization of territorial public self-government held in relation to the objects of civil law regulation, prove its private legal nature. It is suggested to solve the problem of obstacles in the exercise of territorial public self-government by citizens by means of the amendments to Federal Law No. 131-FZ issued on October 6, 2003.

References

1. Presidium of the Council under the President of the Russian Federation for Strategic Development and National Projects, Passport of the national project "National Program" Digital Economy of the Russian Federation" (Minutes of 04.06.2019 N 7) (2019), URL: http://base.garant.ru/72296050/ (date of access 12.01.2020)
2. I. M. Konobeevskaya, Bulletin of Saratov University. New Series. Series Economics. Management. Law **19** **(3)**, 330-334 (2019) https://doi.org/ 10.18500/1994-2540-2019-19-3-330-334
3. S.G. Eremeev, A.B. Mayorov, E.N. Minchenkov, Theory and history of state and law **4 (54)**, 9-17 (2018)
4. E.G. Komissarova, Perm University Bulletin **2 (16)**, 23-29 (2012)
5. S.V. Malyugin, Russian legal journal **3 (108)**, 46-58 (2016)
6. A.V. Fedotov, *Territorial public self-government: constitutional foundations, practice of legal regulation* (Dissertation of Candidate of Legal Sciences) (Moscow State Social University, Moscow, 2004)
7. E.V. Razdyakonova, Modern law **7**, 21-24 (2009)
8. A.A. Larichev, *Collection of articles based on materials of the IV All-Russian Scientific and Practical Conference (December 7, 2018)*, 103-109 (Saint-Petersburg Institute, Saint-Petersburg, 2018)
9. O.I. Bazhenova, Local Law **5**, 77-102 (2015)
10. O.I. Bazhenova, Local Law **1**, 109-110 (2016)
11. E.S. Shugrina, Bulletin of Surgut State University. Issue **1 (7)**, 104-107 (2015)
12. E.A. Bratseva, State power and local government **12**, 50-55 (2019)
13. K.M. Khudoley, D.M. Khudoley, (2017). Electronic supplement to the Russian Law Journal **3**, 68-80.
14. N.A. Korovin, *Collection of scientific papers of the 3rd International Scientific Conference (May 15, 2020),* 103-109 (South Western State University, Kursk, 2020).

Concession as a state investment policy tool

Inna Shor [1*], *Viktor* Pushkarev[2], *Ekaterina* Tokareva [3], and *Olga* Shepeleva [4]

[1] Volgograd State University, Prospect Universitetsky, 100, Volgograd, Russia, ShorIM@volsu.ru
[2] Plekhanov Russian University of Economics, Stremyanny lane, 36, Moscow, Russia, vvp77r@rambler.ru
[3] Leningrad state University named by A. S. Pushkin, St. Petersburg sh., 10, St. Petersburg, Russia, tokareva.e.v@bk.ru
[4] Leningrad state University named by A. S. Pushkin, St. Petersburg sh., 10, St. Petersburg, Russia, shepeleva.o.r@mail.ru

Abstract. Russian Federation's competitiveness is largely interconnected with the state investment policy's development level. However, the need for investments in high-quality infrastructure, on the one hand, and corresponding budget funds deficit, on the other hand, highlight the problem of the state and business entities' cooperation by means of concession. At the same time there are many barriers, which hinder the concession's consecutive development. That is why the research goal includes the elaboration of measures, aimed at the effective application of a concession as a tool of state investment policy. The systematic approach and a set of scientific measures, including observation, expert assessment, statistical data processing, analysis and comparison have been applied in the study. As a result, the research authors define the following growth areas of a concession: legal normative (normative consolidation of the concession's development strategies and programmes, quality improvement of the methodological and legal concession projects elaboration), production organizational (quality improvement of territorial planning when proposing concessional agreements, the application of a complex approach to concession's facilities' modernization and etc.) and economic (broadening the spectrum of budgetary and non-budgetary tools used for financing concessional projects, elaboration of reasonable requirements to guarantee exposure according to the concessional agreement and etc.).

1 Introduction

These days it is essential to ensure effective cooperation between the state and business entities with the help of the public-private partnership toolset in order to implement socially useful projects, primarily connected with building various infrastructure facilities. At the same time it is worth mentioning that private investments are of key significance, while public funding is auxiliary and promotes the attraction and effective treatment of the business entities' investments. [1].

Only of the most widespread form of public-private partnership is a concession, which allows funding infrastructure facilities at the expense of private investments in terms of the insufficient state budget [2]. One of the reasons, why business entities participate in a concession is that they get an opportunity to increase their capital by funding socially useful projects, and they also enhance their competitiveness [3].

2 Problem Statement

The analysis of concessions use has elicited a number of barriers, hindering their development, which are legal-normative, organizational, production and economic [3]. But still the elaboration of practical measures aimed at creating an effective concession remains debatable [4]. The cooperation between the state and business entities is insufficiently scientifically substantiated, that is why it is worth mentioning the topicality of studying a concession, as a tool of state investment policy .

3 Research Questions

The keys to understanding scientific and practical aspects of a concession as a tool of state investment policy are as follows:

- What is the substance of a concession as a state investment policy tool?
- Which features are characteristic of a concession if speaking about the public utility infrastructure system, which belongs to the municipality urban district of Volgograd?
- What are the challenges and growth points in the development of a concession as a state investment policy tool?

4 Purpose of the Study

The research purpose is to elaborate the measures aimed at ensuring the effective application of a concession as a state investment policy tool.

* Corresponding author: ShorIM@volsu.ru

5 Research Methods

The study's methodology has been formed according to the scientific theory in the field of public-private partnership in general and concession in particular. The systematic approach and a set of scientific measures, including observation, expert assessment, statistical data processing, generalization, analysis and comparison have been applied in the study.

6 Findings

The creation and reconstruction of social, cultural and other state facilities are the top-priority goals of the Russian socio-economic development. However, in times of state budget deficit there are certain problems in achieving these goals. That is why the issues of applying a concession, as a public and private partnership priority form, are highly urgent [5].

The core of a concession is the relation between the state (the concession grantor) and a legal or individual entity concessionaire, which arises in the course of ensuring by the concession grantor to the concessionaire the rights for using the state property under an agreement on a paid and return basis as well as the rights for carrying out the activities, comprising the state monopoly [6]. Correspondingly, the following features characterize a concession:

- By the goal – state property reconstruction; construction of a concession agreement facility, which will belong to the state;
- By the investment object – state property and monopoly activities; intended purpose alternation and pledge ban; investments object's return character; laying the risk of damage or the investments object's accidental destruction on the concessionaire unless otherwise specified;
- By the investment subjects and their interrelations – a concession grantor and concessionaire; concession grantor's compensation of a concessionaire's part of expenses; a concessionaire's commission fee payment to the state's budget; distribution of risks between a concession grantor and concessionaire [7].

Thus, it can be mentioned that a concession's advantages for the state deal with solving socio-economic tasks at the business entities' expense, with preserving the initial purpose and provision of the state property's effective use, as well as with the opportunity of getting the rights for the state's movables and immovables, created by the concessionaire for ensuring the activity stated in the concession agreement [8]. Although, there some drawbacks in a concession from the state's side, which include the state's movables and immovables (which have been acquired during the realization of a concession agreement and in case the agreement does not imply this property's transfer to the state's ownership) transfer into the concessionaire's ownership. Another drawback is the concessionaire's acquisition of rights for the production and profit, received in the course of business activity, stipulated by the concession agreement. In return, from the investment subjects' side, one of the concession's main strengths is stimulating the investment activity, which leads to the profitability increase and competitiveness and sustainability enhancement. However, there are some drawbacks as well, which deal with high control pressure from the state under the concession agreement [9].

Taking into account the data of the National Association of Concessionaires and Long-Term Infrastructure Investors, it is worth noting that from 2009 till 2019, 277 concession agreements with a total volume of investment obligations of 1.754 trillion rubles, were concluded in Russia. Out of this investment amount 476 billion rubles is the volume of direct government obligations intended for creating and modernizing the infrastructure and 1.278 trillion rubles is the volume of investment obligations of the concessionaire. The largest numbers of concession agreements were concluded at the municipal level, which is 183 projects or 66 percent out of the total. 81 concession agreements, or 29 percent out of the total, were concluded at the sub-federal level, whereas 13 concession agreements or 5 percent out of the total were concluded at the federal level. At the same time, in terms of the investment obligations volume, concession agreements concluded at the sub-federal and federal levels occupy approximately equal shares - 41 percent each. The volume of investment commitments for municipal projects is 353.6 billion rubles or 18 percent of the total volume of market. The largest number of concession agreements were concluded in the field of utilities and communal services - 162 projects or 59 percent out of the total number of concession agreements. In terms of the investment obligations volume, the transport sector occupies the leading position - which is 1,457.2 billion rubles, or 77 percent out of the total investment obligations volume under concession agreements.

An example of a unique experience in the implementation of a concession not only on the national but also on the world level is the concession agreement, regarding the facilities' infrastructure system, owned by the municipality Urban District of Hero City Volgograd. The concession agreement is concluded between the Volgograd City Administration, Municipal Unitary Enterprise "Volgograd City Vodokanal" and "Water Supply Concessions" LLC. Within the framework of the concession agreement, the "Water Supply Concessions" LLC concessionaire ensures the design, construction, reconstruction and commissioning of water supply and sewerage facilities and the achievement of the planned performance indicators established in the concession agreement. In turn, the concession grantor must establish a tariff taking into account the concessionaire's reasonable expenses, long-term regulation parameters established by the agreement, investment measures and the execution period. Municipal Unitary Enterprise "Volgograd City Vodokanal" continues its activities as an operator of water supply and wastewater disposal facilities. In July 2015, the "Water Supply Concessions" LLC received the status of a guaranteed supplier in the field of water supply and wastewater disposal on the territory of the municipality Urban District of Hero City

Volgograd. And at the end of 2015, this concession agreement was recognized as the largest investment deal in Europe in the field of utilities and communal services.

This concession agreement has been implemented for 30 years. The volume of investments in the project reaches 58 billion rubles. This is the largest concession agreement in the field of utilities and communal services in Russia. The project model implemented involves attracting institutional investors' funds, i.e. non-state pension funds, through the mechanism of concession obligations. The "Water Supply Concessions" LLC represents the interests of investors, providing the unconditional return on investment by achieving the investment program implementation effect, as well as by reducing costs. The investment program includes several sets of activities:

- construction, modernization and reconstruction of facilities of centralized water supply and (or) water disposal systems in order to connect capital construction facilities;
- construction of new facilities for centralized water supply and sewerage systems not related to the connection of new capital construction facilities;
- construction of new facilities for centralized water supply and sewerage systems, not related to the connection of new capital construction facilities (measures aimed at enhancing the environmental efficiency, achieving the planned reliability values, quality and energy efficiency of centralized water supply and (or) sewerage facilities;
- shutdown, conservation and dismantling of objects of centralized water supply and (or) sewerage systems.

The main reasons why the "Water Supply Concessions" LLC concluded a concession agreement with the municipality Urban District of Hero City Volgograd, regarding the communal infrastructure facilities, were the state of the utilities infrastructure system (centralized cold water supply and sewerage) on the territory municipality Urban District of Hero City Volgograd at the beginning (in 2015 the utilities infrastructure system was 80.4 percent deteriorated), low rates and volumes of network reconstruction and equipment modernization, a high level of network losses and a low degree of energy saving and energy efficiency. In the second half of 2015, about 720 thousand cubic metres of untreated wastewater was discharged into the soil and surface water, which caused great harm to the ecology of the region. At the same time, by the second half of 2016, the "Water Supply Concessions" LLC had completely eliminated the discharge of untreated wastewater into the soil and surface water. In addition, the loss of clean water was 46.2 percent, which was 94,557,054 cubic metres of the entire mass of water raised from the Volga River. At the same time, in the first half of 2017 the "Water Supply Concessions" LLC reduced the loss of clean water to 24.9 percent. In general, the actions taken by the "Water Supply Concessions" LLC to reconstruct and modernize the centralized water supply and sewerage networks positively influenced a number of key indicators reflecting the quality of water supply and sewerage.

Without diminishing the significance of the above presented results of the concession agreement implementation, it is necessary to highlight the regulatory, organizational, production and economic barriers that hinder further development of the concession both in the field of utilities and communal services and other infrastructural sectors of Russia. The legal-normative barriers are associated with imperfect legislation in the field of concession activities, low business entities' awareness about a concession's legal structures, business entities' low level of legal protection in the implementation of concession projects, and other [10]. Organizational and production barriers are reflected in an insufficient level of competence and experience in the sphere of implementing concession agreements, low quality or the absence of territorial planning documents when forming the proposals for concluding concession agreements, the lack of an integrated approach to modernizing concession facilities, and other [11]. Economic barriers are characterized by the economic instability, increased requirements to the amount of guarantees, insufficient development of the deferred payments mechanism under a concession agreement, and others [12].

7 Conclusion

The regulatory, organizational, production and economic nature can be traced not only in the barriers, but also in the growth points in the development of a concession as a tool of state investment policy. These growth points include:

- regulatory and legal growth points - normative consolidation of strategies and programs of the concession's development, methodological and legal development of concession projects quality improvement, increase of awareness of the legal structures of the concession and legal protection of concession activities' subjects;
- organizational and production growth points - quality improvement of territorial planning in the formation of proposals for concluding concession agreements, application of an integrated approach to modernizing concession facilities, provision of organizational, technical and other kinds of support from the state, facilitation of measures promoting experience exchange in the sphere of concession projects implementation, personnel training and retraining in the field of concession activity;
- economic growth points – expansion of the toolset of budgetary funding (government guarantees, budget loans, targeted grants) and off-budget funding (loans, bonded loans, pension funds finance, insurance companies and other) of concession projects, development of reasonable requirements to the amount of guarantees and the mechanism of deferred payments under a concession agreement.

In the end, it should be noted that the formation of these growth points requires, firstly, the application of an integrated approach to creating an effective risk management system and ensuring high quality state support to concession projects. At the same time, the study emphasizes that while creating a concession agreement object, it is necessary to: assess all potential risks, draw up and update the matrix of risks and plans for resolving extraordinary situations and deviations from the initial plan in the process of implementing a concession project; create a multi-level system for the concession project implementation qualitative and quantitative parameters control [13]. Consequently, during the exploitation of the concession agreement object, it is important to create long-term partnership relations with contractors in all areas of activity; strict observance of technical regulations for the machinery and equipment operation, as well as safety and labor protection requirements [14], [15]. And, secondly, given that concession projects are long-term, which promotes business entities' stable development and predictable profitability, a high level of confidence in the concession system is crucial for making decisions on such long-term investments.

References

1. V. G. Varnavskii, *The partnership between the government and the private sector* (Nauka, Moscow, 2019)
2. S.S. Chernov, M.V. Filchenkova, Business. Education. Law **3 (40)**, 82-89 (2017)
3. O.V. Tarasova, Bulletin of the Novosibirsk state University of Economics and management **2**, 71-84 (2019)
4. N.N. Kosinova, Bulletin of the Volgograd state agricultural University. Series 3, Economics. Ecology **3**, 87-98 (2017)
5. V.V. Baranov, Property relations in the Russian Federation **9**, 74-81 (2019)
6. V.A. Kabashkin, *Public-private partnership: international experience and Russian perspectives* (LLC "MIZ", Moscow, 2020)
7. I.M. Shor, Financial Analytics: problems and solutions **42**, 38-48
8. A.A. Postnikova, Economics and business: theory and practice **1**, 87-91 (2017)
9. D.M. Shor, I.M. Shor, Financial Analytics: problems and solutions **46 (328)**, 18-31 (2016)
10. N.L. Platonova, Law and Economics **1**, 16-21 (2019)
11. N.N. Semenova, Economy. Taxes. Law **13 (3)**, 59-69 (2020)
12. N.V. Kalinin, T.V. Medvedeva, G.A. Romitsyna, Economic Sciences **2**, 53-56 (2018)
13. V.V. Karginova, Economics and management **1**, 51-62 (2017)
14. V.A. Fursov, N.V. Lazareva, Bulletin of the North Caucasus Federal University **1 (52)**, 103-108 (2016)
15. A.G. Zeldner, *Public-private partnership: theory, methodology and practice*, (Institute of Economics of the Russian Academy of Sciences, Moscow, 2019)

Electronic document flow as evidence in civil proceeding: evaluation features and application problems

G.V. Stankevich [1*], *I.M.* Vilgonenko [2], *Y.N.* Slepenok[4] and *O.M.* Litvishko [3]

[1] Pyatigorsk State University, Pyatigorsk, Russia; Stavropol Branch of Moscow State Pedagogical University, Stavropol, Russia, stankevichg@rambler.ru
[2] Pyatigorsk State University, Pyatigorsk, Russia, vil-irina-m@yandex.ru
[3] Institute of Service, Tourism and Design (Branch) of North-Caucasus Federal University in Pyatigorsk, Pyatigorsk, Russia, slep.80@mail.ru
[4] Pyatigorsk State University, Pyatigorsk, Russia, olitvishko@yandex.ru

Abstract. Due to the development of information technologies, electronic document flow is actively becoming a part of modern life, and the activity of courts is not an exception. Electronic documents are increasingly considered by courts as evidence in civil cases. Judicial practice faces certain problems when evaluating such evidence due to the insufficiency of legal regulation of such kind of evidence. The article examines the theoretical and practical aspects of the existence of an electronic document as a type of written evidence, attempts to reveal the essence of an electronic document, define its features as compared to a traditional written document, and analyzes the approaches to the definition of the concept of 'electronic document', which have developed in Russian legal science. The analysis of the practice in the application of an electronic document as evidence with regard to relevance, admissibility and adequacy is carried out. Certain problems which judges have to face when using an electronic document as evidence have been identified. The authors' understanding of an electronic document is submitted, the framework of problems and gaps under current civil procedure legislation is defined, and methods of solution are proposed. The need to define an independent type of evidence – electronic evidence is substantiated, with it further being enshrined in the Civil Procedural Code of the Russian Federation, as well as to formalize in legislation the concept of an 'electronic information medium'.

1 Introduction

The era of digitalization of the Russian society and state should begin to run from the date of enactment of the Executive Order of the RF Government of 28.07.2017 No. 1632-R that approved the Program of "The Digital Economy of the Russian Federation"[1], which is part of the Strategy of the development of information society in the Russian Federation for 2017-2030, approved by Russian Federation Presidential Decree of 09.05.2017 No. 203 [2]. The ensuing amendments affected all areas, including the judicial system. The amendments made to the current civil procedure legislation of the Russian Federation in 2016, which came into legal force in 2017, enshrined the use of electronic document flow when applying to judicial bodies. However, the legislator did not extend the electronic (digital) document flow to the institute of evidence in civil proceedings, confining itself to mentioning in Part 1 Article 71 of the Civil Procedural Code of the RF (hereinafter CPC RF) that written evidence is represented by documents and materials executed in the form of digital, graphical record, including received via facsimile, electronic or other means of communication with the use of information and telecommunication network "Internet", as well as documents signed by an electronic signature. Thereby, the legislator has recognized the existence of such evidence as an electronic (digital) document, but does not regard it as the original, since Part 2 of Article 71 of the CPC RF binds litigants to submit copies of electronic documents in the form of an original, despite the fact that a copy of an electronic document is essentially the original of an electronic document. Thus, the legislator regards an electronic (digital) document not as an independent type of evidence, but as a type of written evidence, defying its specific character and features of electronic (digital) information exchange. This fact demands analyzing recent developments and substantiates the relevance of the conducted research [3].

2 Problem Statement

The main problem of the research consists in the fact that the current civil procedure legislation considers both a written document in its traditional form and an electronic

* Corresponding author: stankevichg@rambler.ru

(digital) document as single-type phenomena. However, electronic document flow operates through electronic interaction of litigants in civil proceedings by means of information and telecommunication network "Internet" and the official website of the court. Following the implementation of the provisions of the Strategy of the development of information society in the Russian Federation, a system of such interaction has been established. One of the elements of this system, for instance, is information resource Gosulugi.ru (Integrated portal of state services), on the website of which various state services are provided online serving as an alternative to the MFC (Multifunctional centre), and thus performing electronic (digital) document flow. Should any civil disputes arise electronic (digital) documents sent and received within the framework of such interaction represent nothing more than electronic evidence of certain legal facts. With no registration on the website mentioned above, it is impossible to use the resources of the official site of the court when submitting such documents as a statement of claim and its appendices.

As a result, we witness the following situation: electronic (digital) document flow operates successfully, but the documents circulating within its framework and prepared in accordance with the requirements of the current legislation, do not have sufficient legal force to be considered as an independent form of evidence in a civil dispute, as providing their being submitted in electronic form, the judge may request them in the original, which complies with the norms of civil procedural legislation (Paragraph 2 of Article 71 of the code of CPC RF). Obviously, the legislator expresses a reservation in regard to the situations which demand the submission of original documents to the court alongside with those submitted in electronic form, but in general, we observe a situation in which civil procedural norms do not make allowance for the features of electronic (digital) information exchange and do not take into account the essence of an electronic document, especially the fact that in some civil cases an electronic document can act as the only evidence.

3 Research Questions

Having regard to the above, it is necessary to inquire into the essence of an electronic (digital) document, its evidential force and place in the system of evidence in civil disputes, to reveal the features of the assessment by the court the submitted electronic documents and to identify the problems that arise in the framework of law enforcement, which are faced by litigants in civil proceedings.

4 Purpose of the Study

Research of electronic (digital) documents as evidence in civil proceedings, identification of features of their assessment and problems of application in the practice of Russian courts in civil proceedings.

5 Research Methods

Research methods include general scientific methods of cognition (dialectics, synthesis; analogy; functional analysis) and specific legal methods: formal-legal, formal-dogmatic, comparative-legal, technical-legal, etc.

6 Findings

In order to achieve the goal set, it is necessary to refer to the legal framework of electronic (digital) document flow in the system of justice. In addition to the above-mentioned CPC RF, the framework also includes: Federal law No. 149-FZ of 27.07.2006 "On information, information technologies and information protection" [4], in which Paragraph 11.1 of Article 2 provides the following definition of an electronic document – documented information presented in electronic form, i.e. in a form suitable for human perception with the use of electronic computers, as well as for transmission via information and telecommunications networks or processing in information systems; Order of the Judicial Department of the Supreme Court of the Russian Federation No. 251 of 27.12.2016 "On approval of Procedure for submitting documents in electronic form to Federal courts of general jurisdiction, including in the form of an electronic document" [5], which establishes the official definition of an electronic document – a document created in electronic form with no prior documentation as hard copy, signed with an electronic signature in compliance with the legislation of the Russian Federation. In Paragraph 1.4 of the Procedure, the understanding of an electronic image of a document is also formulated, which is regarded as "an electronically-scanned copy of a paper document certified, in compliance with the Procedure for submitting documents, by a basic electronic signature or an enhanced encrypted and certified digital signature". Similarly, Order of the Judicial Department of the Supreme Court of the Russian Federation No. 252 of 28.12.2016 "On approval of Procedure for submitting documents in electronic form to the arbitration courts of the Russian Federation, including in the form of an electronic document" [6] and the Order of the Judicial Department of the Supreme Court of the Russian Federation No. 168 of 11.09.2017 "On approval of the Procedure of submitting documents in electronic form to magistrate judges, including in the form of an electronic document" both make distinction between an electronic document and an electronic image of a document [7].

The analysis of the aforementioned norms of the Russian law leads us to the conclusion that the current legislation and regulatory enforcement presume from the interpretation of an electronic document as written evidence created by a person and representing certain information medium relevant for the civil dispute under consideration, created with the use of modern information technologies, but capable of existing as hard copy, which in turn acquired the image of an electronic document by means of scanning.

As we can see, the legislator regards an electronic document as written evidence, but the science of procedural law has developed the following approaches to understanding the essence and meaning of an electronic document in the system of evidence.

Thus, A.T. Bonner in the monograph "Traditional and non-traditional means of evidence in civil and arbitration proceedings" concludes that "it is unlikely that electronic documents can be automatically considered equivalent to traditional written documents, neither other modern sources of information – to material evidence" [8]. As an electronic document, the researcher regards a body of data in computer memory intended for human perception with the use of appropriate software and hardware, as well as the absence of copies of an electronic document, all copies of which are originals and have equal legal force [8].

A.P. Vershinin in his work "Electronic document: legal form and evidence in court" asserts that an electronic document does not possess one of the main attributes of written evidence, namely an external written form [9]. In his opinion, the material form of the medium of written evidence is secondary. The primary thing is that a written document transfers information by means of written signs which are voluntarily perceived by a person through reading or by ear [9]. This fact is also pointed out by Article 181 of the CPC RF, according to which written evidence is pronounced before the court in session and presented to parties to a case for review.

An electronic document is substantially different from a traditional document in the following ways:

- an electronic document is placed on an electronic medium (hard disk, flash card), it should be borne in mind that the same document can be placed on different media, while preserving the equivalence of the content and in this case remaining the original, not a copy;
- in order to be used as evidence, it must be converted to a format suitable for reproduction;
- it is generated, altered, transferred, stored owing to electronic document flow;
- it is certified by an electronic signature in the manner prescribed by the law;
- its contents must be identified.

For the purpose of using an electronic document as evidence, it must be converted into a form that allows litigants in the process to study its contents, namely, it should be printed out. From this point on, it ceases to be an electronic document, being transformed into a traditional document, but loses many of its properties. A traditional document does not contain information about who created the electronic document and when, what amendments and additions were made to it, etc. Thus, the printed version of an electronic document transfers only its contents in the latest version. An electronic document as a medium of information once encased on an electronic medium and reproduced with the use of a computer carries the full version of its existence, even if it was later deleted, but this data are rarely accessed, particularly as for these purposes it demands to involve specialists and use special means and software technologies. In practice, the final version of an electronic document serves most often as the means of evaluation, and it is this version in the traditional form that is taken cognizance of by the court. However, there may be situations which require the use of specialized methods to evaluate an electronic document, for instance, in cases of consumer protection when conducting cashless payments by means of electronic money, disputes with credit organizations related to opening transaction accounts, deposits, obtaining loans, etc.

An electronic document is a distinctive type of evidence which has its own merits and demerits. Like any media, it may contain false information, in particular, if it is entered negligently. Therefore, it must be evaluated as one of the facts in evidence with no preference over other facts in evidence.

In the practice of courts with regard to an electronic document, difficulties may arise related to the application of provisional measures, for example, when protecting copyrights on created computer programmes or databases.

In particular cases, the evaluation of an electronic document may demand conducting technical expert examination on the motion of the parties, since such electronic medium can also be falsified using special software. An example of such disputes is the protection of patients' rights – a medical report signed by an unauthorized person, laboratory tests are not factually accurate and the patient suffers injuries to health, etc.

One of qualificatory features of an electronic document is its authentication and identification. Only an electronic signature can be used in the process of authentication and identification of an electronic document [10]. In court practice, situations may arise related to confirming the existence of a digital signature, since only a text is visualized in the electronic document flow. By Order of Rospatent No. 57 of 13.04.2016 (ed. 28.07.2016) "On establishing[11]. The rules and Form of visual display of an electronic signature mark on documents in electronic form" this problem is solved as follows: the rules for visualizing the electronic signature on documents have been developed and the form for visual display of the electronic signature mark has been approved, still some authors believe that the legal effect of such document is void [12].

Therefore, when evaluating an electronic document, the court must first of all proceed on the basis of the triad of evidence evaluation: relevance, admissibility and adequacy of evidence. Consequently, electronic documents being served in violation of the current legislation cannot be accepted for consideration, that is, they are recognized as inadmissible.

8 Conclusion

The conducted analysis allows us to formulate the following conclusions.

It is necessary to agree with the conclusions of scholars who conducted research into features of electronic justice in Russian judicial system. They state that the introduction of modern information technologies will increase the standards of legal proceedings, their quality, speed up the process of resolving disputes on the merits, simplify document flow and clerical correspondence in courts, and increase citizens' confidence in justice [3].

We propose to understand an electronic document as a document that appeared in the process of electronic document flow, and certified by an electronic signature.

The features of an electronic document do not allow it to be recognized as identical to a traditional document, and it is considered inexpedient to extend the legal regime of the latter to an electronic document.

So far, an electronic document has not been governed by the civil procedural legislation of the Russian Federation.

An electronic document should be regarded as a type of written evidence.

Judicial practice on the assessment of an electronic document as evidence is based on the content of Article 6 of the Federal law No. 63-FZ of 06.04.2011 "On electronic signature" [14] and recognizes it as equivalent to a hard copy document signed with a handwritten signature.

A range of urgent issues related to the use of an electronic document as evidence in a civil case should include the following:

- Its assessment may demand using special methods, software and involvement of experts, as well as conducting technical expert examination.
- It is difficult to provide interim measures.
- Problems may arise in confirming the existence of an electronic signature or its validity, since the visual image of an electronic signature does not represent the electronic signature itself.
- To overcome the highlighted problems, we consider it reasonable to make the make the following amendments to the current legislation:
- To amend Chapter 6 of the Civil Procedural Code of the Russian Federation by enshrining electronic evidence as a separate type of evidence.
- To supplement the provisions of Article 2 of Federal law No. 149-FZ of 27.07.2006 "On information, information technologies and information protection" with the concept of "electronic information medium"[15].

References

1. Decree of the Government of the Russian Federation No. 1632-R of 28.07.2017 "On approval of the Program of digital economy of the Russian Federation" / Expired.
2. Decree of the President of the Russian Federation No. 203 of 09.05.2017 "On the Strategy of the development of information society in the Russian Federation for 2017-2030".
3. Procedure for submitting documents to the arbitration courts of the Russian Federation in electronic form, including in the form of an electronic document".
4. Federal law No. 149-FZ of 27.07.2006 "On information, information technologies and information protection
5. Order of the Judicial Department of the Supreme Court of the Russian Federation No. 251 of 27.12.2016 "On approval of the Procedure for submitting documents to Federal courts of General jurisdiction in electronic form, including in the form of an electronic document".
6. Order of the Judicial Department of the Supreme Court of the Russian Federation No. 252 of 28.12.2016 "On approval of Procedure for submitting documents in electronic form to the arbitration courts of the Russian Federation, including in the form of an electronic document".
7. Order of the Judicial Department of the Supreme Court of the Russian Federation No. 168 of 11.09.2017 "On approval of the Procedure for submitting documents to magistrate judges in electronic form, including in the form of an electronic document".
8. A.T. Bonner, *Selected works: in 7 v. V. 5. Problems of the theory of forensic evidence* (Moscow, Russia: Prospekt, 2017).
9. A.P Vershinin, *Electronic document: legal form and evidence in court* (Moscow, Russia: JSC "Gorodets-Izdat", 2000)
10. Zubov, V.V. (2020). Electronic signature in the digital economy. DOI: 10.15405 / epsbs.2020.03.89.
11. Rospatent Order No. 57 of 13.04.2016 (ed. of 28.07.2016) "On approval of the Rules and Form for visual display of the electronic signature mark on documents in electronic form".
12. N.A. Hramtsovskaya, Electronic signature visualization: domestic and foreign approaches. Office and document management at enterprise, 12. (2014), URL: https://www.eos.ru/upload/analitica/Delo_2014-12%2002.pdf. (date of access: 16.03.21).
13. A.V. Azarkhin, D.A. Karev, A.A. Petrogradskaya, *Electronic justice in the courts of the Russian Federation* (2020)
14. Federal law No. 63-FZ of 06.04.2011 "On electronic signature".
15. Federal law No. 149-FZ of 27.07.2006 "On information, information technologies and information protection

Russian legal experience in using information technologies in diagnostics of reflux esophagitis

Anna Voevodina [1,*]

[1]Sechenov First Moscow State Medical University of the Ministry of Healthcare of the Russian Federation (Sechenov University), Moscow, Russia, anny1608@mail.ru

Abstract. This publication focuses on a theoretical investigation of the issues of Russian experience in the application of digital technologies in the provision of medical care in the diagnosis of oesophageal hernia complicated by reflux esophagitis. The relevance of the research lies not only in the increasing number of patients diagnosed with reflux esophagitis, but also in the current digital paradigm of the entire Russian healthcare system. Despite domestic legislative attempts to legally regulate the use of digital technologies in health care through the regulatory introduction of telemedicine technology, as is most common, certain issues in the use of this digital "platform" in health care remain unresolved. The methodological basis of this study is the methods of comparative analysis, system-structural analysis and synthesis. The author has assessed the possibility of using telemedicine technologies in the context of the doctor-patient relationship, implemented in a remote format, when diagnosing a patient, including from the position of significant territorial remoteness of the patient. There is a focus on the need for a more specific legal regulation in terms of legally establishing the definition for guaranteeing the security of personal data received by participants in the consultation, processed by them when applying telemedicine technologies.

1 Introduction

The use of digital technology in various spheres of public life in the Russian Federation is becoming more widespread and more relevant every year. The Russian healthcare system is no exception. It consists of state, municipal and private healthcare systems. The prerequisites for the "digital" transformation of health care are not only the use of highly informative techniques, electronic medical devices and treatments, including for surgical interventions, but also the possibility of using telemedical technologies for consultation (case conference) even at the stage of primary health care. The possibility of applying telemedicine technologies in various fields of clinical medicine is one of the significant directions of digital development in health care. The doctrine notes the non-specificity for Russian law enforcement to resolve issues relating to the settlement of electronic processes [1]. On the domestic legislative level, the issue of the possibility of using telemedicine technologies was regulated since 2017 in the Federal Law No. 323-FZ of 21 November 2011 "On the Fundamentals of Citizens' Health Protection in the Russian Federation" (hereinafter referred to as the Law on Fundamentals of Health Protection), which was amended by introducing a new legal norm (Article 36.2) devoted to the specifics of medical care provided using telemedicine technologies. Certainly, the legal regulation of the issue of telemedicine technologies is not limited to the mentioned legal act because there are other legal acts, including those of sectoral agencies, adopted to implement and enforce it. In this case, we are referring to Order No. 965n of the Ministry of Health of the Russian Federation of 30 November 2017 "On Approval of the Procedure for Organizing and Providing Medical Care Using Telemedicine Technologies" (hereinafter, Order No. 965n). It is valid to suggest that there are consequences of the spread of digital technology, such as the creation of new modes and means of regulation [2].

2 Problem Statement

There is a variation in the content of the relationship regulated by and arising from the use of telemedicine technologies, as it involves two areas: the interaction between health care professionals, primarily related to the rapid transmission of information and exchange of medical documentation regarding the patient, and the interaction through the doctor-patient system. The latter involves resolving a rather significant range of issues related to patient diagnosis, determining the applicable treatment methods, issuing referrals to highly specialized medical institutions, monitoring and analysing data obtained from laboratory, instrumental and functional tests. However, regardless of the chosen direction of the relationship based on the use of telemedicine technologies, the actual basis for this relationship is the

* Corresponding author: anny1608@mail.ru

remote interaction realized "at a distance". In the context of digital doctor-patient relationships, the main form of their implementation is the counselling services (consultation, case conference) provided online, i.e. in real time, through the use of videoconferencing, and involving direct interaction between the patient (or his legal representative) and the attending physician, medical workers (participants of the case conference), who determine the presence (absence) of any pathological abnormality in the patient based on the medical data provided electronically. Direct interaction between the participants in the consultation and the patient (or their legal representative) is possible not only through the exchange of voice information, but also through visual information. The doctrine mentions the possibility of collecting and transmitting patient information using modern telemedicine complexes [3]. Moreover, there is a position on the current expansion of telemedicine to all health services and on physician telemedicine consultations as the most well-known and widespread type of telemedicine services, while further noting the correlation of current advances in telemedicine with the advent of digital information transmission channels [4]. Despite the legal regulation of the use of telemedicine technologies in the provision of medical care based on using a "digital platform" of high-speed transmission of various information, the actualization of this method of providing medical care is certainly becoming increasingly important in view of the social processes that are emerging and taking place in society, often contributing to the remote ("distant") interaction between patient and treating physician (e.g. coronavirus infection COVID 2019). The scientific literature points to the unresolved legal issues related to the implementation of telemedicine services, which include issues of licensing and accreditation of doctors practising in different subjects of the state, requirements for diagnostics of patients, their treatment, if personal contact of doctor and patient is required, issues of legal liability [5].

3 Research Questions

The use of telemedicine technology in the diagnosis of oesophageal hernia complicated by reflux esophagitis is significant because of the high prevalence of these diseases in the structure of digestive diseases [6]. The doctrine has also suggested an increase in the number of patients with reflux esophagitis due to an axial hernia of the oesophageal orifice [7]. In addition, there is also a growing subject interest in investigating the dysfunctional relationship between the gastrointestinal tract and the cardiovascular system [8]. The consequences of failing to diagnose and detect this disease in time can be very negative, involving complications that put the patient's life at risk. However, timely diagnosis and detection of oesophageal hernia and associated symptoms is not possible without the use of information and technical diagnostic tools. One of the most informative and basic methods in diagnosing diseases of the oesophagus and stomach is the radiological method, which allows the most reliable examination of the topography, the anatomical structure, the functional state of the oesophagus and stomach, and gastro-esophageal reflux. There is evidence that indirect signs of oesophageal diaphragmatic hernia can be detected by noncontact examination, but with radiological data [9]. Radiological examination not only allows a diagnosis of a herniated oesophagus but also reveals cardiac insufficiency and the presence of reflux esophagitis symptoms of various kinds, while helping to visualize the oesophagus and draw conclusions about the state of the underlying gastrointestinal tract. Of course, this examination is not the only means of diagnosing the patient. The choice and tactics of diagnosis are case-specific, which in fact also predetermines the prospects of specific treatment technologies in each situation [10] However, using radiology data, a "non-contact" consultation with the patient is possible through the use of telemedicine technology.

3.1 Evaluation of radiological data using telemedicine technology: practical relevance

The practical implementation of a remote "digital" doctor-patient interaction when advising a patient about pathological changes and abnormalities suspected to exist in the gastrointestinal tract involves the patient first undergoing a "face-to-face" (in-person) radiological examination. Subsequently, it is interlinked with the electronic transmission of the results to the participants of the consultation for the evaluation of the medical findings reflected in the relevant documentation. The geographical features of the Russian state make it impossible to provide qualified medical care everywhere [11]. The doctrine mentions the existence of pilot telemedicine projects in the Russian Federation for patients living in remote areas and experiencing difficulties in organizing regular visits to specialists [12]. There is also a positive dynamics in the interaction between the federal research centre of the healthcare system and the clinical hospital of the subject (Cheboksary city), expressed in the provision of telemedical consultations for 38 patients and (or) medical documents of these patients in three calendar years [13].

3.2 Safeguarding patient confidentiality in telemedicine: is it guaranteed?

The legal basis for the implementation of telemedicine technologies, as noted, is set out in the provisions of Article 36.2 of the Law on the Fundamentals of Health Care, Order No. 965n, establishing general legal requirements on the need to ensure the confidentiality of patient data included in the patient's personal data and medical secrecy. However, it is an open question as to whether the high-speed channels for transmitting patient information are sufficiently secure, even though participants have limited remote access to video conferencing for a specific consultation and the patient is aware of the name and contacts of the operator of the information system used in the consultation. It is also an unresolved issue as regards the preservation of patient data in the relevant information system from the point of view of preventing "unregulated" adjustments to any information about the patient and his or her medical parameters.

4 Purpose of the Study

The purpose of the study is to examine the prospects of using telemedicine technologies in the diagnosis of patients with hiatal hernia of the diaphragm, complicated reflux esophagitis, and determine the further tactics of their treatment, including the position of a significant territorial remoteness of the patient, located in a sparsely populated point (administrative-territorial unit) of the Russian Federation.

5 Research Methods

The methodological basis of the study is the methods of comparative analysis, system-structural analysis and synthesis.

6 Findings

6.1 Evaluation of radiological data using telemedicine technology: practical relevance

There is a particular interest in optimising the diagnosis of oesophageal hernia complicated by reflux esophagitis due to the increasing prevalence of this nosological condition. The incidence of this disease is typical both for patients residing in large cities (constituent entities) of the Russian Federation and for patients residing and staying in remote (sparsely populated) areas of the Russian Federation. In the latter case, however, the level of quality and accessibility of the medical care provided (this case also requires "highly specialized" care) often does not meet the evolving needs of society and therefore its provision is sometimes extremely problematic. However, it cannot be made dependent on the patient's place of residence (location) to provide quality and timely medical care by the specialists whose expertise they need to heal. The investigated digital paradigm of remote health care delivery, implemented from the primary health care stage, represents an entirely new format of health care, enabling the use of the necessary software and hardware "kit" for patient-physician interaction. The use of telemedical technologies in the form of consultations (case conferences) based on medical data about a patient with specific symptomatology will make it possible to reasonably assess the need for, for example, surgical intervention and also determine the tactics of the patient's treatment. In this case, such actions and activities aimed at assessing the patient's condition (and possibly issuing a referral for surgery), using this digital format of interaction with a patient located in a locality remote from major cities in the Russian Federation, where the level of healthcare development is certainly higher, will be implemented by competent specialists. Thus, the practical application of telemedicine services will actually reduce the number of patient visits to outpatient clinics, minimizing the need for personal visits to the attending doctor, provided that all data on the patient's medical parameters are available electronically, if they are sufficient to make a diagnosis without the patient being present in person. It is fair to note the positive dynamics of using telemedicine technologies in the form of reducing the time a patient spends in hospital for preoperative examinations if they are needed [14], improving the quality of life of patients and reducing emergency admissions and hospitalizations [15] and allowing for consultations with competent specialists [16]). The statement about relieving the patient of the need to travel to a treatment facility that is significantly distant from their place of residence is similarly valid [17]. The administrative-territorial division of the Russian Federation and the significant geographical scale of the entire country, which implies the presence of quite remote (from large cities, including federal cities) settlements (often sparsely populated) in which the provision of specialized (and highly specialized) medical care is not always possible, justifies the incorporation of telemedical technologies in the practical medical activity.

6.2 Safeguarding patient confidentiality in telemedicine: is it guaranteed?

The information sent electronically about a patient, including medical records of any pathological changes, falls into the category of confidential, personal information about the patient, the free dissemination of which is prohibited by Russian law at the domestic level. The use of a digital doctor-patient system should strictly adhere to the principle of information protection of personal data on the patient. The implementation of such a digital approach is not possible without the use of the Internet telecommunication network, the availability of which gives both parties to the relationship access to a videoconference for the consultation. It seems insufficient to have a separate legal provision, referring to the prohibition of dissemination of this information without further specification of the organizational procedure for ensuring the information security of high-speed transmission channels used in telemedicine technologies. Moreover, the absence of a legal provision for the preservation of patient data in uncorrected form, uploaded to an appropriate information system to store such data, does not in fact provide the consumer of a health service with a sufficient level of guarantee of its availability in an unchanged format. It is possible that access to patients' personal data, including information that constitutes medical secrecy, may be open to a wide range of people (e.g. in the event of a technical failure resulting in an unauthorized data breach). To minimize such situations, it would be advisable to store patient information in an uncorrected form on the electronically secure servers of the information processor, using current digital security and encryption tools.

7 Conclusion

The study confirms the increasing importance of digital transformation in healthcare, while demonstrating the feasibility of using telemedicine technologies, implemented in a remote format, to provide medical care to patients with oesophageal hernia complicated by reflux esophagitis, living (staying, residing) in a locality remote from major cities of the Russian Federation. This normatively introduced approach to organising the digital paradigm of the doctor-patient relationship seems to be cost-effective and financially feasible, reducing and reducing the level of monetary costs. The use of a digital procedure for assessing patient data and medical parameters offers the possibility of extending the targeting of highly informative and high-tech radiological diagnostic techniques, while allowing correct and appropriate interpretation of results from specialists - radiologists who are constantly faced with assessing the symptomatology of such diseases. The main form of digital doctor-patient interaction is video-conferencing for consultation (case conference) to examine the patient's medical parameters recorded in electronic documentation, a conversation with the patient to clarify the nature of the individual pathological symptomatology. The use of this form of interaction makes it possible not only to determine the correct diagnosis in good time, but also to choose the best treatment tactics for the patient, including an assessment of the need for surgical intervention, the timeliness and promptness of which, in general, affects the patient's standard of living.

References

1. V.N Grigoryev, A.P. Sukhodolov, S.S. Ovanesyan, M.G. Spasennikova, V.V. Tyunkov, All-Russian Journal of Criminology 13(6), 873-883 (2019)
2. E. Gromova. T. Ivanc, BRICS Law Journal 7(2), 10-36 (2020)
3. O.V. Kamenskaya, I.Y. Loginova, A.S. Klinkova, A.R. Tarkova, R.A. Naydenov, E.I. Kretov, V.V. Lomivorotov, Russian Journal of Cardiology 25(6), 154-160 (2020)
4. T.S. Smal, V.D. Zavadovskaya, I.A. Deev, Bulletin of Siberian Medicine, 15(1), 79-88 (2016)
5. M.S. Zhuravlev, *Law and information: issues of theory and practice: collected works of international scientific and practical conf*, 151-154 (2017)
6. E.L. Nikonov, Surgery. Journal named after N.I. Pirogov **5**, 96-105 (2018)
7. N.V. Gibadulin, I.O. Gibadulina, S.N. Trinov, Surgery. Journal named after N.I. Pirogov **11**, 34-37 (2009)
8. O.N. Antropova, N.V. Pyrikova, I.V. Osipova, Russian Journal of Cardiology **7**, 103-109 (2019)
9. G.A. Zhurbenko, A.S. Karpitsky, S.V. Panko, A.M. Shestyuk, Chest and Cardiovascular Surgery **57 (1)**, 36-42 (2015)
10. V.O. Shchepin, A.I. Bravve, A.I. Babenk, A.L. Tomchuk, Problems of Social Hygiene, Public Health and History of Medicine **5**, 39-42 (2013)
11. M.D. Podolsky, A.A. Strokov, V.I. Kuznetsov, V.V. Shapovalov, Cardiovascular Therapy and Prevention **15 (5)**, 92-96 (2016)
12. V. A. Bondarenko, D.D. Kostoglodov, *Modern problems of economics, law and business of post-coronavirus crisis. Proceedings of the International Scientific and Practical Online Conference*, 114-119 (2020)
13. I.B. Bashkova, T.V. Prokopyeva, A.V. Arkhipova, B.F. Ryapolov, Scientific and Practical Rheumatology **54 (1)**, 108 (2016)
14. M.V. Ionov, N.E. Zvartau, I.V. Emelianov, A.O. Konradi, Telemedical, Arterial Hypertension **25(4)**, 337-356 (2019)
15. D.N. Laptev, V.A Peterkova, Diabetes Mellitus **6**, 420-426 (2017)
16. V.K. Fedyaeva, N.I. Zhuravlev, J.A. Galeeva, Diabetes Mellitus **4**, 328-335 (2019)
17. Yu.N. Gorozhantsev, S.G. Sergienko, E.A. Vorotintseva, Yu.V. Emelianenko, Russian Journal of Cardiology **24**, 39a-39b (2019)

Aspects of administrative responsibility in a digital environment: problems and prospects

Vyacheslav Volkov [1], *Yelena* Yevsikova [2*], *Natalia* Kravchenko[3], *Aleksandr* Ponomarev[4] and *Svetlana* Rusanova [5]

[1] the Crimean branch of the Federal State Budget-Funded Educational Institution of Higher Education "The Russian State University of Justice", Simferopol, Russia, volkow@yandex.ru
[2] the Crimean branch of the Federal State Budget-Funded Educational Institution of Higher Education "The Russian State University of Justice", Simferopol, Russia, helen_yevsikova@mail.ru
[3] the Crimean branch of the Federal State Budget-Funded Educational Institution of Higher Education "The Russian State University of Justice", Simferopol, Russia, nataliakravchenko85@gmail.com
[4] the Crimean branch of the Federal State Budget-Funded Educational Institution of Higher Education "The Russian State University of Justice", Simferopol, Russia, alexandr.ark@mail.ru
[5] the Crimean branch of the Federal State Budget-Funded Educational Institution of Higher Education "The Russian State University of Justice", Simferopol, Russia, sveto4ka_r@mail.ru

Abstract. The article examines the features of administrative liability under the administrative proceedings in the context of the trend towards the widespread digitization of different types of judicial proceedings. The authors research current Administrative-tort legislation of the Russian Federation and elements of the electronic justice system typical for such administrative procedures. It is revealed the features of digitalization during the procedure for handling cases on administrative offenses provided by the Draft of Procedural Code of the Russian Federation on administrative offenses. Researching the peculiarities of the procedure for bringing to administrative responsibility, the authors come to the conclusion that a number of significant problems and violations of the rights, freedoms and legitimate interests of persons participating in the case are inevitable, which will arise as a result of the introduction of an electronic justice system in cases of administrative offenses. As a result of the study, the authors come to the conclusion that there is a need for further scientific and practical understanding of all aspects of the implementation of the electronic justice system in cases of administrative offenses in order to focus it on ensuring high-quality and effectiveimplementation by the participants in proceedings in cases of administrative offenses of their rights, freedoms and legal interests.

1 Introduction

Recently, the transition of Russian society to a digital model of formation and development requires a search for new ways of regulating administrative sphere of legal relations based on modern scientific developments and studies of this problem and analysis of digital technologies usage in law enforcement practice, including within the framework of bringing to administrative responsibility.

At the same time, however, among the general markers of law transformation, one should be taken into account – the emergence of new legal categories: subjects of law (virtual personality, robots, etc.), requiring doctrinal development of possible models for recognizing a legal personality for highly developed robotics; artificial intelligence that determines the legal regime; new digital rights [1].

The relevance of the topic is confirmed by the increased attention to this problem from the top officials of the state. So, in the message of the President of the Russian Federation V.V. Putin of March 1, 2018 and February 20, 2019 is emphasized the fact that as soon as possible we need to create an advanced legislative framework, remove all barriers to the development and widespread use of robotics, artificial intelligence, unmanned vehicles, e-commerce, processing technologies of big data processing. Moreover, such a regulatory framework should be constantly updated, based on a flexible approach to each area and technology [2-3].

Also, Zorkin V.D. noted that one of the tasks of the state is to recognize and protect the digital rights of citizens from all kinds of violations. However, the current legislation of the Russian Federation does not fully meet modern needs, since many laws are weakly related both to the specified basic law and to each other. In this connection, information legislation needs to be systematized, get rid of repetitions and bring its conceptual apparatus into a harmonious, consistent state [4].

These statements generally apply to the administrative-tort legislation, which is currently experiencing the third «wave» of codification [5] and needs to be rethought taking into account the general trend of digitalization and the development of the e-

* Corresponding author: helen_yevsikova@mail.ru

justice system in the Russian Federation.

2 Problem Statement

The transition of Russian society to a digital model of development is carried out in accordance with the Strategy for the Development of the Information Society in the Russian Federation for 2017–2030 [6], which provides that the information society is a society in which information and the level of its application and accessibility radically affect economic and sociocultural living conditions of citizens.

The priority of ensuring national interests in the development of the information society is: a) the formation of the information space, taking into account the needs of citizens and society in obtaining high-quality and reliable information; b) development of the information and communication infrastructure of the Russian Federation; c) creation and application of Russian information and communication technologies, ensuring their competitiveness at the international level; d) formation of a new technological basis for the development of the economy and social sphere; e) ensuring national interests in the field of the digital economy.

At the same time, there is an acute imbalance in the digitalization processes of various types of the administrative process. So, if administrative court proceedings are characterized by the widespread use of digital methods of information exchange, then in proceedings on cases of administrative offenses such use is still not very widespread.

Therefore, carrying out scientific research of the problems and prospects of bringing to administrative responsibility in the context of digitalization of the administrative process as a whole, which affects the rights, freedoms and interests of individuals and legal entities, is relevant and in demand today. It can also serve as a basis for further research and development in this area.

3 Research Questions

Since the adoption in 2001 of the Code of Administrative Offenses of the Russian Federation (hereinafter referred to as the Code of Administrative Offenses) [7], the legal regulation of the procedure for bringing to administrative responsibility has undergone many changes and additions in order to refine and improve. So, for 19 years since the adoption of the Code of Administrative Offenses, in total, more than 530 federal laws have been adopted, which have already introduced over 5.5 thousand changes to the Code of Administrative Offenses, which once again confirms the objective need for scientific and practical comprehension and analysis of the practice of Code of Administrative Offenses application that has developed over these years and critical assessment of the effectiveness of its norms, taking into account the numerous changes introduced [8].

In accordance with the Resolution of the Plenum of the Supreme Court of the Russian Federation of March 24, 2005 No. 5 "On some issues arising from the courts when applying the Code of Administrative Offenses of the Russian Federation" [9], to the legislation on administrative offenses, which should be followed when considering this category cases, belongs to the Code of Administrative Offenses, which determines the conditions and grounds for administrative responsibility, types of administrative punishments, the procedure for proceedings in cases of administrative offenses, including the jurisdiction over cases and judicial jurisdiction of these cases, as well as the laws of the constituent entities of the Russian Federation adopted in accordance with the Code of Administrative Offenses on issues related to the competence of the constituent entities of the Russian Federation.

Thus, the provisions of clause "k" of art. 72 of the Constitution of the Russian Federation on the joint jurisdiction of the Russian Federation and the constituent entities of the Russian Federation of administrative and administrative-procedural legislation.

At the same time, the procedure for proceedings in cases of administrative offenses belongs to the jurisdiction of the Russian Federation, and the establishment by the laws of the constituent entities of the Russian Federation for the organization of proceedings in cases of administrative offenses provided for by the laws of the constituent entities of the Russian Federation belongs to the jurisdiction of the constituent entities of the Russian Federation. However, the elements of digitalization of the procedure for considering cases of administrative offenses by the current administrative-tort legislation of the Russian Federation and the subjects of the Russian Federation is provided for today rather fragmentarily.

It is interesting that the draft Code of the Russian Federation on Administrative Offenses prepared by the Ministry of Justice of the Russian Federation [10] is a more perfect and well-developed legislative act, however, it does not pay due attention to the digitalization of the procedure for bringing to administrative responsibility.

At the same time, the organization of the procedure for carrying out proceedings in cases of administrative offenses, including those provided for by the laws of the constituent entities of the Russian Federation, in accordance with the Draft Code of Administrative Offenses, refers only to the jurisdiction of the Russian Federation, and not to the constituent entities of the Russian Federation, which seems to be quite logical and correct from the point of view of a single the procedure for carrying out proceedings on cases of administrative offenses throughout the territory of the Russian Federation.

It should be borne in mind that in accordance with the Draft Procedural Code of the Russian Federation on Administrative Offenses (hereinafter referred to as the Draft Procedural Code of Administrative Offenses) [11] the procedure for proceeding in cases of administrative offenses in the initiation, consideration and revision of cases on administrative offenses, as well as the procedure for executing decisions on cases of administrative offenses in the territory of the Russian Federation are regulated by the Draft Procedural Code of Administrative Offenses.

At the same time, this Draft Code of Administrative Offenses and the Draft Procedural Code of Administrative Offenses do not provide for the procedure for the implementation of digital production, as such, in cases of administrative offenses.

However, art. 1.17 of the Draft Procedural Code of Administrative Offenses provides that an application, petition, complaint, protest, the presentation can be filed by participants in proceedings in an administrative offense case to a court considering (reviewing) an administrative offense case in the order and terms established by this Code, in electronic form by filling out the form of documents posted on the official website of the court in the information and telecommunication network "Internet" in the manner established by the Supreme Court of the Russian Federation and the Judicial Department under the Supreme Court of the Russian Federation.

And also, a statement, petition, complaint, protest, the presentation can be submitted by the participants in the proceedings in the case of an administrative offense to the body, the official carrying out the proceedings in this case in the order and terms established by this Code, in the form of an electronic document signed by the person who sent such a document, with an electronic signature in accordance with the legislation of the Russian Federation, through the infrastructure providing information and technological interaction of information systems used for the provision of state and municipal services and the performance of state and municipal functions in electronic form.

At the same time, the materials attached to the application, petition, complaint, protest, submission are also submitted in the form of electronic documents. And electronic documents produced by other persons, bodies, organizations in free form or in the form established for these documents by the legislation of the Russian Federation must be signed by them with an electronic signature in accordance with the requirements of the legislation of the Russian Federation.

In addition, the participants in the proceedings in the case of an administrative offense have the right to receive copies of the acts issued in the case, executed in the form of electronic documents, notifications, summons and other documents (their copies) in electronic form, with the exception of documents containing information, access to which, in accordance with the legislation of the Russian Federation is limited.

At the same time, documents drawn up in electronic form, notifications, summons in the case of an administrative offense are sent to interested persons through electronic means of communication, infrastructure that ensures information and technological interaction of information systems used for the provision of state and municipal services and the performance of state and municipal functions in electronic form, including through the Unified portal of state and municipal services.

However, it should be taken into account that the provisions of this article should be applied only if it is technically feasible. It should be noted that the current Code of Administrative Offenses also contains certain elements of digital proceedings in cases of administrative offenses. So, for example, the legislator has provided for the use of special technical means working in automatic mode that have the functions of photography and filming, video recording for the detection and fixation of administrative offenses in the field of road safety or in the field of landscaping using a vehicle which, on the one hand, allows to optimize the process of bringing to administrative responsibility, to make it simpler and more dynamic, however, on the other hand, it can lead to a violation of the principle of the presumption of innocence of a person. So, within the framework of bringing to administrative responsibility in this category of cases, a number of difficulties arise related to the fact that at the time of the offense, another person was driving the vehicle and this vehicle was in the possession or use of another person.

At the same time, one should take into account the legal position of the Constitutional Court of the Russian Federation, which in its Resolution of January 18, 2019 No. 5-P "In the case of checking the constitutionality of art. 2.6.1 and parts 1, 2, 3 and 6 of art. 12.21.1 of the Code of the Russian Federation on administrative offenses in connection with the request of the Kostroma Regional Court and the complaints of citizens A.I. Dumilin and A.B. Sharov" [12], which indicated that the imposition of responsibility in this case on the owner (owners) of heavy and (or) large vehicles is due to the peculiarities of fixing the corresponding offenses, in which it is established that the offense was committed using a specific vehicle, and on the basis of data from the state registration of vehicles, its owner (owners) can be determined. This method of fixing violations of the prescribed rules - with the provision of vehicle owner (owners) with the opportunity to prove their innocence - in itself does not contradict the Constitution of the Russian Federation its introduction into the legislation on administrative offenses belongs to the discretion of the federal legislator, who, in order to increase the efficiency of protecting the values protected by the Constitution of the Russian Federation from unlawful encroachments (which is part of the scope of this legislation – art. 1.2 of the Code of Administrative Offenses), has the right to take into account the current level of development technical means of detecting prohibited acts.

At the same time, the ruling on requesting the information is signed by the official who issued such ruling, reinforced with a qualified electronic signature. In the event that the requested information (information) or notification of the impossibility of submitting such evidences (information) is sent in electronic form using a unified system of interdepartmental electronic interaction and regional systems of interdepartmental electronic interaction connected to it, they are signed by an authorized official to whom such a determination was received, reinforced with a qualified electronic signature (art. 26.10 of the Code of Administrative Offenses).

In addition, art. 29.10 of the Code of Administrative Offenses provides that a decision in a case of an administrative offense can be issued and sent for execution in the form of an electronic document (including using a unified system of interdepartmental electronic interaction and regional systems of interdepartmental electronic interaction connected to it) signed by a judge, a person presiding at a meeting of a collegial body, or an official who issued a resolution, reinforced with a qualified electronic signature in the manner prescribed by the legislation of the Russian Federation.

Also, a decision in a case on an administrative offense with the attachment of materials obtained using special technical means working in an automatic mode, having the functions of photography and filming, video recording, or means of photographing and filming, video recording, is drawn up in the form of an electronic document signed by an official who issued a resolution, reinforced with a qualified electronic signature in the manner prescribed by the legislation of the Russian Federation.

In order to send for execution a resolution in a case on an administrative offense, issued in the form of a paper document, a copy of the said resolution may be made in the form of an electronic document signed by a judge, a person presiding at a meeting of a collegial body, or an official, who is issued a resolution in a case of an administrative offense, reinforced with a qualified electronic signature in the manner prescribed by the legislation of the Russian Federation.

It should be noted that the Draft Procedural Code of Administrative Offenses provided for a similar possibility of signing various procedural documents with an enhanced qualified electronic signature by an official considering a case of an administrative offense:

1)A ruling in a case of an administrative offense (for example, a ruling on the reclamation of information (para. 6 art. 3.5); in order to send for the execution of a ruling in a case of an administrative offense, issued in the form of a paper document, a copy of the said ruling may be made in the form of an electronic document signed by a judge, enhanced with a qualified electronic signature in the manner prescribed by the legislation of the Russian Federation (para. 5 art. 8.45);

2)Resolution in a case of an administrative offense (a resolution in a case of an administrative offense can be issued and sent for execution in the form of an electronic document (including using a unified system of interdepartmental electronic interaction and regional systems of interdepartmental electronic interaction connected to it) signed by the person presiding at the meeting of the collegial body, or by the official who issued the resolution, reinforced with a qualified electronic signature in the manner prescribed by the legislation of the Russian Federation (para. 6 of art. 6.8);

3)Resolution in a case on an administrative offense with the attachment of materials obtained using special technical means operating in an automatic mode, having the functions of photography and filming, video recording, or means of photographing and filming, video recording, is drawn up in the form of an electronic document signed by an official who issued a resolution, reinforced with a qualified electronic signature in the manner prescribed by the legislation of the Russian Federation (para. 7 art. 6.8).

At the same time, the Draft Procedural Code of Administrative Offenses of the Russian Federation is partially focused on digitalization processes, just like the current Code of Administrative Offenses of the Russian Federation, and provided that the person has the appropriate technical capabilities and an electronic signature which, in accordance with Federal Law No. 63-FZ dated 06.04.2011 "On Electronic Signature" [13], is information in electronic form, which is attached to other information in electronic form (signed information) or otherwise associated with such information and which is used to identify the person signing the information.

At the same time, the current legislation does not contain the concept of an enhanced qualified electronic signature, but contains the concept of a qualified certificate of an electronic signature verification key - a certificate of an electronic signature verification key that meets the requirements established by the Federal Law "On Electronic Signature" and other normative legal acts adopted in accordance with it, created by an accredited certifying center or a federal executive body authorized in the field of using the electronic signature, and is in this connection an official document (para. 3 of art. 2).

So, despite the fact that the Federal Law "On Electronic Signature" was adopted in 2011, the administrative-tort legislation of the Russian Federation is very limited in agreement and interacts with it, is fragmentarily focused on digitalization processes and the possibility of participants in proceedings in cases of administrative violations of their rights and interests using electronic document management, however, if technically feasible.

As you can see, the current legislation in this area requires significant improvement and unification, first of all, the categorical apparatus and the establishment of a unified approach to the digitalization of various spheres of public life, which today allows us to talk about the improvement of legal regulation and an increase in certain aspects of the effectiveness of justice. However, there is still no evidence of a transition to e-justice [14].

For example, the Federal Republic of Germany Law on Administrative Procedures [15] stipulates that a condition for the implementation of electronic document circulation is the opening by the recipient of an electronic document of appropriate access, which means not only the availability of technical conditions for electronic document management, but also the desire of a person (citizen) to communicate in electronic form. At the same time, to open access to a person, it is enough to indicate an email address. In turn, all public authorities [16] are obliged to open access to receive electronic documents signed with an electronic signature. In addition, the obligation to open access to receive electronic documents is valid for municipalities [1].

Similarly to the Draft Procedural Code of Administrative Offenses, the Federal Republic of

Germany Law on Administrative Procedures provides for written and electronic forms of documents for the implementation by persons participating in the case of their rights and interests and for the performance of duties provided for by law, provided that the electronic documents are signed with an electronic digital signature.

In addition, the Law of the Federal Republic of Germany on Administrative Procedures provides for the issuance of an administrative act in an automated mode, if such a possibility is provided for by the rules of law, and the issuance of such an administrative act is not associated either with the implementation of administrative discretion or with any other possibility of assessing the circumstances of the case. Thus, administrative acts or other actions that are fully or partially issued or carried out not by a person are also qualified as actions of public authorities, and human will, in this case, ceases to be an essential element of an administrative act, and an automated administrative act does not not fully correspond to such a classical doctrinal feature of an administrative act as «regulated» [1].

As you can see, on the one hand, digital technologies make it possible to expand opportunities for citizens to exercise their rights and interests, however, such implementation is made dependent on the availability of technical capabilities, and also imposes on the state the obligation to ensure this right. Thus, the use of digital technologies in the activities of public authorities makes it possible to increase the transparency of their activities, make them more open and at the same time accessible to citizens [17].

An example of the automation of the judicial process in cases of administrative offenses provided for by the Code of Administrative Offenses of the Russian Federation is a special procedure for bringing to administrative responsibility for administrative offenses in the field of road traffic when they are fixed by special technical means working in the automatic mode having the functions of photographing and filming, video recording, or working in automatic mode by means of photographing and filming, video recording (in these cases, a protocol of an administrative offense is not drawn up, a decision on an administrative offense case is issued without the participation of the owner (owners) of the vehicle and is drawn up in the procedure provided for in art. 29.10 of the Code of Administrative Offenses).

In this case, the automatic mode should be understood as the operation of the corresponding technical device without any direct human impact on it, when such a device is placed in the prescribed manner in a stationary position or on a vehicle moving along an approved route makes fixation in the zone of its review of all administrative offenses for the detection of which it is intended, regardless of the discretion of this or that person. Taking into account the fact that the event of an administrative offense is characterized, among other things, by the place and time of its commission, the materials generated by technical means operating in automatic mode must contain the specified information.

At the same time, technical means operating in automatic mode must be certified, in particular, as a measuring instrument, have a valid certificate of metrological verification and be used in accordance with the documents regulating the procedure for using these means. In the description of the type of measuring instrument, the metrological characteristics must be determined, the algorithm for the operation of the software for identifying and fixing an administrative offense must be disclosed, and a list of revealed offenses must be defined. If during the consideration of a complaint (protest) against a decision to impose an administrative penalty for an offense identified and recorded by a technical means operating in an automatic mode, doubts about the correct operation of such a technical means, including in connection with the arguments of the complaint (protest), arise, the judge has the right to demand documents containing the above information.

It should be noted that when revising a decision on the imposition of an administrative penalty for an offense detected and recorded by technical means operating in an automatic mode, the arguments of the person against whom the said decision was issued are subject to verification about the impossibility, after fixing an administrative offense with a technical means operating in an automatic mode, to stop illegal actions in connection with the organization of traffic on a specific section of the road. So, the frequency of placement of technical equipment operating in automatic mode, which did not allow the driver, after fixing an administrative offense, to reduce the speed of the vehicle without creating an emergency situation or leave the lane intended, for example, for the movement of route vehicles, without crossing the road markings, may indicate the absence of his fault in committing a subsequent administrative offense provided for, respectively, one of the parts of art. 12.9, part 1 of art. 12.16 or part 1.1 of art. 12.17 of the Code of Administrative Offenses [18].

Analysis of the provisions of the current Code of Administrative Offenses of the Russian Federation shows that in this case the legislator abandoned the important principle of proceedings in cases of administrative offenses - the principle of individualization of punishment, giving artificial intelligence the ability to choose a punishment option for a person which shows the possibility of introducing information technologies into proceedings in cases of administrative offenses, which, in turn, may grossly violate the inalienable basic principles of any type of legal proceedings - the principle of the presumption of innocence, the principle of inevitability of administrative responsibility for the guilty person, the principle of proportionality and individualization of punishment, which, in our opinion, does not justify the simplicity, efficiency and efficiency of the execution of proceedings in this category of cases of administrative offenses.

In addition, within the meaning of Ch. 30 of the Code of Administrative Offenses, a complaint against acts held in a case of an administrative offense must be submitted to a higher court on paper, since the Code of

https://doi.org/10.1051/shsconf/202110901041

Administrative Offenses does not provide for the possibility of filing documents, appeals, complaints about acts in an administrative offense case in electronic form, including in the form of an electronic document signed with an electronic signature.

However, documents, appeals, complaints in cases of administrative offenses are not subject to filing with a court of general jurisdiction in electronic form, since the proceedings in such cases are carried out according to the rules of the Code of Administrative Offenses, the norms of which do not provide for their filing in electronic form.

We must agree with the opinion of V.D. Zorkin that digital technologies can significantly improve the quality of law enforcement in executive bodies, significantly speed up and simplify the implementation of the public powers assigned to them. And also, about the role of artificial intelligence in legal proceedings: in the cognitive and evidence base (digital traces as electronic evidence; new types of forensic examinations); electronic means of organizing the work of the court (electronic document management, electronic business, intelligent systems for analyzing case materials, legal regulation); systems of electronic participation in the process (video conferencing, electronic summons and SMS notifications, electronic copies of case materials) [4].

As D.A. Pashentsev rightly notes, the introduction of digital technologies into legal practice makes it possible to automate a number of processes that previously could be carried out only with the direct participation of the subject of law (for example, the introduction of electronic justice, in which decisions on typical cases will be made by a robot, not a living judge). However, one should agree with the opinion of E.E. Tonkov that one cannot agree with such an approach to the organization of electronic justice, since judicial discretion plays an important role in every judicial process, i.e. the mental activity of a specific judge in a specific case, the result of which will be a specific decision made on the basis of an analysis of alternative options provided for by law [19].

However, in our opinion, the problem in this case also lies in the fact that with a relatively vague content of the norm, it will be impossible to guarantee its unified application in each case when the norm provides for several options for making a decision in a case of an administrative offense. We must agree with the opinion that the confidence in automated technology has not yet reached the level when it could be provided with a solution to multivariate situations. The institution of an automated administrative act is only suitable for simple structured administrative procedures, since it is within the framework of standard situations that computer algorithms are highly efficient [1].

So, if, in order to prepare the case for consideration, the judge must also establish whether the protocol on an administrative offense has been drawn up correctly in terms of the completeness of the investigation of the event of the offense and information about the person who committed it, as well as compliance with the procedure for drawing up the protocol, then within the framework of electronic justice, this function will be performed by a robot judge, which, given the low quality of drafting protocols on administrative offenses, will contribute to ineffective consideration of cases of administrative offenses and compliance with the principle of inevitability of administrative responsibility but ensuring the termination of proceedings due to the expiration of the deadlines, since the robot judge will return most of such protocols on administrative offenses to the relevant authorities for revision, which, as practice shows, leads to missing deadlines for eliminating deficiencies in administrative material in most cases, or, vice versa will accept protocols with significant flaws in production. So, a significant drawback of the protocol is the lack of data directly listed in part 2 of art. 28.2 of the Code of Administrative Offenses, and other information depending on their significance for a given case on an administrative offense, which should be assessed by a real judge in each specific case (for example, the lack of data on whether the person in respect of whom a case on an administrative offense is initiated, is conversant in the language of the arbitration, as well as data on the provision of an interpreter when drawing up a protocol, etc.) [9].

In addition, the judge must check the authority of the official to draw up a protocol, taking into account the provisions of art. 28.3 of the Code of Administrative Offenses and regulations of the relevant federal executive bodies. At the same time, in the event of the reorganization of federal executive bodies, it is necessary to check whether the relevant officials of these bodies have retained the right to draw up a protocol and consider a case on an administrative offense and whether these functions have been transferred to officials of other federal executive bodies. However, it is not clear how these functions will be performed by a robot judge in the process of implementing electronic justice in cases of administrative offenses.

At the same time, when considering cases on bringing persons to responsibility for an administrative offense, as well as on complaints and protests against decisions in cases of administrative offenses, if necessary, the possibility of summoning the persons who have drawn up a protocol on an administrative offense to the court is not excluded to clarify the issues that have arisen. In this case, it is not clear how the robot judge will resolve this issue. A similar situation will arise when deciding on the appointment of an examination, when, in the course of proceedings on an administrative offense, it becomes necessary to use special knowledge in science, technology, art or craft, a judge on the basis of art. 26.4 of the Code of Administrative Offenses issues a ruling on the appointment of an examination.

In addition, in practice it is quite difficult to determine the degree of guilt of a legal entity in committing an administrative offense. When determining the degree of responsibility of an official for committing an administrative offense, which resulted from the implementation of the decision of the collegial body of a legal entity it is necessary to find out whether the official took measures in order to draw the attention of the collegial body or the administration to the impossibility of executing this decision due to the fact that this may lead to the commission of an administrative offense. Since the Code of Administrative Offenses does not provide for any restrictions in the specified case when imposing an administrative penalty, the judge has the right to apply any punishment to a legal entity and an official within the sanction of the relevant article including the maximum, taking into account mitigating, aggravating and other circumstances affecting the degree of responsibility of each of these persons, which when considering an administrative offense by a robotic judge will be impossible to implement in practice and will contribute to the violation of the principles of justice, proportionality, individualization administrative punishment [20].

At the same time, in order to comply with the established art. 29.6 of the Code of Administrative Offenses of the Russian Federation, the time frame for considering cases of administrative offenses, the judge must take measures to promptly notify the persons involved in the case about the time and place of the trial. However, the Code of Administrative Offenses of the Russian Federation does not contain any restrictions related to such a notification; depending on the specific circumstances of the case, it can be made by using any available means of communication that will allow you to control the receipt of information by the person to whom it is sent (a subpoena, telegram, telephone message, facsimile, etc.), by means of an SMS message, if the person agrees to the notification in this way and when the fact of sending and delivery of the SMS notification to the addressee is recorded, which in the context of the digitalization of justice should be carried out immediately in automatic mode.

So, para. 1 of art. 2.15 of the Draft Procedural Code of Administrative Offenses stipulates that participants in proceedings in cases of administrative offenses and other persons authorized to participate in proceedings in cases of administrative offenses are notified or summoned to the court, body or official in whose proceedings the case is, by registered mail with delivery receipt, a summons with a receipt acknowledgment, a telephone message or a telegram, by facsimile, using the Unified Portal of State and Municipal Services, or using other means of communication and delivery that ensure that notice or a call is recorded and delivered to the addressee. In this case, the procedure for sending and delivering notifications (notifications) using the Single Portal of State and Municipal Services is established by the Government of the Russian Federation.

In our opinion, in this case, with the notification of the persons participating in the case of an administrative offense, the digitalization of justice will just contribute to their quick and reliable notification, if there are appropriate technical capabilities.

4 Purpose of the Study

The purpose of the research is to study administrative legal relations in the field of bringing to administrative responsibility in the framework of the implementation of proceedings on cases of administrative offenses in the context of the widespread digitization of different types of judicial proceedings.

5 Research Methods

To achieve this goal, the work used general scientific, special research andpartial scientific methods:

1) general scientific methods: logical method (analysis and synthesis provided an accurate and comprehensive consideration of the subject of research; the method of system-structural analysis made it possible to consider the investigated legal phenomena as elements of the system;

2) special methods: the use of the statistical method made it possible to determine the state and dynamics of the administrative-tort situation;

3) partial -scientific (legal) methods: the using of the formal dogmatic (legal and technical) method made it possible to analyze the current federal and regional legislation in the field of researched legal relations.

6 Findings

Thus, the trend of digitalization of all types of legal proceedings, including proceedings in cases of administrative offenses, has significant prerequisites and a basis for its development, but, today, the prospect of total digitalization of all proceedings in cases of administrative offenses seems to be very doubtful, since, on the one hand, digitalization will help to improve the quality of law enforcement, simplify and optimize proceedings in cases of administrative offenses, on the other hand, it can lead to significant violations of basic legal principles - the presumption of innocence, inevitability of punishment, proportionality, fairness and individualization of punishment.

So, the proceedings in cases of administrative offenses to date have not received comprehensive digitalization as the development of the electronic justice system at all stages, at the same time, proper legal regulation and improvement of legal regulation of such constituent elements of electronic justice as: electronic evidence; new types of forensic examinations; electronic means of organizing the work of the court (electronic document management, electronic business, intelligent systems for analyzing case materials, legal regulation); system of electronic participation in the process (video conferencing, electronic summons and SMS notifications, electronic copies of case materials), since the administrative and tort legislation provides for the above elements of e-justice, but does not disclose the specific procedure for their implementation and the requirements that are imposed on them, which significantly complicates the implementation of such elements of digital proceedings in cases of administrative offenses in practice, which can lead to a violation of the relevant rights of citizens, especially in the absence of appropriate technical capabilities.

As you can see, the current legislation in this area requires significant improvement and unification, first of all, the categorical apparatus and the establishment of a unified approach to the digitalization of various spheres of public life, which today allows us to talk about the improvement of legal regulation and an increase in certain aspects of the effectiveness of justice. However, it does not yet indicate a transition to electronic law.

7 Conclusion

In our opinion, today the digitalization of the procedure for proceedings in cases of administrative offenses is undergoing only its formation and development and requires significant improvements in the legal regulation and the development of a unified concept of digitalization of those elements of the justice system which really need such improvement in order to ensure the implementation of the rights, freedoms and legitimate interests of participants in proceedings in cases of administrative offenses, and not create additional opportunities for violation of their rights when considering this category of cases, that when considering an administrative offense by a robotic judge, it will entail the occurrence of many errors and violations, both procedural and substantive law. For this reason, it should be noted that the confidence in automated equipment has not yet reached the level when it could be provided with a solution to multivariate situations.

References

1. U. Kramer, L.A., Mitskevich & A.F. Vasilieva, Bulletin of St. Petersburg University. Pravo, **4,** 756-780 (2019)
2. President of the Russian Federation Message from the President of the Russian Federation (On the situation in the country and the main directions of domestic and foreign policy of the state). 20.02.2019Moscow, Russia (2019), URL:http://www.kremlin.ru/acts/bank/44032 (date of access: 20.02.2021)
3. President of the Russian Federation President's message to the Federal Assembly (01.03.2018) Moscow, Russia. (2018), URL:http://www.consultant.ru/document/cons_doc_LAW_291976/ (date of access: 20.02.2021)
4. V.D. Zorkin, Rossiyskaya Gazeta – Stolichny, **115 (7578)** (2018) URL: https://rg.ru/2018/05/29/zorkin-zadacha-gosudarstva-priznavat-i-zashchis.html (date of access: 20.02.2021)
5. E.V. Evsikova Scientific notes of the Crimean Federal University named after V.I. Vernadsky. Legal Sciences **3 (69),** 151-165 (2017)
6. President of the Russian Federation Decree of the President of the Russian Federation *"On the Strategy for the Development of the Information Society in the Russian Federation for 2017 - 2030".* (09.05.2017, No. 203) Moscow, Russia. (2017), URL: https://www.garant.ru/products/ipo/prime/doc/71570570/(date of access: 20.02.2021)
7. Code of the Russian Federation on Administrative Offenses dated 30.12.2001 №195-ФЗ // Collected Legislation of the Russian Federation of year 2002, No.1, Art. 1(2001)
8. Government of the Russian Federation *The concept of the new Code of Administrative Offenses of the Russian Federation was prepared by an interdepartmental working group formed in pursuance of the instructions of the Prime Minister of the Russian Federation Dmitry Medvedev and in accordance with the order of the Government of the Russian Federation* dated April 4, 2019 No. 631-r. Moscow, Russia. (2019), URL: http://government.ru/news/36971/ (date of access: 20.02.2021)
9. Supreme Court of the Russian Federation *Resolution of the Plenum of the Supreme Court of the Russian Federation "On some issues arising in courts when applying the Code of the Russian Federation on Administrative Offenses".* (24.03.2005, No. 5) Moscow, Russia. (2005), URL: https://base.garant.ru/12139487/ (date of access: 20.02.2021)
10. Ministry of Justice of the Russian Federation (2020). Draft Federal Law "Code of the Russian Federation on Administrative Offenses" (01/30/2020). Moscow, Russia. URL:https://base.garant.ru/56821718/
11. Ministry of Justice of the Russian Federation Draft Federal Law "Procedural Code of the Russian Federation on Administrative Offenses" (30.01.2020). Moscow, Russia. (2020), URL: https://base.garant.ru/56821726/ (date of access: 20.02.2021)
12. Constitutional Court of the Russian Federation Resolution of the Constitutional Court of the Russian Federation "In the case of checking the constitutionality of Article 2.6.1 and parts 1, 2, 3 and 6 of Article 12.21.1 of the Code of Administrative Offenses of the Russian Federation in connection with the request of the Kostroma Regional Court and the complaints of citizens A.I. Dumilin and A. . B. Sharova ". (18.01.2019, No. 5-P) Moscow, Russia. (2019), URL:http://www.consultant.ru/document// (date of access: 20.02.2021)
13. Federal Assembly Federal Law "On Electronic Signatures". (06.04.2011, No. 63-FZ) Moscow, Russia. (2011), URL: http://www.consultant.ru/document/cons_doc_LAW_112701/ (date of access: 20.02.2021)
14. E.E. Tonkov & L.A. Pozharova Scientific statements of the Belgorod State University. Series: Philosophy. Sociology. Pravo **44(3),** 490-497 (2019)
15. Ausfertigungsdatum Neugefasst durch Bek. v. 23.1.2003 I 102; zuletzt geändert durch Art. 5 Abs. 25 G v. 21.6.2019 I 846.(25.05.1976). (2019), URL: https://docviewer.yandex.ru/view/0/?*=0lq%2BxnnyWgkX0il5ICH6KOg97Y57InVybCI6Imh0dHLmd (date of access: 20.02.2021)
16. *Das Gesetz zur Förderung der elektronischen Verwaltung sowie zur Änderung weiterer Vorschriften (E-Government-Gesetz).* (2013), URL:

http://rusrim.blogspot.ru/2013/06/blog-pos.html) (date of access: 20.02.2021)

17. B.G. Koibaev & Z.T. Zoloeva, Humanities and Legal Studies, 119-124. (2020).
18. The Supreme Court of the Russian Federation Resolution of the Plenum of the Supreme Court of the Russian Federation "On some issues arising in judicial practice when considering cases of administrative offenses provided for by Chapter 12 of the Code of the Russian Federation on Administrative Offenses". (25.06.2019, No. 20) Moscow, Russia. (2019), URL: http://www.consultant.ru/document/cons_doc_LAW_327611/(date of access: 20.02.2021)
19. E.E. Tonkov, Bulletin of the judicial community of the Belgorod region **10**, 6-10 (2019)
20. A.V. Rudenko Academic Bulletin of the Rostov Branch of the Russian Customs Academy **4 (25),** 77-81 (2016)

AUTHOR INDEX

AUTHOR INDEX

EDP Sciences – Web of Conferences
17, Avenue du Hoggar
Parc d'Activité de Courtabouf
BP 112
F-91944 Les Ulis Cedex A
France

ISBN 978-1-7138-3066-5

CURRAN ASSOCIATES INC.
proceedings.com